Cryobiochemistry

AN INTRODUCTION

Cryobiochemistry

AN INTRODUCTION

PIERRE DOUZOU

Ecole pratique des Hautes Etudes
Institut National de la Santé et de la Recherche Médicale
Institut de Biologie Physico-Chimique, Paris

AP

ACADEMIC PRESS · 1977
LONDON · NEW YORK · SAN FRANCISCO
A Subsidiary of Harcourt Brace Jovanovich, Publishers

ACADEMIC PRESS INC. (LONDON) LTD
24/28 Oval Road
LONDON NW1

United States Edition published by
ACADEMIC PRESS INC.
111 Fifth Avenue
New York, New York 10003

Library of Congress Catalog Number: 77-72085
ISBN: 0-12-221050-6

FILM SET BY COMPOSITION HOUSE, SALISBURY, ENGLAND
PRINTED IN GREAT BRITAIN BY WHITSTABLE LITHO LTD., KENT

Preface

This book sums up ten years' work, carried out mostly in the author's laboratory, reporting data and observations related to low temperature biochemistry.

Such a project offers the opportunity to collate published data and observations, not as a simple compilation, but as a comprehensive study and potential tool for future investigators.

The author is aware of the fact that the methodology can largely be improved upon as can the devices built or adapted to such studies. While the References are by no means comprehensive they should lead the reader to more extensive reviews which contain a fuller list of references to the original literature. Thus, the book presents both the defects and, we hope, the advantages of a personal experience at its very early stage of development, and should be considered merely as an introduction to an emerging methodology.

This book has grown out of a course at the Illinois University during the academic year 1974–5 and owes very much to the stimulating and friendly climate created there by Professor I. C. Gunsalus. The kind hospitality of Dr. E. Zuckerkandl at Saint Guilhem Le Desert provided a wonderful opportunity to try to make a synthesis of the experimental work carried out over years at the Institut de Biologie Physico-Chimique in Paris, a wonderful place to work among distinguished scientists and friends such as Alberte and Bernard Pullman, Sabine and René Wurmser, Marianne Manago, Mike Michelson, Pierre Joliot.

Studies reported and discussed in this book were supported by the Centre National de la Recherche Scientifique, the Institut National de la Santé et de la Recherche Médicale, the Delegation Générale à la Recherche Scientifique et Technique, the Fondation pour la Recherche Médicale Française, and of course by the Fondation Edmond de Rotschild (Institut de Biologie Physico-Chimique, Paris), homeland of our team and of the procedure described in this book. That support is gratefully acknowledged.

I would now extend my warmest thanks to Simon Freed, Edward Reich, Constant Burg, Carlos Chagas, Britton Chance, Martin Kamen, Takashi Shiga, Takashi Yonetani, Hans Frauenfelder, Bert Vallee, Wolland Hastings, David Phillips, Gregory Petsko, Roma Banerjee, Sylvanie Guinand,

Evelyne Begard, Monique Thore and many other colleagues whose own work, cooperation, help, comments and advice were essential for developing the low temperature procedure and writing this book.

Pierre Douzou
February 1977

Acknowledgements

The author wishes to express his sincere thanks to Gregory Petsko for permission to reproduce data and for the loan of plates and diagrams; to Monique Thore and Denise Weckerlé for arduous work in the preparation of the manuscript.

The following have kindly given permission to reproduce the plates and diagrams: *Analytical Biochemistry*, *Advances in Protein Chemistry*, *Biochemistry*, the *Journal of Molecular Biology*, the *Journal of Biological Chemistry*, *Biochimie*, the *National Academy of Sciences*, *l'Academie des Sciences de France*, le *Journal de Chimie Physique* and the publishing companies, Longmans, Wiley and Academic Press.

The author trusts that due acknowledgements have been made in all cases and apologizes for any omission that may have been made inadvertently.

P.D.

To

CLAUDE BALNY, GASTON HUI BON HOA,

PASCALE DEBEY, FRANCK TRAVERS, PATRICK MAUREL

whose efforts made this monograph possible

Contents

1

Introduction

The essence of truth is superior to the terminology of "How"? or "Why"?

Hakim Sanaī XIth century

Why?

One of the most important aspects of biochemical processes is their dynamics and the measure of their overall rate is in fact the measure of the rate of the slowest of the series of successive reaction steps involving structural and catalytic transformations. All the steps should be known to explain any reaction mechanism but most of them are so fast that they go to completion in times going from the millisecond to a fraction of a nanosecond and necessitate rapid kinetic techniques. The fastest biochemical reactions are now accessible to experimental measurements since time resolutions of the order of a nanosecond are attainable, but many of the present rapid techniques give nothing more than a characteristic "relaxation time" with no indication about what this time refers to, since measurements involve optical spectroscopy rather than the more sophisticated techniques essential to determine the structures but which are not adapted to recordings in very short time ranges.

Thus new procedures are needed which would permit all the required structural information about any biochemical system to be obtained as a function of time. Several years ago, we decided to explore in this laboratory the possibility of slowing down reactions by temperature-effect, since almost all biological processes are temperature-dependent and, were it not for the crystallization of water, it is probable that most biochemical systems could be reduced in temperature sufficiently to reduce their reaction rates enormously, thus to improve their analysis, and even to induce complete suspension of their activity.

Moreover, since most biochemical processes consist of several successive steps, the effect of temperature on their overall rates will be the sum of its separate effects on these steps according to their temperature coefficients and activation energies, hence lowering the temperature should consequently lead to a thermal resolution of the individual steps.

In the case of enzyme-catalyzed reactions which consist of at least three consecutive steps

$$E + S \rightleftharpoons E-S \rightleftharpoons E-P \longrightarrow E + P$$

involving at least 6 thermodynamic parameters for each step (namely heat, free energy and entropy of both activation and of the process itself), a very small number of them can be determined and their relationships are unknown because very few separate velocity constants have been obtained and the measurements have been carried out only at one selected temperature. The low temperature procedure should therefore shed a new light and enhance our knowledge of forward and backward reaction pathways in biochemical reactions.

In the field of analytical biochemistry, the eventual adaptation of some of the procedures for the fractionation and concentration of lipid and protein components of biological tissues might open new interesting ways: column chromatography techniques, used at subzero temperatures, should permit the stabilization of some thermodynamically labile enzymes, the concentration and isolation enzyme-substrate complexes, and finally the improvement in resolution.

The introduction of a suitable low temperature procedure could even revive the interest and increase the usefulness of the old methods of salt and organic solvent fractionation which appear to have fallen into disuse, but offer certain advantages when compared to the newer procedures and which can use the solvents employed, as we will see later, to cool our media. Because of the lower density of these mixed solvents, compared to concentrated salt solutions, shorter times or lower speeds of centrifugation would be sufficient to collect the precipitated components and here again new resolution possibilities might be found.

The fact that exposure to organic solvents at room as well as at subzero temperatures can denature proteins is presumably responsible for the neglect of organic solvent fractionation as a technique, since the methods used in analytical biochemistry ought to isolate and concentrate valuable components as nearly as possible in their natural state. Any procedure which retains at low temperature the properties of water as a solvent of proteins and its fluidity would certainly induce a new surge of interest of older fractionation techniques.

The low temperature procedure could be applied to the elucidation of the intermediate steps in metabolic processes. On cell-free extracts, metabolic intermediates can be accumulated by use of specific inhibitors or by inactivation of specific enzymes. Substrates normally in too small a concentration to be detected will accumulate and will be isolated by lowering the temperature enough to interrupt the sequence of reactions at a given step.

Such a method could replace the use of specific "poisons" with greater flexibility and total reversibility.

Another potential application of cryobiochemistry could be cryobiology—a discipline concerned with the singular and yet largely unexplained events associated with life and death at low temperatures, primarily below freezing, of living systems. Cryobiology is almost entirely dominated by its applications, consisting mainly of organ banking; the remarkable developments which have occurred in recent years in surgical "grafting" techniques require methods for prolonging the viability of the organs *in vitro*, awaiting transplant, below zero degrees.

Thus cryobiology is faced with the necessity of finding methods of keeping cells dormant but potentially alive by cooling or freezing at subzero temperatures. This is a challenge since it is known that water does crystallize to ice with a multitude of consequences which are fatal to most of living systems; ice-crystals produce histological injuries both by mechanical effect and dehydration as water is removed to form ice.

Dehydration itself is known to cause the following changes: increase in electrolyte concentration with a resulting increase of ionic strength of the suspending medium and precipitation of proteins from solution, changes in pH, concentration of certain solutes to toxic levels (urea, gases etc.) sufficient removal of water to bring structures into actual physical contact and leading to abnormal cross-linking, removal of structurally essential water, and so on.

Some of these consequences of freezing seem to be avoided by macro addition of "cryoprotective" agents (glycols, polyols, dimethylsulfoxide) but current level of knowledge in cryobiology is still insufficient to explain manifestations of cryoinjury and cryoprotection at any level of biological complexity. Up to now, the approach of such problems have been too empirical and basic information is essential to understand the causes and consequences of subjecting cells to subzero temperatures.

It is within cells, cell-free extracts and biochemical enzyme-dependent relationships that cryoinjury and cryoprotection should be studied. The chief limitations of such observations are due to freezing, so cooling above freezing might open the way to such studies.

Joint *in situ* and *in vitro* experimentation of enzyme-catalyzed reactions and of multi-enzyme processes leading to biological functions could be performed before, during and after cooling, and provide background informations still lacking in cryobiology.

Finally, cosolvents and temperature variations could be used as tools to investigate the macromolecular basis of catalysis and the kinds of rearrangements that proteins undergo both under the influence of substrates and of effectors. In some cases is should be possible to observe such rearrangements by X-ray analysis of enzyme-substrate intermediates stabilized in the crystalline state at subzero temperatures. In other cases these rearrangements could be detected in solution through the reversible perturbations determined by cosolvents as opposed to substrates and effectors, and recorded by various methods.

Thus, there are a number of obvious advantages in carrying out low temperature biochemistry but the question is how to succeed in such an enterprise, a problem which will now be examined and then analyzed at length in chapters 2 and 3.

How?

The requirement for performing low temperature biochemistry is the prevention of water from freezing by the macro addition of miscible organic solvents without altering the capacity of water as a solvent of biomolecules in their ionic environment. This is *a priori* a difficult if not impossible task since water is unique in its structural and physico-chemical properties which determine the structural integrity of biological components and of their systems.

Highly polar water molecules give rise to electrostatic forces of high magnitude and these properties will be affected by any addition of "non-biological" components even when they are highly polar in themselves and give mixed homogeneous solutions retaining many of the properties of pure water. The possibility of special effects on structural and catalytic integrity of proteins due to the addition of miscible organic solvents to water as "antifreeze" has caused caution among biochemists in the application of low temperature procedures and therefore the advancement of cryobiochemistry.

As stated more than a century ago by Claude Bernard, the properties of the internal environment of living systems must be kept constant within very narrow limits, for biomolecules are exceedingly sensitive to small changes in physico-chemical properties of their immediate surroundings. The presence of organic solvents depressing the freezing point of solutions of biomolecules indeed alters the properties of the original purely aqueous medium, these alterations being proportional to the concentration of dissolved solvents.

Thus, by the addition of a solvent on the one hand and lowering the temperature on the other, we will alter the delicate balance maintaining the physico-chemical properties of biological systems at constant values and henceforth their structural and catalytic integrity.

In fact, there are some amendments to this rule since it is found in Nature that some cold-blooded species perform the synthesis of organic substances to "supercool" and then to prevent a substantial portion of their body water from freezing.

Resistance to cold in Nature

Resistance to cold in Nature is a fascinating and encouraging example of the eventual possibility of preventing water from freezing by the presence of organic substances without altering its capacity as a reactive component and as a solvent of biological systems.

It is known that insects hibernating in temperate or cold regions can avoid freezing injury by "supercooling" after accumulation of small molecular weight compounds (polyols and sugars) which depress their body fluid freezing point. Most poïkilotherm organisms (micro-organisms, plants and animals) do not freeze during the winter season, in some cases at temperatures as low as −35°C.

In many insects, polyol formation has been found to occur only during diapause, that is in the larval or pupal stage, but in some adult insects such as the carpenter ant or an Alaskan beetle, seasonal fluctuations of polyol content have been observed which mirrored fluctuations in ambient temperature.

The polyol is glycerol which is formed from glycogen. Sugars such as sorbitol and trehalose are also found in some species. The concentration of glycerol, expressed as a percentage of fresh body weight, can be as high as 25% in the larvae of *Bracon cephi* (hymenoptera), or as low as 1.1% in the eggs of *Bombyx mori* (lepidoptera) where 2.2% sorbitol was also detected.

In high concentrations, glycerol lowers the freezing point of the haemolymph to about −15°C and can determine a supercooling as low as −47°C.

In the adult beetle *Pterostichus brevicornis* (Alaskan), direct correlation between haemolymph glycerol content and supercooling and freezing points have been found in both naturally and laboratory acclimatized forms. Mean glycerol level was less than 1 mg% during summer and increased to 22 mg% during winter, the supercooling points being respectively −4.2°C and −11.5°C, the freezing points −0.6°C and −5°C.

As we mentioned above, glycerol seems to originate from carbohydrate precursors, as indicated for instance by high fructose diphosphatase activity levels in winter and glycerol's rapid conversion to glycogen when winter acclimatized species are taken into the laboratory. On the other hand, the fat content as a percentage of dry body weight declines from about 30% in the fall to about 20% for the winter and spring in the case of Alaskan beetles.

Some insects seem to use sugars to prevent cold injury; diapausing prepupae of the Japanese fly *T. Populi* accumulates sugars at the beginning of the fall; the total sugar levels average 5 to 7% of fresh body weight, of which more than 90% is trehalose. A clear correlation between sugar content and frost resistance is observed in these insects during the period from August to

June. During the cold season, none of the prepupae are killed in liquid nitrogen provided they were previously frozen at −30°C.

Finally, we can say that a slight increase in glycerol or sugar content is always followed by a remarkable increase in frost resistance capacity of most of insect species, and the mechanism of the protective action of glycerol in these cases may well differ from the mechanism operating in the protection of mammalian cells and tissues found in cryobiology, protection which confirms the possibility of the use of organic additives for low temperature biochemistry.

After reconversion of glycerol or sugars to glycogen, insect larvae are able to complete metamorphosis, demonstrating that both organization and function are unaffected by the presence of such solutes.

Prevention of freezing injury: cryobiology

Apart from the abilities of poïkilothermic organisms to avoid freezing and its fatal consequences, it was discovered independently in 1949 that some organic substances behave as protective agents against freezing injury to certain mammalian cells.

Glycerol was the first of these substances to be used successfully and is still the leading cryoprotective agent, presumably owing its comforting virtue of being a normal intermediary product of lipid metabolism, both in mammals and insects.

On the other hand, some "non-biological" substances such as dimethylsulfoxide, ethylene glycol and propylene glycol can be used as cryoprotectives. Nevertheless, in spite of the considerable amount of literature on such additives, the basis of their action still remains largely obscure.

The requirements for use are their nontoxicity, their ability to penetrate the cell membrane freely and their capacity to dissolve electrolytes. All contain hydrogen-bonding groups (OH, NH, =0) and bind strongly with water, preventing it from freezing at 0°C. Their protective action can be understood in the cooled state since it is known that water crystallizes to ice with a multitude of consequences. Let us recall that the production of ice crystals determines histological injuries and also dehydration as water is removed to form ice; dehydration causes several changes including a resulting increase in the ionic strength of the suspending medium hence precipitation of proteins from solutions, changes in pH, abnormal concentration of solutes such as dissolved gases, urea, etc. to toxic levels; finally, sufficient removal of water could bring structures into actual physical contact, leading for instance to abnormal cross-linking.

It is difficult to offer an explanation of how such effects could be avoided during freezing of media containing cryoprotective additives, but here again we have to stress that the presence of high concentrations of organic and eventually "non-biological" solvents does not alter irreversibly the functioning of most of biological systems.

Physical-chemical basis of low temperature biochemistry

Besides the action of organic solvents in cryobiology and their innocuity towards biological systems, it must be remembered that many soluble (purified) enzymes are currently dissolved in water–glycerol mixtures and stored at low temperature were they are found to be "cryoresistant", and also that certain enzymes which consist of subunits are protected against irreversible inactivation during freezing-thawing in the presence of large amounts of polyols. However, storage conditions can be quite different from conditions insuring optimal activity.

Before the beginning of our own work, very few attempts to carry out enzyme catalyzed reactions in mixed solvents at subzero temperatures were reported in the literature (3), (4), (5). Among them the author would especially mention the work of Simon Freed who injected solutions of enzymes in water cooled at 0°C as a fine spray into several cooled alcohol–water mixtures, obtained the progressive dissolution of the small "icebergs", showed that the hydration of the enzymes persisted at low temperature in predominantly alcoholic solvents, and demonstrated the occurrence of reactions in such conditions (6). Freed understood the potentialities of the procedure as well as the absolute necessity to learn more about its basis and was a pioneer in the field of a low temperature biochemistry.

Nevertheless, most purified enzymes and cell-free biological systems might be sensitive to the macro-addition of organic solvents. Such an addition not only decreases the freezing point of aqueous systems but indeed alters physical properties such their as molecular volume, viscosity, dielectric constant, which in turn influence the behavior of structural and catalytic biomolecules, their reaction rates and eventually their mechanisms.

The broad variations in solvent properties, first on addition of "antifreeze" solvents and then on cooling of the mixtures, are at least hypothetically accessible and the documentation of their numerical values is necessary both to predict and explain or even to correct such variations so that the alteration of the structure and function of biomolecules could be minimized.

The physical chemistry of mixed solvents in the range of normal and subzero temperatures has been carried out mainly in this laboratory and will be treated at length in chapter 2.

For the moment, since we are dealing with the "how" of an eventual low temperature biochemistry, we can try to get a bird's eye view of this essential problem.

Among the physical properties mentioned above, the dielectric constants are of particular importance since they reflect the electrostatic interactions which are the basis of most of physico-chemical processes involving dipolar-molecules and ions, that is between proteins, metabolites and electrolytes, in the highly polar media represented by water and also by mixed solvents composed of water and weakly protic solvents (alcohols, polyols, etc).

It is known that any addition of organic solvent decreases the dielectric constant of aqueous solutions and that such a decrease is proportional to the concentration of added solvent. On another hand, since all dipoles develop an orientation which is governed by the Boltzman distribution law and results in the dielectric constant being strongly dependent on the absolute temperature, great attention must be paid to its variation as a function of temperature in mixed solvents having a very low freezing point. We will see that the logarithm of the dielectric constant D plotted against $1/T$ is a straight line over the entire range of subzero temperatures above the freezing point, and the D reaches the value holding for pure water at 20°C (D 80) well before the freezing point of most aqueous–organic mixtures.

It will be shown that the synchronization of the addition of an organic solvent to water with cooling allows D to maintain its original value in water at 20°C and represents a simple and efficient way of preparing cooled solutions of enzymes, avoiding the failures observed when these two operations are consecutive rather than simultaneous. But at the same time, it will be shown that such a procedure does not suppress the solvent effects on acid-base equilibria, solubilities, conformations of biomolecules, reaction rates and mechanisms, thus indicating that factors other than electrostatic ones are pertinent in mixed solvents as a function of temperature.

The relative importance of electrostatics will be illustrated by changes in acid-base equilibria; the pK of ionizable solutes (indicators as well as amino acids and proteins) and the pH, or more precisely the proton activity pa_H, of buffers can vary with the addition of solvent, apparently in conformity—at least qualitatively—with electrostatic expectations of the influence of D on the dissociation of all these solutes; but pKs and pa_Hs will increase during cooling, i.e. in conditions where D is increasing significantly and should theoretically increase the dissociation and therefore *decrease* pKs and pa_Hs.

Thus, the most elaborate electrostatic models available in the literature will present mathematical advantages handling problems of solvent and temperature effects on solutes, but will have physico-chemical deficiencies. It will be obvious that, in spite of the theoretical expressions and empirical rules about electrostatic factors and their physico-chemical consequences, much about the origin of observed effects remains obscure, and that "non-electrostatic" factors, due to the presence of the solvent and to changes occurring in mixtures when the temperature varies, act by themselves and sometimes in opposition. It must be remembered that a mixed solution is not the continuous, structureless medium appearing from our macroscopic measurements but that there is an immense gap between any macroscopic property and its inaccessible microscopic counterpart.

In such conditions the physico-chemical basis of low temperature biochemistry will remain partially empirical, but the data gathered about variations of several accessible parameters will be required to devise a suitable methodology and carry out interesting experiments on solubilized biomolecules in their native conformation.

Once the dangers of denaturation, and other effects incident on changes in solubility and acid-base equilibria, are avoided, work can begin in the test tube on a wide range of subzero temperatures.

Further, it is necessary to record pH-activity profile of reactions which are pH-dependent, both to determine any sequent shift due to solvent and temperature effects on the p*K*s of the ionizing groups at the active site thereby causing a shift in the pH optimum, and to check whether the shapes of the profiles remain unchanged—a prime requirement for the demonstration of an unchanged reaction mechanism.

We will see that after all the above requirements are fulfilled, there is still a noticeable intrinsic cosolvent effect on enzyme activity; activating or inhibiting, it must be reached fairly rapidly, independent of time, and fully reversible by infinite dilution.

Other evidence that reaction mechanisms remain unchanged must be sought by kinetic analysis at selected temperatures and as a function of temperature using the Arrhenius plot. Straight lines and constant values of the activation energy over a broad range of subzero temperatures can demonstrate that reaction mechanisms are unaltered, for it would be a coincidence if the mechanism was changed without a variation of the activation energy.

The goal can then be reached: the rate constants being temperature-dependent according to the expression: $K = A^{-E/RT}$ the slowing down of reactions can be postulated and then observed and applied to the study of a large number of enzyme mechanisms.

References

1 H. T. Meryman. *Cryobiology*. Academic Press, London and New York (1966).
2 E. Asahina. "Frost Resistance in Insects." In *Advances in Insect Physiology*, Vol. 6. Academic Press, London and New York (1969).
3 G. K. Strother and E. Ackerman. *Biochem. Biophys. Acta* **47**, 317 (1961).
4 S. M. Siegel, T. Speitel, and R. Stocker. *Cryobiology* **3**, 160 (1969).
5 B. H. J. Bielky and S. Freed. *Biochem. Biophys. Acta* **89**, 314 (1964).
6 S. Freed. *Science* **150**, 576 (1965).

2

Properties of Mixed Solvents as a Function of Temperature

Structural aspects of mixed solvents

WATER

Any addition of a miscible and in most of cases weakly protic organic solvent to water, and then lowering the temperature of the mixture disturb the distribution of water molecules and therefore the physico-chemical properties of liquid water.

Water molecules

Let us recall that the shape of the water molecule is that of an isosceles triangle—roughly speaking a "V"—and each molecule is capable of forming hydrogen bonds with four others acting as hydrogen donor twice and as hydrogen acceptor twice, see Fig. 1.

This molecule is a highly polar structure since the oxygen nucleus attracts electrons away from the hydrogens, thus leaving the region around them with a net positive charge. The two pairs of unshared electrons of the oxygen atom tend to concentrate in directions pointing away from the O–H bonds. Such a polarity manifests itself through the formation of hydrogen bonds, a positively charged region in one molecule orienting itself towards a negatively charged region in one of its neighbors.

The number of hydrogen ions which form the positive ends of the hydrogen bonds around any given oxygen atom will be equal to the number of unshared electron pairs which form the negative ends.

Thus each molecule tends to have four nearest neighbors and the tetrahedral arrangement of the bonds determines the formation of three-dimensional structures. This type of pattern is found in ice and, in a more imperfect form, in liquid water.

In fact, the shape of each water molecule as well as its hydrogen bonding and three-dimensional structure shows a peculiar behavior which is not quite completely described by the simple picture of dipoles, but depends on the distribution of charges within the water molecule.

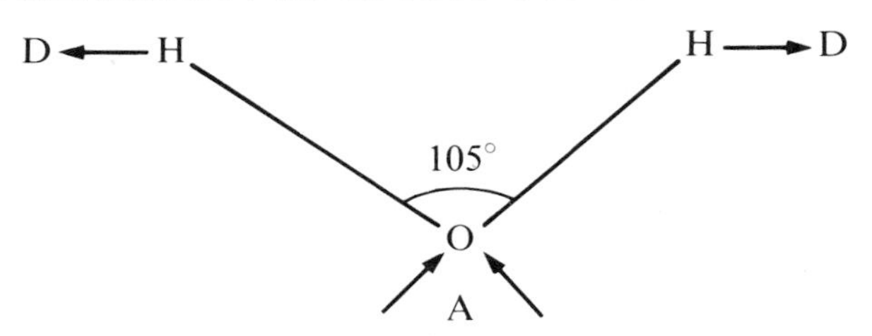

FIG. 1. Shape of the water molecule.

The oxygen atom contains the orbits 1s and 2s filled by four electrons, and the three 2p orbits denoted by (2x), (2y) and (2z), containing altogether four more electrons.

Thus, at least one of these functions must hold two unshared electrons (let us suppose (2x) is that function); it is no longer available for the formation of a valence-orbital function, the function (2y) and (2z) being used for the formation of valences.

The (2y) function has maximum probability densities in the $\pm y$ direction, has a node in the $y = 0$ plane, and one must add an atomic wave function of hydrogen in such a way that the function should overlap as much as possible. To get the greatest overlap it is best to locate the hydrogen-atomic function and with it the hydrogen atom itself in the $+y$ and the $-y$ direction. The valence-orbital function so arrived at will be filled by one of the oxygen electrons and the hydrogen electron.

Once a hydrogen atom is placed in one of these directions, and the valence-orbital function made up from the hydrogen function, and the (2y) function is filled with two electrons, the (2y) function is no longer available for valence formation. Other functions which could be constructed would have an anti-bonding character and lead to repulsion.

Thus, a hydrogen atom approaching from the $-y$ direction would be repelled, since it would strongly overlap a wave function already filled with two electrons.

Therefore a second hydrogen atom can be bound only by interaction with the (2z) function which as yet contains only one oxygen electron. The smoothest valence-orbital function can be formed from the hydrogen-atomic function and from (2z) if the second hydrogen atom is located along the plus or minus z direction. Then 90° is obtained for the stable H–O–H angle.

It is a consequence of the orthogonality of the functions that the two H–O bonds tend to be perpendicular, which has nothing to do with the concept of perpendicular lines in space.

The (2y) proper function and the function of the hydrogen atom situated on the y axis have cylindrical symmetry arounb the y axis, that is, around the direction of the valence bond, and this orbital can be called a σ function.

However, the influence of the other hydrogen atom will destroy the strict cylindrical symmetry and therefore the σ notation must be taken as an approximation.

As soon as a σ valence orbital is filled by an electron pair, it behaves like a complete shell, and it will repel other filled valence orbitals; the two O–H bonds will tend to spread, making, as a result, an angle greater than 90°, and such a tendency will be enhanced by the polar character of the O–H bonds since the positive ends of the dipoles repel each other. As seen above, the observed angle is actually 105°.

Distribution of charges within the water molecule has a decisive influence on the most stable configuration of these molecules with respect to each other. Taking into consideration the existence of two distinct centers of positive charge in H_2O, it appears that the most stable configuration of water molecules would be the following:

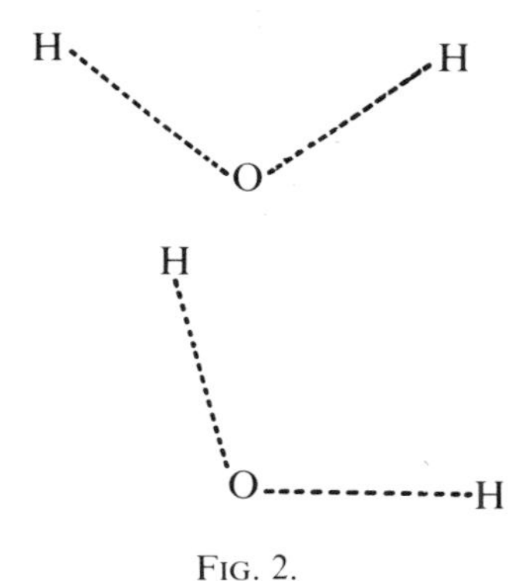

FIG. 2.

instead of the configuration predicted from the electrical properties of the molecules by attaching to them one dipole moment bisecting the H–O–H angle:

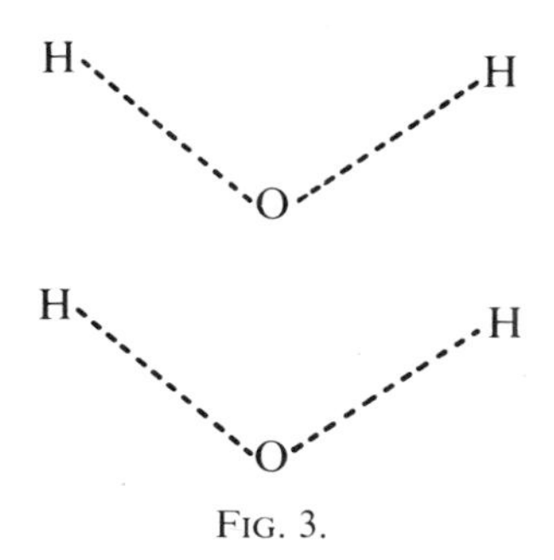

FIG. 3.

The resulting tetrahedral coordination of water molecule is sketched below:

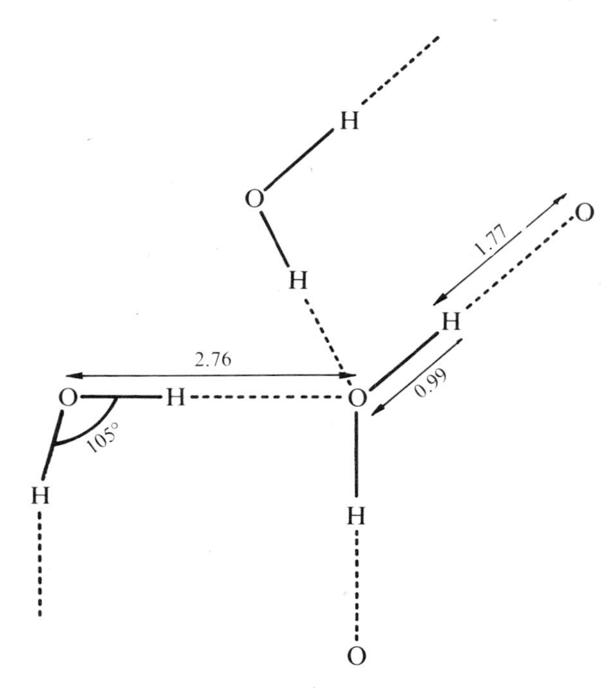

FIG. 4. Sketch of tetrahedral coordination of water molecules.

As we said previously, such an arrangement is found in ice and, in a more imperfect form, in liquid water. The arrangement has been established by X-ray diffraction of ice, as well as by X-ray scattering of liquid water. It has been found that, in ice, the O–H · · · O distance is 2.76 Å, and thus the nearest distance from a proton to the oxygen in the nearest neighboring molecule is 1.77 Å. The energy of the H · · · O bond in water has been estimated as approximately 4.5–5.0 kcal/mol.

When ice melts, the highly coordinated crystalline structure keeps the same pattern, except that some of the bonds break and form, so that there is no fixed permanent crystalline structure.

High melting and boiling points and the high heat of vaporization of water are indicative of strong intermolecular forces in liquid water, requiring high energies to separate the molecules from one another.

The most important property of liquid water is its capacity as a solvent of salts and ionic compounds, and also of non-ionic compound containing hydroxyl, carboxyl, amino, or keto groups, which can interact with water molecules via hydrogen bonds.

It is known that water is capable of dissolving a far greater variety of ionic compounds than any other medium, because its high dielectric constant causes a weakening of the attractive forces of ions of opposite charge and permitting them to remain in solution.

Liquid water

There is no established molecular description of liquid water. All the current models and theories are both incomplete and oversimplified and the water problem in biochemistry is yet unsolved.

There are two major schools of thought on liquid water structure, viz. that of the continuum model as opposed to that of a mixture model in which water consists of distinct species having different densities. In the latter model, water could exist in a limited number of distinct species in equilibrium as proposed by Franck and Evans (1), or even in two structures composed of monomeric water molecules and "ice-like" clusters of water molecules, as assumed by Ben Naïm (2).

For the purpose of our interest in cryobiochemistry, that is in cooled mixed solutions of biopolymers, we are primarily concerned with any eventual shift of the equilibrium of distinct species of liquid water after addition of a partially non-polar solvent and the cooling of the resultant mixture. We will, therefore, briefly summarize some of the conclusions drawn from a large body of experimental evidence and theoretical explanations in terms of the mixture model.

If liquid water exists in a limited number of distinct species in equilibrium, addition of a partially non-polar solvent would drive the equilibrium towards the species of highest density and therefore stabilize the structure of water. In Eley's model (3), assuming that water consists of a definite number of natural holes, as well as in the treatment of Franck and Evans (1), the water builds a microscopic "iceberg" around the non-polar molecule. That sort of quasi-solid structure does not imply that its structure is exactly ice-like. The assumption that non-polar molecules, and also rare gases, form icebergs has received some support from thermodynamics, as well as from the existence of crystalline hydrates of these substances, and that picture has been widely used in physical biochemistry. Klotz (4) has postulated that the amino acid residues of identical structure may also do this, and other authors maintain that icebergs can be formed around the non-polar moieties of dissolved proteins.

The molecular picture of water behavior at protein surfaces is still a matter of controversy; the idea of a stable hydration shell with either polar or ice-like character, depending on the estimates of hydration water which represents approximately 30% of the weight of the protein, is now challenged by new considerations based on X-ray diffraction, NMR studies and buoyant density measurements.

X-ray diffraction demonstrated that there are large interstices of hemoglobin molecules which must be filled by water molecules (5,6). Some of the dimensions of these interstices are such that sequested water molecules should be highly structured and quite different from normal fluid water. Sequested water must be considered as part of the protein and should form an essential role in the subunit reorganization occurring on oxygen binding (7). Thus a new molecular model of water is emerging and has to be extended to other protein systems, including the most highly organized ones.

On the other hand, clusters of structured water molecules have been found at a number of points on the surface of carboxypeptidase A in the crystal. The same studies have shown that many of the water molecules in the interstices of the crystal are also in fixed positions (8).

Studies of buoyant density on a number of protein systems have shown some effects which are interpreted as indicating relatively large amounts of water which are not penetrable by ions and other small molecules. Such a form of water might be sequestered in some interstices, still ice-like in nature, or on outer surfaces (9).

Thus the conventional earlier models of protein-water interacting are progressively replaced by new pictures, much more dynamic in essence, opening new perspectives in the understanding of protein behavior under the influence of a number of solutes including cofactors, as well as under the influence of cosolvents. The reader interested in such developments should read the papers by Lumry and Rajender (9), Nemethy and Sheraga (10,11) and references therein.

NON AQUEOUS SOLVENTS

Organic solvents proposed as "antifreeze" are alcohols (methanol), polyols (methyl pentanediol, ethylene glycol, propylene glycol, glycerol) and finally esters and nitrogen compounds (dimethylsulfoxide, dimethylformamide) (12).

They are miscible with water and weakly protic, i.e. only weak proton donors or acceptors, or both—their presence in water at concentration 1 M being characterized by $6 < pH < 8$. In this respect they resemble water; some of them most nearly resemble water in structure and tend to combine by hydrogen bonding to polymers.

Certain essential requirements regarding chemical inertness and stability must be met and demonstrated; all the solvents must be purified and used after recent distillation; their choice will vary according to the type of molecules (solutes) under investigation.

Alcohols

Alcohols associate to an indefinite extent by hydrogen bonding, forming zig-zag chains:

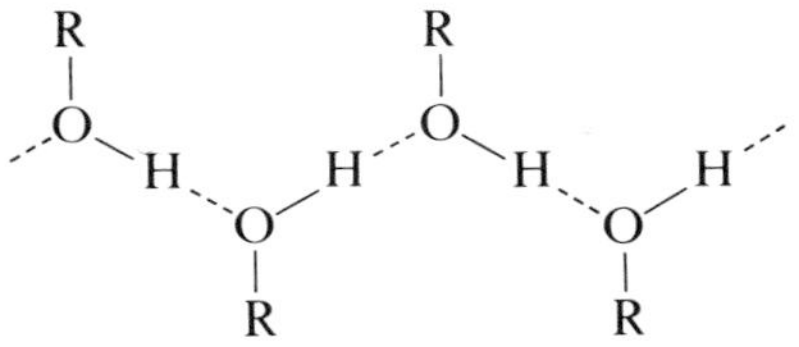

FIG. 5. Alcohol association by hydrogen bonding.

It was found that the O–H ··· O distance in methanol crystals is 2.75 Å (at −110°C).

If only a few bonds link the molecules, the crystals melt and for this reason the heat of fusion and the melting point are only slightly abnormal, whereas the heats of vaporization and boiling point show the effect of the hydrogen bonds strongly, and, in consequence, the liquid state is stable over a wide temperature range. This is the case for most alcohols and polyols; their freezing points are given in Table 1.

TABLE 1. Freezing points of usual organic solvents

Organic solvent	Freezing point (°C)
Methanol	−97.68
Ethylene glycol*	−13
Propylene glycol	−60
Glycerol	18.18
Dimethylsulfoxide	18.54
N-*N*-dimethylformamide	−60.43

* Freezing point doubtful as the tendency of ethylene glycol is to supercool and form a glass.

It is instructive to compare the heats of vaporization of water and alcohol: the amount of heat required to vaporize a fixed quantity of each liquid decreases significantly from water (heat of vaporization at 0°C = 0.595 kcal/g) to alcohols; in this respect, methanol which resembles water more than does any other solvent shows a heat of vaporization of 0.289 kcal/g.

On the other hand, approximate values of the energies of the O–H ··· O bonds in crystalline alcohols can be obtained comparing their melting and freezing points as well as the molal enthalpy of sublimation with their methyl-esters.

Structurally one can visualize the bonding between hydroxy solvents and water, see Fig. 6.

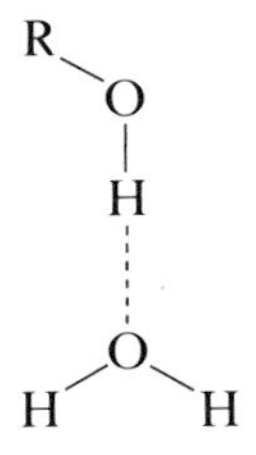

FIG. 6. Alcohol–water association by hydrogen bonding.

Nitrogen and sulfur compounds

Dimethylsulfoxide and dimethylformamide can be used as additives "anti-freeze". Their freezing points are respectively 18.54°C and −60.43°C (see Fig. 7, for their formulae).

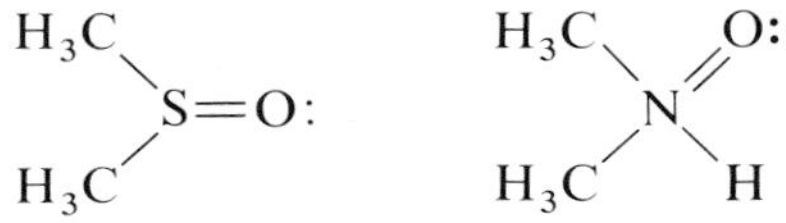

H_3C H_3C S=O: H_3C H_3C N O: H

FIG. 7. Dimethylsulfoxide and dimethylformamide formulae.

It can be seen that the ability to form hydrogen bonds in the absence of water is likely for both dimethylsulfoxide and dimethylformamide; these observations can be extended to mixtures with water, where the bonds shown in Fig. 8 can be predicted.

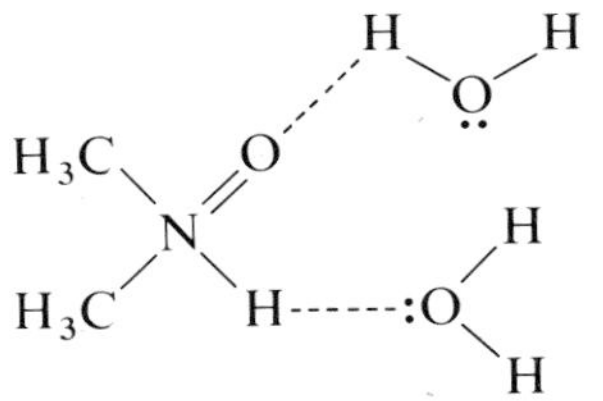

H_3C H_3C N O H------:O H H H O H

FIG. 8. Dimethylformamide association to water by hydrogen bonding.

It is evident from Fig. 7 that dimethylsulfoxide is strongly basic, behaving only as hydrogen acceptor.

MIXED SOLVENTS

The physical properties of mixed solvents have not been studied extensively and often a particular property has been determined at only one or a few selected volume-ratios and temperatures.

Most physical properties which are a function of organic solvent concentration are also temperature dependent, and their numerical values must be known for both the execution and interpretation of experimental results.

The first property which has to be established is the depression of the freezing point of mixed solvents upon increasing addition of the organic moiety. Several binary and ternary mixtures are listed in Tables 2 and 3. Some of them show the interesting property of supercooling (termed SC) and can be used eventually as transparent glasses.

Other physical properties of interest to us are the broad variations of volume (mostly volume contraction), viscosity and dielectric constant; these will be examined separately.

Apart from the atomic properties the structure of mixed solvents is largely unknown and in most cases X-ray diffraction of frozen samples would be of little interest as the two solvents separate into two different phases during freezing.

In fact, mixed solvents become even more complex systems in the presence of solutes because of the selective attraction of one of the components of these solvents by the solutes. The molecules of the more polar component (water) tend to cluster around the solutes (ions and dipolar ions), forcing away the molecules of the less polar component. Thus the composition of the solvent in the neighborhood of charged solutes does not correspond to that in the bulk of the solution but varies from point to point as one moves away from the surface of the charged solute.

It has been widely admitted that, even in mixtures containing a very small amount of water, the shell of solvent immediately around the ions will be made up entirely of water molecules.

We will see later that Debye proposed a theory of the distribution of the two kinds of solvent molecules around ions which gives a general molecular picture much closer to the actual molecular situation that is the picture of the mixed solvent as a continuous, structureless medium of uniform physical properties. This does not qualitatively alter the general electrostatic picture of the solvents but explains quantitative discrepancies between observed properties and calculations derived from the simple continuous picture.

Selective (or preferential) solvation of charged solutes by the more polar solvent molecules will thus play an essential role on the physical and chemical properties of mixed solvents as well as on the behavior of solutes.

Another important property of mixed solvents is that alcohols, polyols, dimethylsulfoxide and dimethylformamide are sufficiently more basic in

TABLE 2. Freezing points, in °C, of binary mixtures

Organic solvent	% solvent volume/water volume									
	10	20	30	40	50	60	70	80	90	100
Ethylene glycol (EGOH)	−4	−10	−17	−26	−44	−69	< −100	−83	−50	−12.5
Methanol (MetOH)				−40	−49	−67	−85	< −100	< −100	−96
2-Methyl-2-4-pentanediol (MPD)	−1.5	−5	−10.5	−26	−48	SC*	SC	SC	SC	SC
Propylene glycol (1-2-propanediol) (PrOH)	−3.5	−8	−17	−38	SC	SC	SC	SC	SC	SC
Glycerol (GlOH)	−3	−8	−14.5	−29	SC	SC	SC	SC	SC	+18
Dimethylsulfoxide (DMSO)	−3	−12	−19	−41	SC	SC	SC	−38	−7	18.5
N-N-dimethylformamide (DMF)	−2.5	−7	−13	−25	−40	−62	−83	−100	−90	−62

TABLE 3. Freezing points, in °C, of ternary mixtures

% EGOH–% MetOH–% water	40–20–40	25–25–50	10–50–40	10–60–30
Freezing point	−71	−50	−69	SC

* SC: Supercooling

character than water to behave as hydrogen acceptors rather than as amphoteric compounds.

The above properties, as well as other physical characteristics such as expansion or contraction, the viscosity and the dielectric constant, represent a largely uncharted area and we will now examine the evolution of these last properties in the various mixtures we intend to use, as a function of temperature.

Density and viscosity

Up to now these physical properties and the others considered in the following sections have been studied extensively for relatively few mixtures. Often a particular property has been determined at one or a few selected volume ratios and temperatures only.

Most physical properties of water and miscible organic solvents are dependent from the volume ratio of the mixtures and are temperature dependent. Numerical data were gathered about the evolution of the physical properties, primarily of density and viscosity of mixtures.

DENSITY

The density ρ, of a solvent is defined as the mass per unit of volume, and is expressed in grams per cubic centimeter in the CGS system. It is a useful physical property and is a factor both in any eventual correction of spectrophotometric data and in the calculation of many properties, including the viscosity.

It is well known that the addition of increasing percentages of organic solvent to aqueous solutions produces a volume contraction and that the lowering of temperature enhances this process. Nevertheless, no data about such a volume contraction are available for the range of subzero temperatures, although this knowledge is essential to make corrections to any spectroscopic recording.

Density measurements are performed with a cylinderical piece of quartz (radius, 7 mm, weight, 23 g) suspended under the plate of a Bunge precision balance and plunged into the liquid to be studied. The sample itself is inside a double-walled cryostat, in the first jacket of which circulates a thermal regulating fluid (nitrogen), the external jacket being under vacuum to avoid convection losses. The temperature control device will be described in chapter 4; the temperature is measured by a thermocouple and kept constant within 1°C. The reference liquid is water at +21°C (density ρ_0 = 0.998 g/ml). If M_0 is the weight of displaced water and M the weight of displaced sample, the density (ρ) of the latter is given by the expression

TABLE 4. Density of some hydro-organic solvents as a function of temperature

Volume ratio (%)	Temperature (°C)								
	+20	+10	0	−10	−20	−30	−40	−50	−60
20 EGOH–80 H_2O	1.026	1.030	1.034	1.037	—	—	—	—	—
30 EGOH–70 H_2O	1.041	1.044	1.048	1.052	—	—	—	—	—
50 EGOH–50 H_2O	1.067	1.063	1.079	1.085	1.092	1.098	1.104	—	—
50 MetOH–50 H_2O	0.928	0.935	0.943	0.951	0.959	0.967	0.974	—	—
70 MetOH–30 H_2O	0.888	0.897	0.906	0.915	0.923	0.932	0.941	0.950	0.959
40 EGOH–20 MetOH–40 H_2O	1.025	1.032	1.038	1.044	1.051	1.057	1.064	1.070	1.077
25 EGOH–25 MetOH–50 H_2O	0.994	1.003	1.010	1.017	1.023	1.030	1.037	1.043	—
10 EGOH–50 MetOH–40 H_2O	0.934	0.942	0.950	0.958	0.967	0.975	0.983	0.992	1.000
10 EGOH–60 MetOH–30 H_2O	0.916	0.924	0.933	0.942	0.951	0.960	0.969	0.977	0.986
50 MPD–50 H_2O	0.990	0.996	1.003	1.009	1.016	1.022	1.029	—	—
50 PrOH–50 H_2O	1.039	1.045	1.051	1.057	1.062	1.068	1.074	1.079	1.085
50 GlOH–50 H_2O	1.136	1.142	1.147	1.153	1.158	1.163	1.168	1.174	1.179
50 DMSO–50 H_2O	1.077	1.084	1.091	1.097	1.104	1.110	1.117	1.124	1.130

$M/\rho = M_0/\rho_0$, neglecting the volume change of the quartz with temperature. The precision obtained is 0.1 %.

Table 4 and Fig. 9 give ρ values as a function of temperature for various binary and tertiary mixtures. It can be seen from Fig. 9 that the variations of ρ are linear as a function of temperature so that corrections are easy to make on spectroscopic recordings.

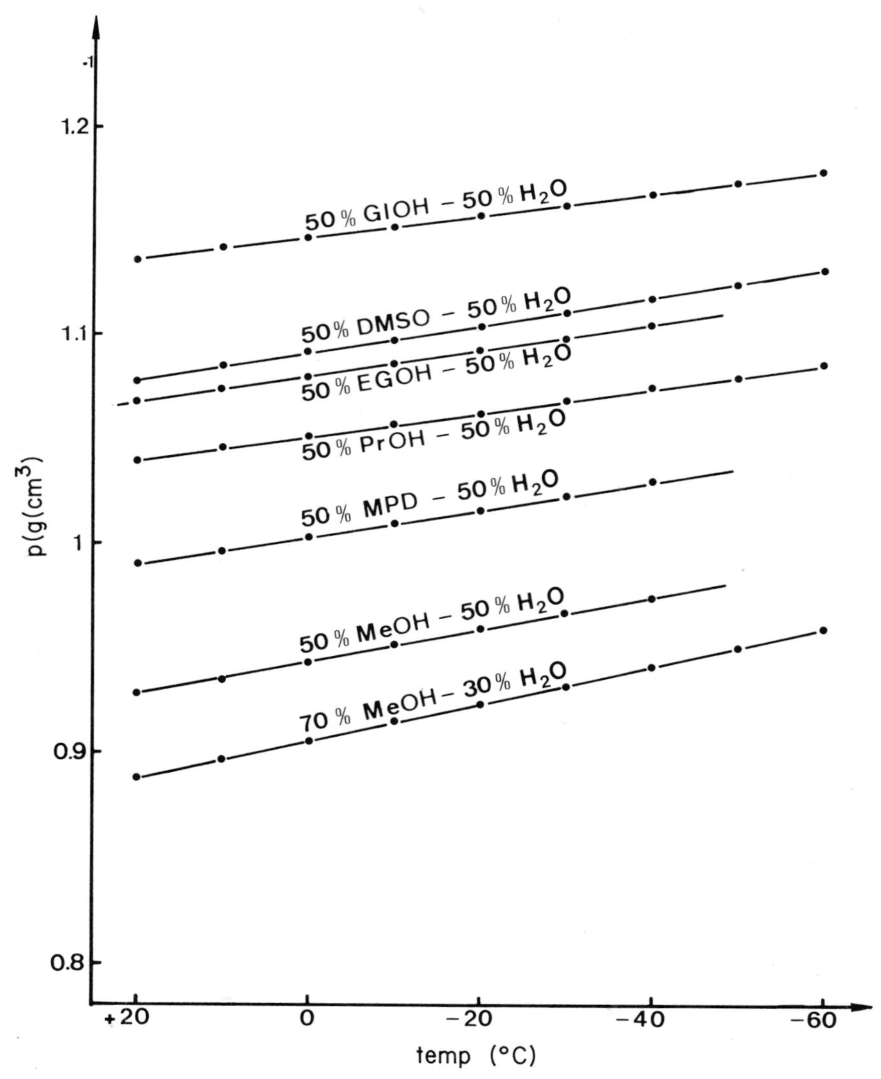

FIG. 9. Density (ρ) of some hydro-organic mixtures as a function of temperature.

VISCOSITY

The dynamic viscosity, η, is defined as the force per unit area necessary to maintain a unit velocity gradient between two parallel planes a unit distance apart. The unit of viscosity is the poise (dyne-s/cm^2), but numerical values are usually given in smaller subsidiary units such as the millipoise (10^{-3} poise) or the centipoise, cP, (10^{-2} poise). The kinematic viscosity is defined as the ratio of the dynamic viscosity to the density of a fluid. It is used in fluid mechanics and is the only parameter characteristic of the fluid itself to appear in the equations of hydrodynamics. The kinematic viscosity is directly proportional to the time required for the liquid to flow through a capillary-tube viscosimeter under its own hydrostatic head. The CGS units are called stokes (cm^2/s) and may be converted to poises by means of the equation, poises = stokes × density. Viscosities in this book will be expressed in centipoises, unless otherwise indicated.

Organic solvents added to water to depress its freezing point also change the network of hydrogen bonds and therefore the viscosity. Increasing this network will produce higher viscosities such as those obtained for the glycerol–water mixtures. As the temperature is decreased, the number of hydrogen bonds increases and accordingly the energy of activation of viscosity also increases, as does the viscosity value. The viscosities of common liquids and mixtures, including aqueous–organic solvents, have been determined in the range of normal temperatures, usually around +20°C. We needed data on the variations of viscosity as a function of temperature down to the freezing points of our mixtures.

We first had to choose an accurate method. Among the numerous methods for viscosity measurements which have been developed over the years, we choose one of the earliest and least precise, but the only one which could be adapted to low temperature measurements, namely the falling-sphere method. The determination of viscosity is made by measuring the resistance to the fall of a sphere in a fluid. Hydrodynamic theory indicates that, if the velocity of fall (v) and the radius (r) of the sphere are sufficiently small, and if the fluid is infinite in extent, the flow of the fluid around the sphere is laminar. In such a case the resistance is due to shearing of the medium, and the viscosity is the dominating factor. Under these conditions Stokes's law will be

$$W = 6\pi\eta r \qquad (2\text{-}1)$$

where W is the resistance to motion of a sphere of radius r moving at uniform velocity. For the steady state at uniform velocity under the action of gravity, the resistance may be equated to the effective weight of the sphere. Solving for the viscosities gives

$$\eta = \frac{2\pi r^2}{9}(y - \rho)\frac{g}{v} \qquad (2\text{-}2)$$

where y and ρ are the densities of the sphere and the fluid, respectively. In practice, the assumptions made in the derivation of this equation are almost realized. When the velocity of a sphere of diameter d is low enough for Stokes's law to apply, the proximity of the walls of a cylindrical container of diameter D may be accounted for by the equation

$$\eta = \eta_s \times F \tag{2-3}$$

in which η_s = the viscosity calculated by Stokes's equation (2-2), and F is a function of d/D. It has been shown experimentally that (2-3) holds over a wide range of viscosities and for d/D values up to 0.32. Since the viscosities measured in our experiments on aqueous-organic mixtures at cryogenic temperatures are high (>10 cP), the present method can be used without any risk of major error.

Another factor that makes the precise experimental use of this method difficult is an unequal temperature distribution in the cylindrical container, causing viscosity variations and convection currents, which make a uniform axial fall hard to achieve. This difficulty has been overcome by strict thermostating and temperature control. The cell used to measure viscosity at low temperatures is shown schematically in Fig. 10.

The viscosimeter was used only for relative measurements, and in each case numerous repetitive measurements under identical conditions were carried out; these showed that the method was both sensitive and relatively precise (the precision was approximately $\pm 8\%$).

Results for various binary and ternary mixtures are shown in Table 6. It can be seen that these mixtures can be classified into two groups:

1. Mixtures with a rather low alcohol content, whose viscosities never exceed 100 cP.
2. Mixtures with a higher alcohol content ($>60\%$), which reach 300 to 500 cP at low temperatures.

The viscosities of the ethylene glycol–water (in the volume ratio, 50:50, most often used) and glycerol–water mixtures are several times those of water at room temperature and attain levels of thousands of centipoises at subzero temperatures. It was already known that the viscosity of the glycerol–water mixture (54:46, v/v) is 6.3 cP at +20°C and is increased to 80 cP at −20°C. Slightly lower values are reported for the ethylene glycol–water mixture (50:50). In both mixtures, however, many rate constants of enzyme catalyzed reactions might become diffusion controlled, and such high viscosities might therefore be responsible for changes in reaction kinetics.

Moreover, it has been checked in this laboratory (13) that the rapid mixing of reactants is impossible when the viscosity exceeds 50 cP, preventing fast

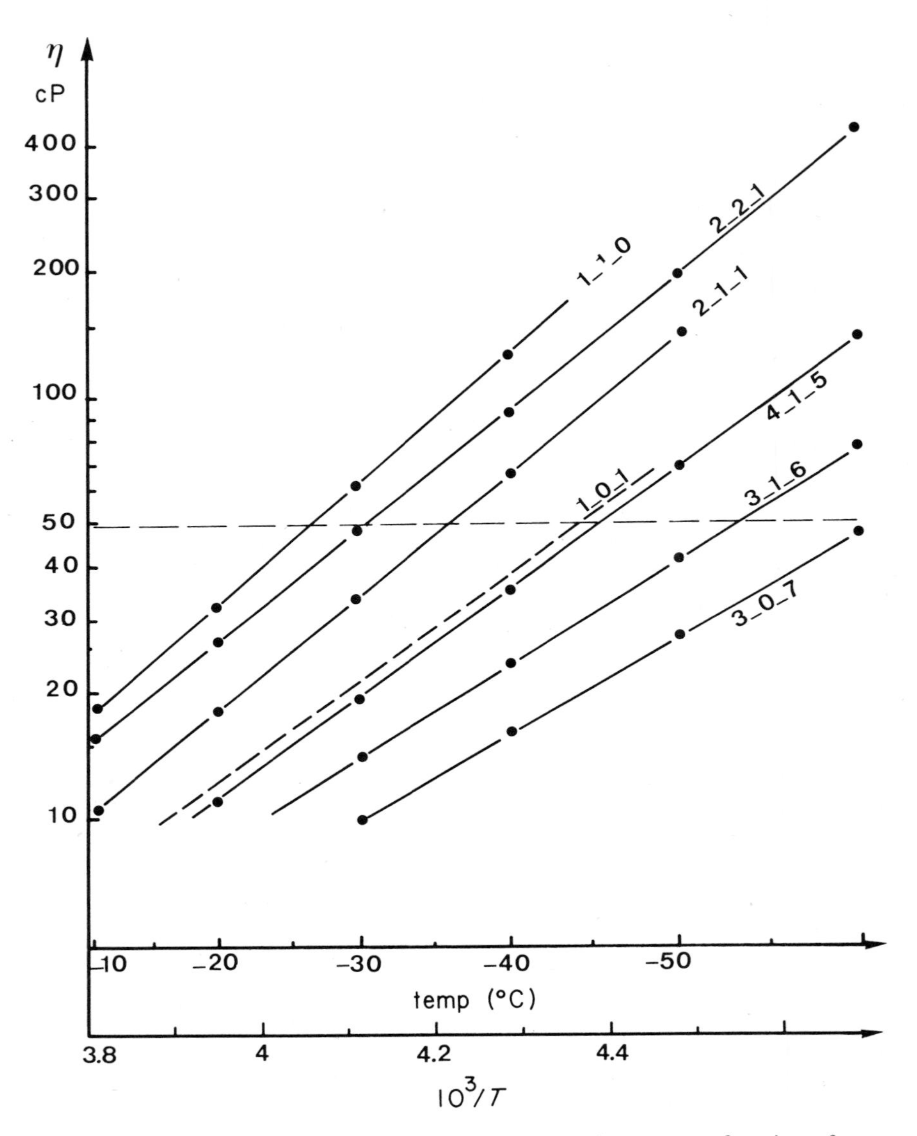

FIG. 10. Viscosity of water–ethylene glycol–methanol mixtures as a function of temperature. Numbers indicate the volume ratio of each constituent.

TABLE 5. Viscosity of ethylene glycol–methanol–water mixtures as a function of temperature

Volume ratio (%)	Temperature (°C)					
	−10	−20	−30	−40	−50	−60
20 EGOH–80 H_2O	11	—	—	—	—	—
30 EGOH–70 H_2O	13	—	—	—	—	—
50 EGOH–50 H_2O	18	32	63	125	—	—
50 MetOH–50 H_2O	—	12	21	38	—	—
70 MetOH–30 H_2O	—	—	10	17	28	48
40 EGOH–20 MetOH–40 H_2O	15	26	48	91	195	420
25 EGOH–25 MetOH–50 H_2O	10	19	34	66	145	—
10 EGOH–50 MetOH–40 H_2O	—	11	19	35	69	140
10 EGOH–60 MetOH–30 H_2O	—	—	14	23	42	76

kinetic measurements in polyol–water mixtures below 0°C. These observations prompted us to try to prepare ternary mixtures involving both ethylene glycol and methanol or dioxane in order to decrease the viscosity at low temperatures as shown in Table 5.

Dielectric constant

GENERAL

The addition of organic solvents to water decreases the dielectric constant (D) and since such a parameter is a fundamental expression of electrostatics which give a qualitative explanation of chemical laws (solubilities, acid-base equilibria, etc.) acting on reaction rates and mechanisms, it is evident that the knowledge of numerical values of D for mixed solvents and as a function of $1/T$ will be essential to understand the properties of cooled media.

Let us recall that the dielectric constant (this term is a poor alternative as the factor is a constant in only very limited way and should be called electrical permittivity) is the electrostatic expression of the interaction of atoms and molecules with macroscopic electric fields produced by classical apparatus, rather than with the exceedingly strong fields of individual atoms and molecules. The interaction between the homogeneous outside field and electrically asymmetrical (polar) molecules will result in a finite interaction since, in these molecules, the effects of positive and negative charges will not cancel.

Water, hydroxy compounds and a number of aqueous–organic mixtures are highly dipolar and in most cases, the effect of the individual dipoles is greatly intensified through their connections by hydrogen bonds to form more extended oriented structures. We know that, in fact, the unique structure of water is expressed by higher value of the dielectric constant and that mixing with any miscible polar solvent will perturbate its oriented structure and lower this value.

The behavior of any molecule under the influence of an outside electrical field is represented by rigidly attaching to the molecule an imaginary vector called its dipole moment; the length of this vector is given by the maximum interaction the molecule can have with a unit electric field. The orientation of the vector has to be fixed in the molecule in such a way that the maximum interaction with the field is obtained when the vector is parallel to the field. If the molecule is rotated through an angle θ around any axis perpendicular to the electric field, then the interaction will be multiplied by $\cos \theta$. For $\theta = 180^\circ$, the interaction will have changed sign.

Let us recall the physical significance of the dipole moment: it is in fact a measure of the separation of the center of the positive charges from the center of the negative charges in the molecule. It can be represented by locating an appropriate positive charge q and the corresponding negative charge $-q$ at the distance r from each other, and it is easy to verify that such a charge distribution will have just the required interaction with the outside homogeneous electric field. It represents a dipole of the magnitude $d = qr$; dE is equal to the maximum interaction energy between the dipole d and on electric field E. The direction of the dipole will be given by the line pointing from the negative towards the positive charge, and in general, the interaction energy between the electric field and the dipole is $-dE \cos \theta$, where θ is the angle between the directions of the dipole and the electric field, the minus sign being due to the fact that the energy of the dipole is at a minimum when it points in the direction of the electric field. Since the magnitude of the dipole moment determines only qr, and not q and r separately, it is customary to define a pure dipole as a special case where q tends to ∞ and r tends to zero, so that their product remain constant. It is permissible to represent the charge distribution in a molecule by the dipole model only in the context of an interaction with an external homogeneous field that is in conditions where the dielectric constant measurements are performed.

The fact that electric fields cause a net dipole within media gives rise to their properties as dielectrics, characterized by the dielectric constant D which, experimentally, is a measure of the relative effect of medium has on the force with which two oppositely-charged plates attract each other, and is expressed as a unitless number.

The dielectric constant D is correlated to dipole moment μ by the Debye expression:

$$\frac{D-1}{D+2} = \tfrac{4}{3}\pi\left(\alpha_0 + \frac{\mu^2}{3kT}\right)n \qquad (2\text{-}4)$$

with α_0, electronic polarizability; n, number of molecules per unit volume. The quantity

$$P = \tfrac{4}{3}\pi N\left(\alpha_0 + \frac{\mu^2}{3kT}\right) = \frac{D-1}{D+2}\frac{N}{n} \qquad (2\text{-}5)$$

where N is Avogadro's number, is the molar polarizability which has the dimension of volume.

The molar dielectric polarizability of a mixture of two solvents is expressed P_{12} and is given by the equation

$$P_{12} = \frac{D-1}{D+2}\frac{f_1 M_1 + f_2 M_2}{d} \qquad (2\text{-}6)$$

where f_1 and f_2 denote mole fractions, M_1 and M_2 molar weights and d density of the mixture. This equation is supposed to be used only for media consisting of non-associated compounds in non-polar solvents, and not for molecules such as water and the orginic solvent used as "antifreeze", which are strongly associated owing to hydrogen bonding and which exhibit other abnormalities of behavior.

However, it has been shown by several authors (14,15,16) that there is a linear variation of the polarizability (often termed, less appropriately, polarization) with the mole fraction of one of the polar solvents present. Polarization data for aqueous–organic mixtures used in our low temperature method are now known: the points all lie in a straight line. The calculation of the dielectric constant of such mixtures from the known polarization of the pure solvents would, assuming linear variation with mole fraction, give values of low accuracy. On the contrary, measurement of the dielectric constant is not difficult if one selects a method of measurement whose results would be only marginally influenced by the conductance of mixed solvents.

DIELECTRIC CONSTANT AND TEMPERATURE

Since the solvents are permanent dipoles which develop an orientation under the influence of an extremal electric field in opposition to the disordering influence of thermal agitation, such an orientation process is governed by the Boltzman distribution law and results in the dielectric constant being strongly dependent on the absolute temperature. Thus, as the systems become cooler, the random motion of their molecules decreases and the electric field becomes more effective in orienting them; the dielectric constant should then increase markedly with temperature drop, at constant volume.

Molar polarizability can be evaluated from D measurements by expression (2-5) and a critical test of the Debye theory (17) is provided by a plot of P against $1/T$ which should be a straight line whose slope is given by:

$$\frac{dP}{d\left(\frac{1}{T}\right)} = \frac{4\pi N\mu^2}{9k} \tag{2-6}$$

and the intercept with the ordinate axis by

$$\frac{4\pi}{3} N\alpha_0 = P_0 \tag{2-7}$$

A vast amount of data based on dielectric constant measurements of substances in dilute solution in nonpolar solvents indicate that the theory is correct, but it loses validity and breaks down completely in the case of strongly polar media. It is interesting to notice that such a breakdown can be shown by the concept of the Curie point, introduced in connection with studies of magnetism but directly applicable to the case of dielectric constants. Since the dielectric constant can be related to the molar polarizability P and c the molar concentration by expression

$$D = \frac{2Pc + 1}{1 - Pc} \tag{2-8}$$

and the value of P itself by

$$P = \frac{4\pi N}{3}\left(\alpha_0 + \frac{\mu^2}{3kT}\right) \tag{2-9}$$

it is obvious that P will increase as T diminishes, mainly because of the presence of $1/T$ as a multiplier of μ^2, and at a certain critical temperature, P will become equal to unity as D goes to infinity. This critical temperature, if it exists, is known as the Curie point.

The sudden increase of D towards infinity at a certain critical value of $\mu^2/3kT$ has been referred as the "$4\pi I/3$ catastrophe", arising from the presence of such a term in the expression for the electric field strength ($F = E + 4\pi I/3$).

In fact, no Curie point can be observed with polar liquids as a function of $1/T$ and the Debye theory is therefore not valid for our strongly polar mixtures. A much more refined theory than that of Debye, Kirkwood (18), suffices to account for the general character of the dielectric constant of strongly polar liquids and their evolution as a function of temperature in their cooled state. It is not the purpose of this book to give a full account of this theory.

DIELECTRIC CONSTANT VALUES OF MIXED SOLVENTS

Dielectric constant as a function of solvent concentrations

Measurements carried out at a selected temperature (20°C) as a function of the concentration of the organic solvent from 10 to 100% are shown in Fig. 11.

It can be seen that the dielectric constant decreases markedly as the concentration in organic solvent is increased, and that this effect is very pronounced in the case of methanol and methyl pentanediol, less so for DMSO and polyols.

Adding 50% of most of the selected organic solvents decreases the dielectric constant of water from 10 to 30 units according to the solvent, except in the case of dimethylsulfoxide where the dielectric constant 76 (instead of 80 for water at the same temperature).

Dielectric constant at subzero temperatures

In 1932, Akerlof (16) published experimental values for dielectric constants of some aqueous–organic mixtures as a function of temperature between 0 and 80°C. Curves for the logarithm of D of pure solvents (water, methyl-alcohol, ethylene glycol, glycerol) were plotted against T and gave straight lines obeying without any doubt the following mathematical expression:

$$\log D = a - bT \tag{2-10}$$

where a and b are empirical constants and T is absolute temperature. This relation has a great value for interpolation purposes and therefore is well worth testing over as large a temperature range as possible. It was tested using a number of previous measurements for ethylbromide and chloro-benzene from -52 to $+126$°C, and for dimethylpentane from -120 to $+80$°C and was found valid within experimental errors over a temperature range of at least 150°C. Every straight line obtained in these conditions seems to indicate that the above equation is valid.

In this laboratory, Travers (19,20) carried out measurements on various aqueous–organic mixtures, as a function of the volume-ratio at constant temperature and then for selected volume-ratios as a function of temperature, between room temperature and the freezing point of mixtures. Results are shown in Tables 6–13 and on Fig. 12.

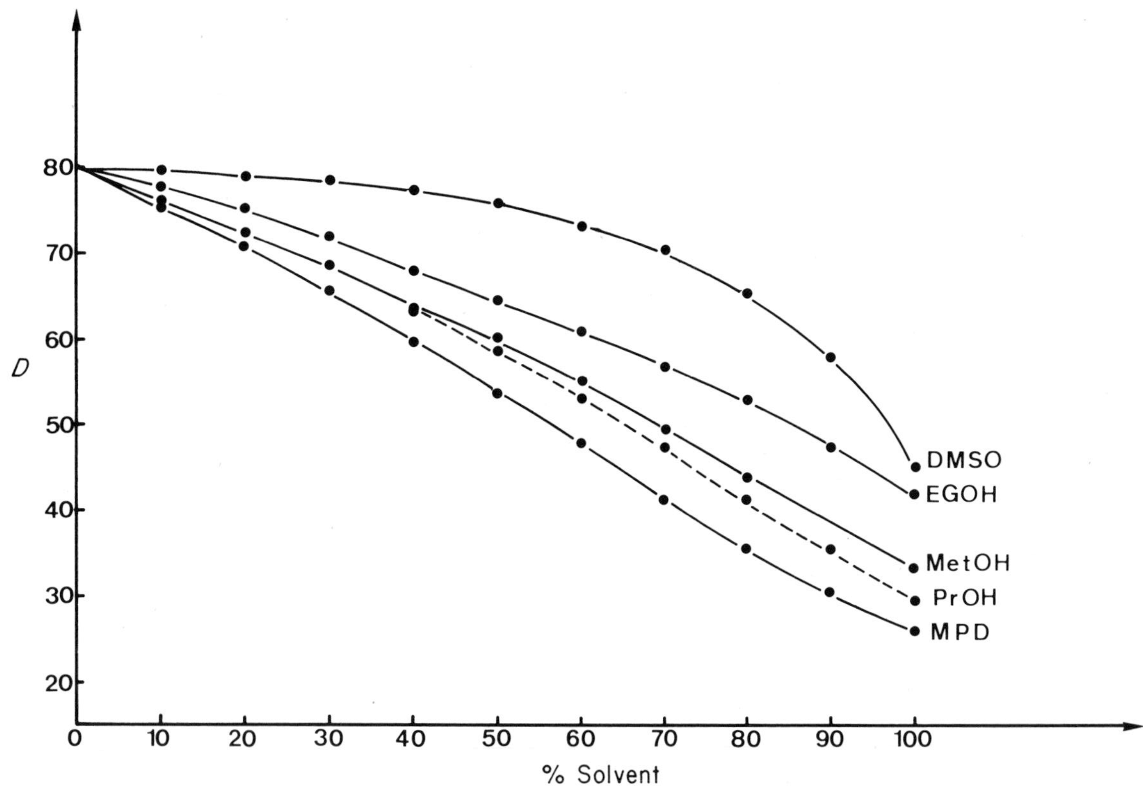

FIG. 11. Dielectric constant (*D*) of hydro-organic solvents as a function of percentage in volume. Temperature, 20°C.

TABLE 6. Dielectric constant of ethylene glycol–water mixtures

Solvent (%)	Temperature (°C)													a	b
	+20	+10	0	−10	−20	−30	−40	−50	−60	−70	−80	−90	−100	(10^3)	
0	80.4	84.2	88.1											1.945	2.00
10	77.7	81.4	85.3											1.931	2.02
20	75.1	78.4	82.5	86.9										1.916	2.15
30	72.0	75.7	79.5	84.0										1.912	2.20
40	68.1	72.1	76.3	80.2	84.4									1.880	2.30
50	64.5	68.4	72.4	76.5	80.7	85.0	89.3							1.860	2.35
60	61.1	64.6	67.9	72.0	76.3	80.8	85.3	90.1	95.7					1.832	2.41
70	56.9	60.0	63.4	67.5	71.3	75.3	79.8	84.5	89.5	94.6	100.0	106.1	112.1	1.803	2.47
80	53.0	55.6	58.8	62.3	66.2	70.0	74.2	78.5	83.2	88.1	93.0			1.770	2.50
90	47.5	50.5	53.5	56.8	60.2	63.8	67.8	72.0						1.728	2.52
100	41.9	44.7	47.6	50.3										1.675	2.54

TABLE 7. Dielectric constant of methanol–water mixtures

Solvent (%)	Temperature (°C)													a	b
	+20	+10	0	−10	−20	−30	−40	−50	−60	−70	−80	−90	−100	(10^3)	
0	80.4	84.2	88.1											1.945	2.00
40	63.8	67.7	71.9	75.6	79.5	83.5	87.9							1.855	2.20
50	60.3	64.0	67.8	71.2	75.5	79.2	83.9							1.830	2.30
60	55.1	58.7	62.5	66.0	70.5	73.8	78.2	82.2	86.7					1.790	2.50
70	46.3	49.4	53.0	56.6	60.1	63.5	66.8	70.9	74.9	79.2	83.5			1.730	2.45
80	43.7	46.4	49.5	52.3	55.4	58.6	61.9	65.7	69.3	73.5	77.8	82.8	88.4	1.695	2.50
100	33.6	35.4	37.9	40.6	42.7	45.4	48.3	51.3	54.6	58.0	62.0	66.5		1.580	2.65

TABLE 8. Dielectric constant of 2-methyl-2-4-pentanediol–water mixtures

Solvent (%)	Temperature (°C)													a	b
	+20	+10	0	−10	−20	−30	−40	−50	−60	−70	−80	−90	−100	(10^3)	
0	80.4	84.2	88.1											1.945	2.00
10	75.4	79.2	83.2											1.920	2.13
20	70.8	74.7	78.8											1.895	2.33
30	65.7	69.7	73.5	78.0										1.866	2.48
40	59.6	63.7	67.3	71.3	75.0									1.829	2.50
50	53.7	57.0	60.5	64.5	68.3	72.7	77.1							1.783	2.60
60	47.7	50.6	53.5	56.9	60.5	64.1	68.1	72.0	76.5	81.2	86.2	91.7	97.5	1.728	2.59
70	41.3	43.8	46.8	49.4	52.6	55.7	59.4	62.6	66.8	70.6	75.6	79.6	84.7	1.668	2.60
80	35.6	37.8	40.3	42.9	45.3	48.4	51.6	54.9	58.9	62.1	66.5	70.9	75.1	1.605	2.71
90	30.6	32.6	34.7	36.9	39.1	41.9	44.6	47.3	50.7	54.3	57.4	60.9	64.7	1.540	2.72
100	26.3	28.0	29.9	31.9	31.0	36.2	38.9	41.0	44.4	46.9	50.3	53.5	57.4	1.476	2.80

TABLE 9. Dielectric constant of 1-2-propanediol–water mixtures

Solvent (%)	Temperature (°C)													a (10³)	b
	+20	+10	0	−10	−20	−30	−40	−50	−60	−70	−80	−90	−100		
0	80.4	84.2	88.1											1.945	2.00
10	76.2	79.8	83.6											1.922	2.02
20	72.0	76.0	79.8											1.902	2.20
30	68.0	71.7	76.6	80.0										1.879	2.38
40	63.6	67.4	71.3	75.2	79.4	83.7								1.852	2.37
50	58.3	61.6	65.4	69.3	73.3	77.5	81.8	86.7	92.0	97.6	103.5	108.1	114.8	1.816	2.47
60	53.2	56.4	59.7	63.3	67.1	70.8	74.7	79.3	83.9	89.0	94.2	99.6	105.0	1.776	2.47
70	47.6	50.4	53.5	56.6	60.1	63.7	67.6	71.6	76.1	80.6	85.7	90.6	96.2	1.728	2.55
80	41.6	44.2	46.9	49.8	52.8	56.2	59.5	63.1	66.8	71.1	75.3	79.6	84.8	1.671	2.58
90	36.2	38.5	40.9	43.5	46.2	49.0	52.3	55.6	59.1	62.8	66.8	70.6	75.5	1.612	2.66
100	29.4	31.3	33.4	35.8	38.0	40.5	43.3	46.0	49.1	52.3	55.9	59.4	63.3	1.524	2.78

TABLE 10. Dielectric constant of glycerol–water mixtures

Solvent (%)	Temperature (°C)													a	b
	+20	+10	0	−10	−20	−30	−40	−50	−60	−70	−80	−90	−100	(10^3)	
0	80.4	84.2	88.1											1.945	2.00
10	77.0	80.8	84.6											1.928	2.09
20	74.3	77.9	81.6											1.913	2.10
30	71.3	75.0	78.9	83.0										1.897	2.15
40	68.6	72.0	75.7	80.0	84.5									1.881	2.23
50	65.0	68.8	72.6	76.6	80.7	85.1	90.2	95.3	99.8	106.0	111.5	117.9	123.0	1.861	2.36
60	61.5	65.1	68.7	72.8	77.0	81.3	86.1	91.0	96.2	102.5	108.5	114.0	121.1	1.837	2.43
70	57.6	60.9	64.2	67.9	71.8	75.9	79.7	84.7	89.6	94.7	100.1	106.0	112.1	1.808	2.40
80	53.5	56.4	59.6	63.0	66.5	70.1	73.9	78.2	82.3	87.2	91.8	97.2	103.0	1.775	2.35
90	48.6	51.3	53.9	57.1	59.8	63.3	66.8	70.2	74.4	78.2	82.4	86.7	91.7	1.732	2.30
100	43.4	45.7	48.2	50.5										1.683	2.24

TABLE 11. Dielectric constant of dimethylsulfoxide–water mixtures

Solvent (%)	Temperature (°C)													a	b
	+20	+10	0	−10	−20	−30	−40	−50	−60	−70	−80	−90	−100	(10^3)	
0	80.4	84.2	88.1											1.945	2.00
10	79.4	82.9	87.0											1.939	2.04
20	78.8	82.7	86.5	90.7										1.937	2.03
30	78.6	82.3	86.0	90.3	94.7									1.934	2.02
40	77.2	80.8	85.1	89.1	93.0	97.3	101.2							1.928	2.01
50	76.0	79.6	83.9	87.7	91.9	96.3	100.5	105.0	110.7	115.1	119.8	125.8	132.4	1.922	2.02
60	73.6	77.4	81.0	83.9	88.7	92.6	96.4	101.0	105.3	110.9	115.1	122.0	127.9	1.908	1.98
70	70.4	73.4	76.8	80.2	83.9	87.4	91.2	95.2	98.9	103.9	109.0	113.7	119.0	1.885	1.88
80	65.4	68.0	71.0	74.1	77.1	80.0								1.851	1.75
90	58.1	60.3	62.4											1.795	1.55
100	45.0													1.703	1.48

TABLE 12. Dielectric constant of *N*-*N*-dimethylformamide–water mixtures

Solvent (%)	Temperature (°C)													a	b
	+20	+10	0	−10	−20	−30	−40	−50	−60	−70	−80	−90	−100	(10^3)	
0	80.4	84.2	88.1											1.945	2.00
10	78.2	82.3	86.7											1.938	2.20
20	76.2	79.8	83.9											1.924	2.07
30	73.9	76.9	81.1	84.6										1.908	1.96
40	70.4	73.9	77.1	80.8	84.3									1.887	1.95
50	66.5	69.9	72.9	76.2	79.4	83.1	86.8							1.863	1.92
60	62.4	65.2	68.1	71.3	74.0	77.2	80.2	84.0	87.7					1.832	1.85
70	57.7	60.1	62.6	65.3	68.0	70.8	73.8	76.8	79.8	83.4	76.9			1.797	1.78
80	51.7	54.0	56.4	58.9	61.4	64.0	66.7	69.7	72.7	76.1	79.3	82.6	86.2	1.751	1.85
90	45.1	47.3	49.6	51.8	54.3	57.9	59.5	62.5	65.2	68.7	72.0	75.6		1.696	2.00
100	38.2	40.2	42.2	44.6	46.9	49.5	51.9	55.0	57.8					1.625	2.25

TABLE 13. Dielectric constant of ethylene glycol–methanol–water mixtures

Solvent (%)	Temperature (°C)										
EGOH–MetOH–H_2O	+20	+10	0	−10	−20	−30	−40	−50	−60	*a* (10^3)	*b*
40–20–40	59.6	63.3	67.0	70.8	74.8	79.1	83.6	88.3	93.8	1.825	2.45
25–25–50	62.1	65.8	69.5	73.6	77.7	82.2	87.1	91.9	97.5	1.842	2.45
10–50–40	57.3	60.5	64.3	68.1	72.1	76.2	80.9	85.5	90.8	1.808	2.50
10–60–30	52.2	55.4	58.7	62.4	66.1	70.2	74.7	79.3	84.0	1.769	2.60

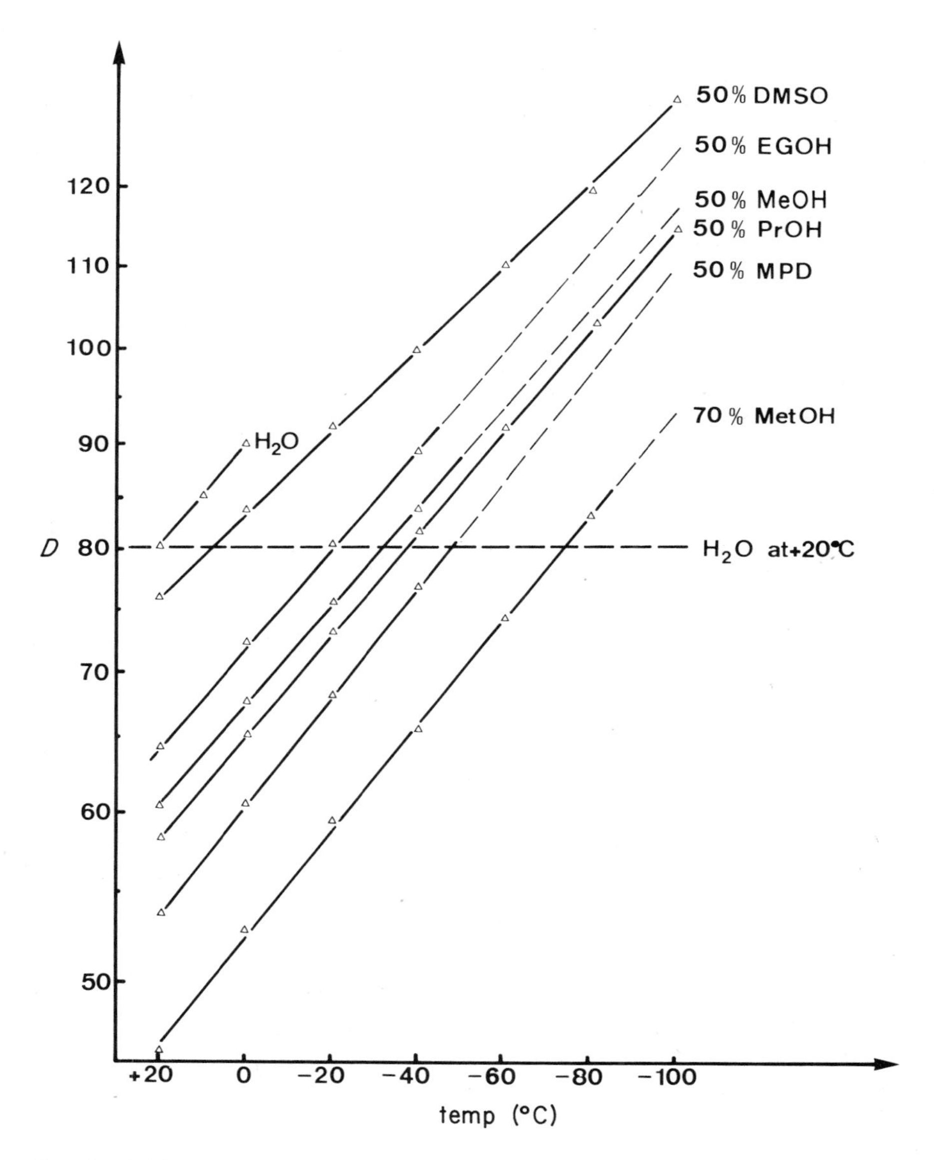

FIG. 12. Dielectric constant (logarithmic scale) of some hydro–organic mixtures as a function of temperature.

AQUEOUS–ORGANIC SOLUTIONS ISODIELECTRIC WITH WATER

It might be useful in some cases to raise the dielectric constant of mixed solvents by addition of suitable substances and it is known that dipolar molecules such as amino acids do this in pure water. These amino acids are virtually insoluble in non-polar solvents, they dissolve readily in salt aqueous solutions and in most mixed solvents, as expected from their highly polar structure.

Most of what is known about their dielectric behavior relates to aqueous solutions where they were studied up to concentrations not far from the limit of their solubility.

If we consider the case of glycine, it is known that the dielectric constant D of water increases rapidly and linearly with the concentration of the amino acid, reaching a value of about 135 at a concentration of 2.5 M at 25°C.

D is given by $D = 78.54 + 22.58\,C$ where 78.54 is the measured value of D for pure water, 22.58 the numerical value of the "dielectric increment" and C is the molar concentration.

This great increase of D with concentration reflects the extremely large moment of glycine as a dipolar ion, and the linearity of the relationship represents the proportionality between D and polarizability characteristic of strongly polar media.

Essentially similar behavior is shown by other α-amino acids, since the numerical value of the slope of the curves dD/dC is very nearly the same for all. It is because all these α-amino acids:

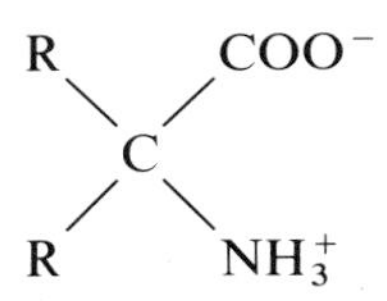

have virtually the same dipole moment, and then the quantity dD/dC is a direct expression of the molecular polarizability of the amino acids, and known as "molar dielectric increment" δ expressed in terms of concentrations in moles per liter $[C]$.

Thus for a given amino acid, the effect of D at a selected concentration is essentially independent of the nature of the solvent and its numerical value of D.

We have said that the quantity dD/dC is a direct reflection of the polarizability of molecules and the polarizability can be expressed by an expression derived from the Kirkwood equation:

$$P_2 = \tfrac{2}{9}(1000\,\delta + D_1 V_2) \qquad (2\text{-}11)$$

where subscript 1 is associated with the solvent and subscript 2 with the amino acid.

For dipolar ions, the predominant term on the right of the above equation is 1000 δ. In the case of glycine in water when $D_1 V_2$ is about 3500, 1000 δ will be about 22 580.

δ varies from the value 22.58 in pure water to 20.4 in the presence of 60% ethanol at 25°C, in the concentration range of glycine from 0 to 0.133 M, and it was confirmed in this laboratory that similar results are obtained for the other mixed solvents considered for cryobiochemical investigations as indicated in Table 14.

TABLE 14. Dielectric increment δD in M^{-1} l for glycine in aqueous–organic mixtures as a function of temperature. *Do* is the dielectric constant of the solvent without glycine: $D = Do + C \cdot \delta D$ where C is the glycine concentration (M)

	+20°C		0°C		−20°C		−40°C	
Solvent (%)	*Do*	δD	*Do*	δD	*Do*	$\delta D\phi$	*Do*	δD
Water 100	80.4	22.5	88.1	23.0	—	—	—	—
Ethylene glycol 50	64.5	22.6	72.4	23.8	80.7	25.4	89.3	27.4
1-2 Propanediol 50	58.3	23.4	65.4	25.2	73.3	26.4	81.8	28.2
Glycerol 50	65	23.0	72.6	24.6	80.7	26.6	90.2	28.0
2-Methyl-2-4-pentanediol 50	53.7	17.0	60.5	18.8	68.3	20.0	77.0	—
Methanol 50	60.3	21.4	67.8	22.4	75.5	24.0	87.9	25.5
Dimethylsulfoxide 50	76	21.0	83.9	22.4	91.9	24.1	100.5	25.6

Equation (2-11) presents the behavior of the dielectric constant of the dipolar ions in polar solutions understandably. It shows that the linear increase of *D* with concentration for changes in the partial molar volumes, only slightly dependent on concentration, can only affect the term $D_1 V_2$; the nearly identical values of *D* for amino acids of the same moment, and the fact that *D* for a given amino acid is insensitive to changes in the dielectric constant of the solvent, since the change of solvent can directly affect δ only through the term $D_1 V_2$.

The addition of an amino acid to mixed solvents at selected temperatures can be a means of compensating even partially for the decrease in dielectric constant due to the solvent addition. Limitations are thus imposed by the solubility of the amino acid in such mixtures; for instance there is a "salting-out" effect in methanol–water 50:50 at 25°C when the concentration in glycine is about 0.5 M ($\delta \simeq 20$).

PREPARATION OF COOLED SOLUTIONS

It can be seen that for solvent concentrations up to 80%, the dielectric constant of any mixture reaches the value 80 (i.e. that of pure water at 20°C) well before freezing point, and that the lower the concentration of organic solvents the higher temperature at which each mixture reaches $D = 80$.

For instance, in the case of ethylene glycol, for the concentrations 10, 20, 30, 40, 50, 60, 70 and 80%, D reaches 80 at 5, 0, −10, −20, −30, −40 and −50°C.

Accordingly, one only needs to coordinate the addition of organic solvent with the lowering of temperature to keep the dielectric constant of mixed cooled solutions at the value 80, i.e. at the original value of the aqueous solution.

There is an easy sampling procedure to obtain cooled solutions of nucleic acids and proteins, starting from aqueous media at 4°C. It will be described in chapter 4. Moreover, some problems of "salting-out", which are encountered on addition of organic solvent at 4°C, can be avoided with the coordinated procedure, media of high dielectric constant retaining their property of ionization. Finally, conformational changes and denaturation due to dielectric constant variations might be avoided by such a procedure. In his early experiments, Freed (see ref. 6 in chapter 1) injected by means of a micropipette a solution of an enzyme in water cooled to 0°C, as a fine spray, into a previously cooled organic–water mixture. The spray froze at once as floating particles of ice in which the enzyme was dissolved, and then the solution became clear within times varying from seconds to hours, depending on the temperature and the concentration of organic solvent. Such a procedure is hazardous and cannot be applied to a large number of enzyme systems or to the study of reaction kinetics. The synchronizing procedure described in chapter 4 appears to be definitely superior.

DIELECTRIC CONSTANT AND THE OTHER PHYSICAL–CHEMICAL PROPERTIES

For atomic dimensions, the electric field originating in a neighboring molecule is strongly inhomogeneous. In order to elucidate the interaction of the homogeneous macroscopic electric field with such molecular fields, the knowledge of the dipole moment of the molecules is not sufficient, but a rather more detailed description of the charge distribution in the molecules is needed. Unfortunately, the dipole moment is the only well-defined quantity which can be determined by direct experiment, and no more exact information about the detailed charge distributions are available. Thus the models depicting ions as charged spheres are far too simple to account for the real phenomena of electrostatics.

The reason why chemical laws are not easily transposed to electrostatics is that electrons behave under the influence of their own or applied electric fields not according to classical mechanics but according to quantum mechanics, and furthermore they obey the Pauli exclusion principle. In fact, electrostatics and dielectric constants are simpler applications of the electrical structure of molecules, and use outside macroscopic homogeneous electric fields interacting with microscopic inhomogeneous fields.

Dielectric constants cannot explain quantitatively most of the physicochemical properties and laws and we will soon see that they can become irrelevant: the molecules of the more polar solvent which tend to cluster around the ions and dipole ions, giving a preferential or "selective" solvation at any selected temperature should be responsible for such an "erasing" effect, which will be observed on examining properties such as solubility, acid-base equilibria and reaction rates.

Also, non-electrostatic effects probably interfere with electrostatic effects to reduce their actual influence: these comprise the basicity of some solvents, their hydrogen bonding capacity, the internal cohesion and the viscosity of mixtures, etc. We have already stated that mixtures of water and non-aqueous solvents are enormously complex systems; their effective microscopic properties can be vastly different from their macroscopic properties and vary with the solute used, because of the selective attraction of one of the solvents by the solute. Consequently accessible, macroscopic parameters such as solvent molecular volume, viscosity, dielectric constant, etc. fail to explain the behavior quantitatively.

For these various reasons, the dielectric constant is not that "critical" value often invoked by biochemists, mainly when they have to explain the effects of organic solvents on protein structure and in enzyme–catalyzed reactions.

We will now examine chemical properties: acid-base equilibria, solubility, effects on reaction rates and mechanisms, and see that it is often difficult to appreciate their underlying physical-chemical principles in terms of electrostatics.

Acid-base equilibria and buffered solutions

THE pK OF ACIDS AND BASES

Definitions

It is known that acid-base equilibria are rather sensitive to the addition of organic solvents and to the corresponding changes in the dielectric constant and in the basicity of the medium. On the other hand, it is admitted that the interpretation of measured pH values are limited to pure water solutions if the pH is to retain its full significance in terms of the hydrogen ion concentration; thus it is necessary to look into the concepts and meaning of acidity and basicity as well as into the role of a solvent in acid-base phenomena as a basis for possible pH scales for mixed solvents, at normal and subzero temperatures.

The solvent plays an essential role in acid-base equilibria because most solvents are themselves acidic (proton donors; protogenic), or basic (proton acceptors; protophilic). Water and the alcohols are called amphiprotic because they are capable of accepting and donating protons. These solvents can be subdivided into acidic and basic amphiprotic solvents.

Pure water is perfectly amphiprotic: exchanges of protons between water molecules proceed constantly and this process may be visualized as the jumping of a proton from the oxygen to which it is attached by a covalent bond to an adjoining oxygen which is attached by a hydrogen bond.

The water molecule acts as both acid and base:

$$H_2O + H_2O \rightleftharpoons H_3O^+ + OH^- \qquad (2\text{-}12)$$

This reaction involves a large increase in electrostatic energy since an anion and a cation are produced from two neutral molecules. Consequently, the H_3O^+ ion has a strong tendency to lose a H^+ whereas the OH^- ion has a strong tendency to accept a proton from one of its neighbors.

At equilibrium, the concentration of H_2O^+ and OH^- are very small. In pure water, they are necessarily equal:

$$[H_3O^+] = [OH^-] = 1.00 \times 10^{-7} \text{ at } 25^\circ\text{C}.$$

The equilibrium constant of reaction (2-12) may be written;

$$\frac{[H_3O^+][OH^-]}{[H_2O]^2} = K$$

In fact, the ionization constant of water, K_w, is written:

$$[H_3O^+][OH^-] = K[H_2O]^2 = K_w = 10^{-14} \text{ at } 25^\circ\text{C}$$

It is customary to denote the negative log of an equilibrium constant K by the symbol pK. Thus pK_w = 14.00 at 25°C.

The value of pK_w decreases—that is K_w increases—with rising temperature i.e. pK = 14.94 at 0°C and 13.59 at 37°C, thus it can be predicted that the pK of polar solvents will increase with lowering temperature.

We said already that the weakly protic organic solvents nearly resemble water in structure but they are not perfectly amphiprotic. Methanol, glycols and glycerol (hydroxy compounds) are sufficiently more basic in character to behave as hydrogen acceptors in aqueous solutions. Esters such as dimethylsulfoxide and amides such as dimethylformamide are strong bases, because of the electron-donating properties of the oxygen and nitrogen atoms, with and would affect the acid-base equilibria of these solvents when added to water.

If A and B denote an acid and its conjugate base:

$$A + H_2O \rightleftharpoons B + H_3O^+$$

In fact this expression is more often expressed as

$$A \rightleftharpoons B + H^+$$

In an aqueous–organic solvent the equation is

$$K^* = \frac{a_H \times a_B}{a_A} \tag{2-13}$$

a_i is the activity of the "i" species which is expressed as:

$$a_i = \gamma_i C_i$$

where γ_i is the activity coefficient and C_i concentration. From (2-13) we get

$$-\log a_H = -\log \gamma_H H = -\log K^* + \log \frac{[B]}{[A]} + \log \frac{\gamma_B}{\gamma_A} \tag{2-14}$$

$-\log a_H$ is noted as pa_H (or pH*) which is equal to pH when the medium is pure water and $-\log K^*$ is pK^*.

$$pa_H = pK^* + \log \frac{[B]}{[A]} + \log \frac{\gamma_B}{\gamma_A} \tag{2-15}$$

In the case of an ionic species, γ_i in a given solvent can be calculated by the Debye–Hückel equation, which takes into account the electrostatic interactions:

$$-\log \gamma_i = \frac{(1.82455 \times 10^6) Z_i^2 \sqrt{I\rho}}{(DT)^{3/2}(1 + 50.2904(DT)^{-1/2} a\sqrt{I\rho})} \tag{2-16}$$

where Z_i is the charge number of the i ion; I, the ionic strength; a, the ionic radius of the species i; ρ, the density of the solvent and D, the dielectric constant of the solvent.

Solvent effect

Variations in dielectric constant will alter the relative strength of acids of different charge types since the amount of electrical energy involved in a proton transfer reaction must vary with the dielectric constant of the medium. Since a lowering in dielectric constant value increases the energy required to separate the ions, for instance to ionize an uncharged acid creating an anion and a cation, any addition of organic solvent should determine an increase of pK_A values.

Calculations have been carried out by Edsall and Wyman (21) for dioxane–water solutions (70:30, v/v) the dielectric constant of which is 18. The increase in pK_A^* between this medium and water should correspond to 5.3 units at 25°C. The observed pK_A values for acetic acid are 4.756 in water and 8.321 in the above medium, i.e. a difference in pK_A^* of 3.565.

Thus this value is of the same order of magnitude as that calculated from the electrical effects using a very crude model (the anion and the cation resulting from the ionization of acetic acid are treated as two charged spheres). The difference between the values 5.3 and 3.565 could be readily adjusted by a moderate increase in the assumed values of the radii of the ions. However when the pK_A^* values in different mixed solvents such as dioxane–water and methanol–water are plotted against $1/D$, the resulting graph is not linear as the theory would predict, but concave to the $1/D$ axis.

Thus the model of charged spheres is too crude, and non-electrostatic effects due to the solvent itself are presumably also involved. It must be reiterated that the composition of the solution around the ions does not correspond to that in the bulk of the solution, so that pK_A^* values do not depend entirely of the recorded macroscopic dielectric constant. With these observations in mind, we must try to gather as much as possible data of the pK values as a function of the concentration of each miscible organic solvent and, for every mixture, as a function of temperature.

Temperature effect

The temperature coefficients of $\log K$ and $\log \gamma$, in terms in which the temperature coefficient of pa_H is expressed, are readily estimated for many buffers from data in the literature.

Some general conclusions can be drawn regarding the effect of temperature on pH values. In dilute solutions the variation of the activity coefficient with temperature is negative and uniform over the whole pH range and tends to lower $\delta pa_H/\delta T$ for strong bases, salts and weak acid buffers with negative Z, and to raise it for strong acids and buffers containing weak bases with positive Z.

However, the effect of the altered activity coefficient is small and most often overshadowed by $\delta \log K/\delta T$, and then changes in concentration are frequently without noticeable effect on temperature coefficients. For these reasons, pa_H vs temperature curves are of much the same type as the curves of pK vs temperature.

At constant pressure, we have from the Van't Hoff equation

$$\frac{\delta \log K}{\delta T} = \frac{\Delta H^\circ}{RT^2} \qquad (2\text{-}17)$$

where R is the gas constant and ΔH° is the change of enthalpy for the dissociation of one molecule of the acid or base in the standard state.

The plots of $-\log K(\mathrm{p}K)$ as a function of temperature are roughly parabolic in form, and pK for many weak electrolytes has a minimum value in the range 0 to 60°C. Hence $\delta \log K/\delta T$ may be either positive or negative at room temperature.

In general, the characteristic minimum is shifted towards higher temperatures as the acid strength decreases, so that $\mathrm{p}K_A^*$ increases with temperature for the stronger acids and higher temperatures, and decreases with temperature for weaker acids and with rising temperatures.

$\mathrm{p}a_H$, like pK, often passes through a minimum within the experimental range and the effect of the term $\delta \log \gamma/\delta T$ is to shift the maximum to a temperature lower or higher than that characteristic of the dissociation constant alone. Thus the temperature coefficient of pH may appear small at temperatures in the vicinity of the minimum.

SPECTROPHOTOMETRIC DETERMINATION OF $\mathrm{p}a_H$

The colored indicators

A colored indicator is an ionizable molecule whose protoned InH and unprotoned form In have different spectra.

The equilibrium $\mathrm{InH} \rightleftharpoons \mathrm{In} + \mathrm{H}^+$ is dependent on the $\mathrm{p}a_H$ of the solution from:

$$K_I^* = \frac{a_{In} \times a_H}{a_{InH}}$$

which gives:

$$-\log a_H = -\log K_I^* + \log \frac{a_{In}}{a_{InH}}$$

and

$$\mathrm{p}a_H = \mathrm{p}K_I^* + \log \frac{\gamma_{In}}{\gamma_{InH}} + \log \frac{[\mathrm{In}]}{[\mathrm{InH}]} \tag{2-18}$$

In the usual conditions of ionic strength ($\sim 10^{-2}$)log γ, as determined by the Debye–Hückel equation (2-16), can be neglected due to the limitations of detection.

$r = [\mathrm{In}]/[\mathrm{InH}]$ is spectrophotometrically determined. As an example a series of spectra of 2-5-dinitrophenol indicator whose pK is 5.87 in water, is given in Fig. 13. C_0 is the total concentration of the indicator.

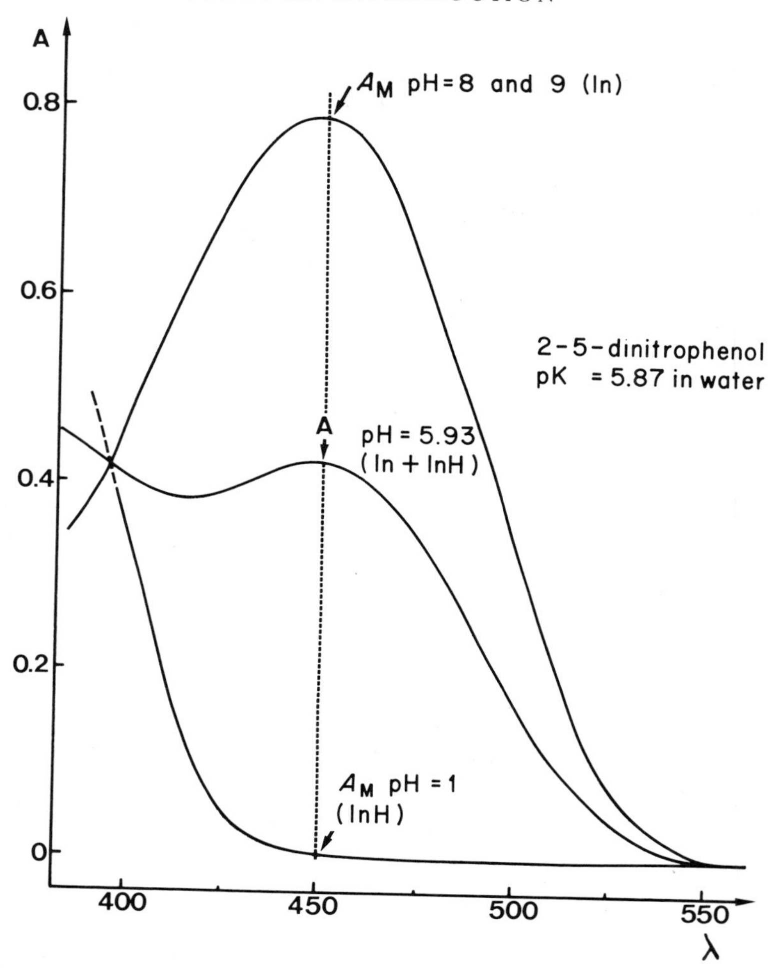

FIG. 13. Absorption spectra of 2-5-dinitrophenol in water as a function of pH.

At $\mathrm{p}a_{\mathrm{H}} \ll \mathrm{p}K_{\mathrm{I}}^*$, i.e. pH = 1, the absorbance is

$$A_{\mathrm{m}} = \varepsilon_{\mathrm{InH}} C_0 \tag{2-19}$$

in this case [In] = 0 and [InH] = C_0.

At $\mathrm{p}a_{\mathrm{H}} \gg \mathrm{p}K_{\mathrm{I}}^*$, i.e. pH = 9, [InH] = 0 and the absorbance is

$$A_{\mathrm{M}} = \varepsilon_{\mathrm{In}} C_0 \tag{2-20}$$

for an intermediary $\mathrm{p}a_{\mathrm{H}}$ near pK absorbance is:

$$A = \varepsilon_{\mathrm{InH}}[\mathrm{InH}] + \varepsilon_{\mathrm{In}}[\mathrm{In}]$$

from (2-19) and (2-20) we get

$$A = \frac{A_m}{C_0}[\text{InH}] + \frac{A_M}{C_0}[\text{In}]$$

since $[\text{In}] + [\text{InH}] = C_0$ we have

$$r = \frac{[\text{In}]}{[\text{InH}]} = \frac{A - A_m}{A_M - A}$$

and

$$\text{p}a_H \simeq \text{p}K_I^* + \log r = \text{p}K_I + \log \frac{A - A_m}{A_M - A} \qquad (2\text{-}21)$$

From experimental values of r, this expression permits one either to determine $\text{p}a_H$ if indicator $\text{p}K_I$ is known or $\text{p}K_I$ is $\text{p}a_H$ is known.

Definition and determination of a $\text{p}a_H$ *scale in mixed solvents*

It has been shown (22) that, in the presence of organic solvents whose concentration does not exceed 70%, 10^{-2} M HCl is fully dissociated whatever the temperature. In these conditions, the activity coefficient of the hydrogen ion γ_H, $\text{p}a_H = -\log \gamma_H[\text{H}^+]$, can be calculated by the Debye–Hückel formula (2-16), for which the parameters D and ρ have been measured in the mixed solvents used. In this formula, temperature is only involved in the term DT, which does not change with temperature so that γ_H, and therefore $\text{p}a_H$ of the HCl solution do not vary with temperature. $\text{p}a_H$ of a 10^{-2} M HCl solution, denoted as $\text{p}a_{H_0}$, provides the starting point reference of the $\text{p}a_H$ scale to be established in the mixed solvent considered. A first indicator is selected whose $\text{p}K_1$ is slightly higher than $\text{p}a_{H_0}$; r is determined as a function of T and $\text{p}K_1$ can then be measured for each temperature using (2-21). In a second step, a buffer (I) is chosen, the $\text{pH}_{(I)}$ of which is slightly higher than $\text{p}K_1$. Here again, r is determined as a function of temperature so that (2-21) allows $\text{p}a_{H(I)}$ to be measured at any temperature. This procedure is then followed step by step, alternating buffers and indicators of increasing basicity. Accordingly a choice of convenient buffers and indicators enable a protonic activity scale to be established in mixed solvents at room and subzero temperatures between $\text{p}a_H$ 2 and 11.

Some experimental aspects

Buffers currently used in biochemistry have been selected. They are prepared by appropriate dilution of an aqueous solution, contraction upon mixing being taken into account. Their ionic strength is adjusted to 10^{-2}, or in some exceptional cases 5×10^{-3}, to avoid precipitation on cooling. Routinely

TABLE 15. Usual colored indicators with their pK value in water

Indicator	pK_a in water t (20°C)
Metanitroaniline	1.59
Thymosulfone phtaleine	2.16
α-Naphtylamine	3.40
2-6-Dinitrophenol	4.42
2-5-Dinitrophenol	5.87
Chloronitrophenol	6.35
Paranitrophenol	7.95
Metanitrophenol	9.48
Parachlorophenol	10.53

used colored indicators are given in Table 15 as well as their pK in water at 20°C.

Some of them possess more than one pK, so that care must be taken in order to avoid overlapping of the various zones. A criterion of good experimental results is the recording of the isobestic point as pa_H is varied. Moreover, to obtain the greatest accuracy, r has to be kept between 0.1 and 9. Such a limitation requires a careful choice of indicators. With this procedure of spectrophotometric determination of pa_H an absolute error of ± 0.1 is obtained on pa_H.

Results

Since weak electrolytes show a variation of their dissociation constants as a function of temperature, buffers containing such electrolytes should be similarly influenced in the same conditions, retaining their buffer capacity at each temperature. This prediction has been verified in buffers containing equimolar concentrations of a weak acid and its salt in which the protonic activity is close to the pK_a of the weak electrolyte component.

The values of pa_H as a function of temperature for various buffers in different hydro–organic mixtures have been measured by the indicator method (13,23,24). Results are given in both Tables 16–26 and Figures 14–20. The graphs between 0 and -50°C (pa_H as a function of $1/T$) are linear for each solvent, and the same behavior is found with non-equimolar buffers (dotted lines). We will discuss later the thermodynamic implications of these results. For the moment, it is sufficient to say that from the slope of Δpa_H for a given equimolar buffer, it is possible, extrapolating from a pa_H value at room temperature, to predict the value of pa_H in a supercooled solution. This should permit any desired value to be adjusted from the data given in

TABLE 16

Buffers 10^{-2}M	pH water	pa_H in 50% EGOH–50% H_2O					
	+20°C	+20°C	0°C	−10°C	−20°C	−30°C	−40°C
Chloracetate	2.35	2.90	2.95	2.95	3.00	3.05	3.05
	2.55	3.25	3.30	3.30	3.35	3.40	3.40
	2.85	3.55	3.60	3.60	3.65	3.70	3.70
Acetate	3.65	4.25	4.35	4.40	4.45	4.50	4.55
	4.10	4.65	4.75	4.80	4.85	4.90	4.95
	4.40	4.90	5.00	5.05	5.10	5.15	5.20
	4.55	5.05	5.15	5.20	5.25	5.30	5.35
	4.75	5.25	5.35	5.40	5.45	5.50	5.55
Cacodylate	5.4	5.90	6.00	6.05	6.10	6.15	6.25
	5.8	6.30	6.40	6.45	6.50	6.55	6.65
	6.2	6.65	6.75	6.80	6.85	6.90	7.00
	6.4	6.85	6.95	7.00	7.05	7.10	7.20
	7.0	7.30	7.40	7.45	7.50	7.55	7.65
Phosphate	6.0	6.60	6.75	6.80	6.90	7.00	7.10
	6.5	7.10	7.25	7.30	7.40	7.50	7.60
	7.0	7.60	7.75	7.80	7.90	8.00	8.10
	7.5	8.10	8.25	8.30	8.40	8.50	8.60
Tris	7.0	7.20	7.95	8.35	8.75	9.20	9.70
	7.5	7.65	8.40	8.80	9.20	9.65	10.15
	8.0	7.95	8.70	9.10	9.90	9.95	10.45
	8.5	8.45	9.20	9.60	10.00	10.45	10.95

Tables 16–26. Another interesting result concerns the variation of pa_H in itself. We can see that the protonic activity of buffers such as oxalate, acetate, cacodylate and phosphate does not vary widely as a function of temperature and that in some cases one could repeat cooling ⇌ warming cycles (over several degrees) without changing their protonic activity.

On the other hand, Tris buffer should enable us to couple a change in temperature with a change of protonic activity in a pa_H range spanning 3 to 4 units. Finally, we must deal with the important practical problem of the solubility of phosphate and Tris buffer as a function of temperature. At a concentration of 10^{-2} M, phosphate buffer precipitates at −60°C in solutions containing 70% methanol, and at −40° when the solutions contain 50% methanol.

Under the same conditions, Tris buffer precipitates at −70°C and −50°C in solutions containing 70% methanol or dimethylformamide and 50% methanol. In the presence of a high concentration of proteins or other solutes, the precipitation can occur even at higher temperatures. The same buffers do

TABLE 17

Buffers 10^{-2}M	pH water	pa_H in 50% MeOH–50% H_2O					
	+20°C	+20°C	0°C	−10°C	−20°C	−30°C	−40°C
Chloracetate	2.35	2.90	2.95	2.95	3.00	3.05	3.10
	2.55	3.25	3.30	3.30	3.35	3.40	3.45
	2.85	3.60	3.65	3.65	3.70	3.75	3.80
Acetate	3.65	4.50	4.55	4.55	4.60	4.60	4.65
	4.10	4.85	4.90	4.90	4.95	4.95	5.00
	4.40	5.10	5.15	5.15	5.20	5.20	5.25
	4.55	5.30	5.35	5.35	5.40	5.40	5.45
	4.75	5.45	5.50	5.50	5.55	5.55	5.60
Cacodylate	5.4	6.30	6.30	6.35	6.35	6.40	6.40
	5.8	6.60	6.60	6.65	6.65	6.70	6.70
	6.2	6.95	6.95	7.00	7.00	7.05	7.05
	6.4	7.10	7.10	7.15	7.15	7.20	7.20
	7.0	7.55	7.55	7.60	7.60	7.65	7.65
Phosphate	6.0	7.20	7.25	7.25	7.30	7.30	7.35
	6.5	7.65	7.70	7.70	7.75	7.75	7.80
	7.0	8.10	8.15	8.15	8.20	8.20	8.25
	7.5	8.50	8.55	8.55	8.60	8.60	8.65
	8.0	8.90	8.95	8.95	9.00	9.00	9.05
Tris	7.0	7.25	7.80	8.15	8.55	8.90	9.35
	7.5	7.70	8.25	8.60	9.00	9.35	9.80
	8.0	7.95	8.50	8.85	9.25	9.60	10.05
	8.5	8.40	8.95	9.30	9.70	10.05	10.50
	9.0	8.85	9.40	9.75	10.15	10.50	10.95

not precipitate at all when used at 10^{-3} M, in 70% methanol or *N*-*N*-dimethylformamide. Moreover, phosphate buffer can eventually be replaced by cacodylate buffer which is soluble over the whole temperature range in all of the mixtures studied up to a concentration of 10^{-1} M, even in the presence of solutes.

Thus the results show that it is possible to choose a suitable buffer system at any required subzero temperature, in the presence of any amount of organic solvent; the properties of the buffer which have been described also allow their use to effect controlled simultaneous "jumps" of temperature and pa_H.

POTENTIOMETRIC DETERMINATION OF pa_H

Method

The potentiometric determinations have been realized by the use of a glass-calomel electrodes assembly (25) as described in Fig. 21 (p. 71). The potential

TABLE 18

Buffers 10^{-2}M	pH water	pa_H in 70% MeOH–30% H_2O					
	+20°C	+20°C	0°C	−10°C	−20°C	−30°C	−40°C
Chloracetate	2.35	3.20	3.20	3.20	3.20	3.20	3.20
	2.55	3.70	3.70	3.70	3.70	3.70	3.70
	2.85	4.05	4.05	4.05	4.05	4.05	4.05
Acetate	3.65	5.0	5.10	5.10	5.15	5.20	5.25
	4.10	5.35	5.45	5.45	5.50	5.55	5.60
	4.40	5.55	5.65	5.65	5.70	5.75	5.80
	4.55	5.75	5.85	5.85	5.90	5.95	6.00
	4.75	5.95	6.05	6.05	6.10	6.15	6.20
Cacodylate	5.4	6.65	6.70	6.70	6.75	6.80	6.80
	5.8	7.0	7.05	7.05	7.10	7.15	7.15
	6.2	7.35	7.40	7.40	7.45	7.50	7.50
	6.4	7.50	7.55	7.55	7.60	7.65	7.65
	7.0	8.0	8.05	8.05	8.10	8.15	8.15
Phosphate	6.0	7.75	7.85	7.90	7.95	8.00	8.10
	6.5	8.25	8.35	8.40	8.45	8.50	8.60
	7.0	8.60	8.70	8.75	8.80	8.85	8.95
	7.5	9.10	9.20	9.25	9.30	9.35	9.45
	8.0	9.45	9.55	9.60	9.65	9.70	9.80
Tris	7.0	7.30	7.95	8.40	8.80	9.25	9.80
	7.5	7.70	8.35	8.80	9.20	9.65	10.20
	8.0	8.00	8.65	9.10	9.50	9.95	10.50
	8.5	8.55	9.10	9.65	10.05	10.50	11.05
	9.0	9.10	9.65	10.20	10.60	11.05	11.60

of a glass electrode reversible by respect to H^+ immersed in a solution whose proton activity is a_H, is given by

$$E_{\text{glass}} = E_{0g} + \frac{RT}{F} \ln a_H$$

Similarly, the calomel electrode, reversible by respect to Cl^- has a potential

$$E_{\text{cal}} = E_{0c} + \frac{RT}{F} \ln a_{Cl}$$

In these relations, E_{0g} and E_{0c} are respectively the standard potentials of the glass and calomel electrodes, F is the Faraday; a_{Cl} is the activity of solvated ion Cl^-.

Let a_H be the protonic activity of a solution to be tested and E the e.m.f. measured between the electrodes

$$E = E_{\text{cal}} - E_{\text{glass}} + E_j$$

TABLE 19

Buffers 10^{-2}M	pH water	pa_H in 50% MPD–50% H_2O			
	+20°C	+20°C	0°C	−20°C	−25°C
Chloracetate	2.2	2.55	2.55	2.55	2.55
	2.6	3.24	3.24	3.24	3.24
	3.0	3.76	3.76	3.76	3.76
	3.4	4.20	4.20	4.20	4.20
Acetate	3.8	4.55	4.59	4.65	4.66
	4.3	5.05	5.09	5.15	5.16
	4.8	5.56	5.60	5.66	5.67
	5.3	6.04	6.08	6.14	6.15
Cacodylate	5.5	6.08	6.09	6.12	6.12
	6.0	6.77	6.78	6.81	6.81
	6.5	7.34	7.35	7.38	7.38
	7.0	7.80	7.81	7.84	7.84
Phosphate	6.5	7.36	7.47	7.78	7.87
	7.0	8.05	8.16	8.47	8.56
	7.5	8.57	8.68	8.99	9.08
	8.0	8.90	9.01	9.32	9.41
Tris	8.0	8.10	8.64	9.39	9.60
	8.5	8.54	9.08	9.83	10.04
	9.0	8.90	9.44	10.19	10.40
	9.5	9.20	9.74	10.49	10.70
Carbonate	9.0	9.70	10.00	10.50	10.65
	9.5	10.67	10.97	11.47	11.62
	10.0	11.16	11.46	11.96	12.11
	10.5	11.61	11.91	12.41	12.56

where E_j is the sum of both asymmetry and junction potentials. We have then

$$pa_H = -\log a_H = \frac{E \times F}{RT \ln 10} - \log a_{Cl} + \frac{(E^\circ - E_j)F}{RT \ln 10} \quad (2\text{-}22)$$

with $E^\circ = E_{0c} - E_{0g}$.

In practice E° cannot be measured and therefore absolute pa_H determinations are not possible with this method. However the term

$$-\log a_{Cl} + \frac{(E^\circ - E_j)F}{RT \ln 10}$$

is a constant for a given temperature, so that relative pa_H measurements can be performed using a standard medium of reference termed as (st) in these conditions.

$$pa_H = pa_H(\text{st}) + \frac{E - E(\text{st})}{RT \ln 10} F \quad (2\text{-}23)$$

TABLE 20

Buffers 10^{-2} M	pH water	pa_H in 50% PrOH–50% H_2O					
	+20°C	+20°C	0°C	−10°C	−20°C	−30°C	−40°C
Chloracetate	2.40	3.35	3.35	3.40	3.40	3.40	3.45
	2.60	3.55	3.55	3.55	3.60	3.60	3.60
	3.00	4.00	4.00	4.00	4.00	4.05	4.05
	3.40	4.40	4.40	4.40	4.40	4.45	4.45
Acetate	4.00	4.90	4.90	4.95	4.95	5.00	5.05
	4.50	5.35	5.35	5.40	5.40	5.45	5.45
	5.00	5.80	5.80	5.85	5.85	5.90	5.95
	5.60	6.30	6.35	6.35	6.40	6.40	6.45
Cacodylate	5.50	6.05	6.15	6.15	6.20	6.25	6.30
	6.00	6.55	6.60	6.65	6.70	6.75	6.80
	6.50	7.00	7.10	7.10	7.15	7.20	7.30
	7.00	7.45	7.50	7.55	7.60	7.65	7.70
Phosphate	6.50	7.30	7.40	7.45	7.55	7.60	7.70
	7.00	7.75	7.85	7.90	7.95	8.05	8.10
	7.50	8.10	8.20	8.25	8.35	8.40	8.45
	8.00	8.55	8.65	8.70	8.80	8.85	8.90
Tris	8.00	7.80	8.40	8.80	9.20	9.65	10.10
	8.50	8.20	8.80	9.20	9.55	10.00	10.45
	9.00	8.60	9.25	9.60	10.00	10.45	10.90
	9.50	8.90	9.55	9.90	10.30	10.75	11.15
Carbonate	9.50	9.70	10.00	10.15	10.35	10.55	10.75
	10.00	10.25	10.55	10.70	10.90	11.10	11.30
	10.50	10.60	10.90	11.10	11.25	11.45	11.65
	11.00	10.75	11.05	11.20	11.35	11.60	11.80

Electrodes and measuring cell (Fig. 21, p. 71)

The glass electrode. To perform pa_H measurements at subzero temperatures, the inner reference solution of the electrode must stay fluid in the whole temperature range investigated. For this purpose we use a solution of 10^{-2} M HCl in a hydro–organic solvent mixture. The organic solvent must be the same as that present in the test solution. Its volumic proportion is arbitrarily choosen as 50%, very low temperatures can thus be reached without freezing. In such conditions, HCl has been shown to be fully dissociated (22) so that the inner solution reference pa_H could be calculated by the Debye–Hückel formula (2-16). An empty glass electrode (Tacussel TB-HS) is filled with such an HCl, hydro–organic solution (one electrode for each particular organic solvent). As the inner reference electrode a Ag:AgCl element (Tacussel) is immersed in this internal liquid and the glass electrode is sealed with "Araldite". When not used, the electrode is stored

TABLE 21

Buffers 10^{-2} M	pH water	pa_H in 50% GlOH–50% H_2O				
	+20°C	+20°C	0°C	−10°C	−20°C	−30°C
Chloracetate	2.60	3.10	3.25	3.35	3.45	3.55
	3.00	3.50	3.65	3.75	3.85	3.95
	3.40	3.90	4.05	4.15	4.25	4.35
Acetate	3.80	4.25	4.45	4.55	4.70	4.80
	4.30	4.70	4.90	5.05	5.15	5.30
	4.80	5.15	5.35	5.45	5.60	5.75
	5.30	5.60	5.80	5.90	6.05	6.20
Cacodylate	5.50	5.70	5.90	6.05	6.15	6.30
	6.00	6.15	6.35	6.45	6.60	6.75
	6.50	6.65	6.85	7.00	7.10	7.25
	7.00	7.15	7.35	7.50	7.60	7.75
Phosphate	6.50	6.70	6.95	7.10	7.30	7.45
	7.00	7.15	7.45	7.60	7.75	7.90
	7.50	7.65	7.90	8.05	8.20	8.40
	8.00	8.10	8.30	8.50	8.65	8.80
Tris	8.00	8.05	8.90	9.35	9.80	10.35
	8.50	8.50	9.35	9.75	10.25	10.85
	9.00	8.85	9.70	10.15	10.65	11.20
	9.50	9.15	9.95	10.45	10.90	11.50
Carbonate	9.50	9.50	9.95	10.20	10.45	10.70
	10.00	9.95	10.40	10.60	10.90	11.20
	10.50	10.25	10.70	10.95	11.20	11.50
	11.00	10.45	10.90	11.10	11.40	11.70

in a 10^{-2} M HCl hydro–organic solution of exactly the same nature and composition as that contained in the electrode. Before measurements are made, the glass electrode is equilibrated with a 10^{-2} M HCl hydro–organic solution of exactly the same composition as that in which measurements are to be made. Such a procedure allows reasonably low equilibrium times.

The calomel reference electrode. Basically the calomel electrode (Fig. 21) consists of mercury, mercurous chloride (calomel) and chloride ion. The concentration of potassium chloride is 0.1 M in a hydro–organic solvent (50/50) of the same nature as that contained in the solution to be investigated. The junction with the test solution is realised either with a capillary or a porous stone. When the capillary is used, a small hydrostatic pressure is maintained inside it in order to avoid any electrode contamination by the test solution. In the main part of our investigation, the porous stone junction was used. Moreover, the calomel electrode is thermostated at +20°C, any temperature variations of this electrode giving appreciable e.m.f. variations

TABLE 22

Buffers 10^{-2} M	pH water	pa_H in 50% DMSO–50% H_2O					
	+20°C	+20°C	0°C	−10°C	−20°C	−30°C	−40°C
Chloracetate	2.60	3.05	3.15	3.20	3.30	3.40	3.45
	3.00	3.60	3.75	3.85	3.90	4.00	4.15
	3.40	3.95	4.10	4.20	4.25	4.35	4.45
Acetate	3.80	4.45	4.65	4.75	4.85	5.00	5.10
	4.30	4.90	5.10	5.20	5.30	5.45	5.60
	4.80	5.40	5.55	5.70	5.80	5.95	6.10
	5.30	5.75	5.95	6.05	6.20	6.30	6.45
Cacodylate	5.50	6.00	6.25	6.40	6.55	6.70	6.90
	6.00	6.45	6.70	6.85	7.00	7.15	7.35
	6.50	7.00	7.20	7.35	7.50	7.70	7.85
	7.00	7.45	7.70	7.85	8.00	8.15	8.35
Phosphate	6.50	7.75	8.10	8.30	8.50	8.80	9.00
	7.00	8.20	8.60	8.80	9.00	9.25	9.50
	7.50	8.60	8.95	9.20	9.40	9.65	9.90
	8.00	9.05	9.40	9.60	9.85	10.10	10.35
Tris	8.00	7.20	7.90	8.35	8.75	9.25	9.80
	8.50	7.55	8.25	8.70	9.15	9.65	10.15
	9.00	7.95	8.70	9.10	9.55	10.05	10.55
	9.50	8.25	9.00	9.40	9.85	10.35	10.85
Carbonate	9.50	10.75	11.30	11.60	11.90	12.30	12.65

involving uncertainty on the pa_H determination of the order of 0.2–0.3 pa_H unit/10°C.

The measuring cell and temperature control. This cell, made from Pyrex, is described in Fig. 21. The operational volume of test solution is 10 ml. The cell is carefully stoppered, and a slow flow of dry N_2 is maintained inside it. Around this cell there is a chamber which controls the temperature of the sample. This chamber is isolated from the surrounding atmosphere by a vacuum jacket. The sample is continuously stirred magnetically and can be thermostated from +50°C to −45°C, ±0.05°C, with a liquid thermostat. The temperature is directly measured in the sample with a chromel–alumel thermocouple.

In all pa_H measurements at very low temperature (< -10°C), in order to avoid thermic shock on the glass electrode which is itself cooled or warmed very slowly, solutions to be tested are previously cooled to the temperature of the electrode. In these conditions, the electrical equilibrium time is reasonably short, of the order of five minutes or less. Finally, the assembly glass–calomel electrodes and measuring cell is placed into a Faraday box.

TABLE 23

Buffers 10^{-2} M	pH water	pa_H in 50% DMF–50% H_2O					
	+20°C	+20°C	0°C	−10°C	−20°C	−30°C	−40°C
Chloracetate	2.35	4.20	4.30	4.35	4.40	4.45	4.50
	2.55	4.40	4.50	4.55	4.60	4.65	4.70
	2.85	4.60	4.70	4.75	4.80	4.85	4.90
Acetate	3.65	5.90	6.05	6.15	6.25	6.35	6.50
	4.10	6.35	6.50	6.60	6.70	6.80	6.95
	4.40	6.55	6.70	6.80	6.90	7.00	7.15
	4.55	6.70	6.85	6.95	7.05	7.15	7.30
	4.75	6.90	7.05	7.15	7.25	7.35	7.50
Phosphate	6.0	8.55	8.55	8.60	8.60	8.65	8.65
	6.5	8.90	8.90	8.95	8.95	9.00	9.00
	7.0	9.20	9.20	9.25	9.25	9.30	9.30
	7.5	9.35	9.35	9.40	9.40	9.45	9.45
	8.0	9.55	9.55	9.60	9.60	9.65	9.65
Tris	7.0	6.90	7.50	7.90	8.20	8.60	9.05
	7.5	7.35	7.95	8.35	8.65	9.05	9.50
	8.0	7.60	8.20	8.60	8.90	9.30	9.75
	8.5	8.20	8.80	9.20	9.50	9.90	10.35
	9.0	8.50	9.10	9.50	9.80	10.20	10.65

The pa_H *response of the glass electrode*

The pa_H response of a glass electrode is defined as:

$$R^e = \frac{\Delta E}{\Delta pa_H}$$

so that the ideal pa_H response is from formula (2-23):

$$R^e = \frac{RT \ln 10}{F} = 0.198\ T \qquad (2\text{-}24)$$

It appears in this formula that R^e is only a function of absolute temperature. In practice, no glass electrode has this theoretical response in all types of samples and over the whole pa_H scale. Nevertheless, if either too acidic or basic pa_H are excluded, it is generally found that the departure from the ideal behavior is negligible.

The pa_H response is tested by the mean of the pa_H values as determined by the indicator method: the electromotive force of the cell immersed in buffer solutions whose pa_H was known, is measured and pa_H spectrophotometrically determined is plotted against E (Fig. 22, p. 72).

TABLE 24

Buffers 10^{-2} M	pH water	pa_H in 40% EGOH–20% MeOH–40% H_2O					
	+20°C	+20°C	0°C	−10°C	−20°C	−30°C	−40°C
Chloracetate	2.2	2.80	2.80	2.80	2.80	2.80	2.80
	2.6	3.40	3.40	3.40	3.40	3.40	3.40
	3.0	3.90	3.90	3.90	3.90	3.90	3.90
	3.4	4.25	4.25	4.25	4.25	4.25	4.25
Acetate	3.8	4.55	4.60	4.65	4.70	4.70	4.75
	4.3	5.10	5.15	5.20	5.25	5.25	5.30
	4.8	5.55	5.60	5.65	5.70	5.70	5.75
	5.3	6.10	6.15	6.20	6.25	6.25	6.30
Cacodylate	5.5	6.10	6.10	6.15	6.15	6.20	6.20
	6.0	6.70	6.70	6.75	6.75	6.80	6.80
	6.5	7.10	7.10	7.15	7.15	7.20	7.20
	7.0	7.60	7.60	7.65	7.65	7.70	7.70
Phosphate	6.5	7.45	7.45	7.45	7.45	7.50	7.50
	7.0	8.00	8.00	8.00	8.00	8.05	8.05
	7.5	8.45	8.45	8.45	8.45	8.50	8.50
	8.0	8.80	8.80	8.80	8.80	8.85	8.85
Tris	8.0	8.05	8.65	9.00	9.30	9.70	10.15
	8.5	8.65	9.25	9.60	9.90	10.30	10.75
	9.0	8.95	9.55	9.90	10.20	10.60	11.05
	9.5	9.20	9.80	10.15	10.45	10.85	11.30
Glycine	9.5	9.50	10.00	10.30	10.65	10.95	11.25
	10.0	9.95	10.45	10.75	11.10	11.40	11.70
	10.5	10.35	10.85	11.15	11.50	11.80	12.10
	11.0	10.60	11.10	11.40	11.75	12.05	12.35

It can be seen that for this ethylene glycol glass electrode the practical response is in good agreement with the theoretical one, between pa_H 2 and 9 and for +21, +1 and −19°C. The reproducibility of the determinations, estimated by the use of two different assemblies of electrodes is better than ±1.0 mV and the error on the pa_H determinations is estimated at ±0.1 pa_H unit. On the other hand, it appears that the results obtained by the two methods are in good agreement.

Practical determinations of pa_H

The results presented above show that, with conveniently modified electrodes, potentiometric determinations of pa_H can be performed in hydro-organic solvents at normal and subzero temperatures. In particular, it must be emphasized that these modified electrodes actually present the theoretically expected pH response.

TABLE 25

Buffers 10^{-2} M	pH water	pa_H in 25% EGOH–25% MeOH–50% H_2O					
	+20°C	+20°C	0°C	−10°C	−20°C	−30°C	−40°C
Chloracetate	2.2	2.80	2.80	2.80	2.80	2.80	2.80
	2.6	3.35	3.35	3.35	3.35	3.35	3.35
	3.0	3.95	3.95	3.95	3.95	3.95	3.95
	3.4	4.35	4.35	4.35	4.35	4.35	4.35
Acetate	3.8	4.80	4.80	4.80	4.80	4.85	4.85
	4.3	5.30	5.30	5.30	5.30	5.35	5.35
	4.8	5.80	5.80	5.80	5.80	5.85	5.85
	5.3	6.30	6.30	6.30	6.30	6.35	6.35
Cacodylate	5.5	6.45	6.45	6.45	6.45	6.45	6.45
	6.0	7.10	7.10	7.10	7.10	7.10	7.10
	6.5	7.55	7.55	7.55	7.55	7.55	7.55
	7.0	8.00	8.00	8.00	8.00	8.00	8.00
Phosphate	6.5	7.85	7.90	7.90	7.95	8.00	8.05
	7.0	8.25	8.30	8.30	8.35	8.40	8.45
	7.5	8.80	8.85	8.85	8.90	8.95	9.00
	8.0	9.30	9.35	9.35	9.40	9.45	9.50
Tris	8.0	8.40	9.05	9.35	9.70	10.05	10.55
	8.5	8.80	9.45	9.75	10.10	10.45	10.95
	9.0	9.25	9.90	10.20	10.55	10.90	11.40
	9.5	9.50	10.15	10.45	10.80	11.15	11.65
Glycine	9.5	9.80	10.30	10.55	10.85	11.10	11.50
	10.0	10.30	10.80	11.05	11.35	11.60	12.00
	10.5	10.65	11.15	11.40	11.70	11.95	12.35
	11.0	11.00	11.50	11.75	12.05	12.30	12.70

Once the electrodes have been prepared for a given hydro–organic solvent, the pa_H determinations can be made at each temperature, either graphically or by direct reading on commercial pH meters calibrated for pa_H measurements.

(*a*) *Graphical determination.* In this procedure the pH meter is used as a millivoltmeter according to the appropriate adjustment of the apparatus. A solution (A): 10^{-2} M HCl in the hydro–organic solvent considered, is selected as the standard reference solution, its pa_H being calculated for any temperature according to the Debye–Hückel formula as previously indicated. After the electrodes have been immersed in this solution, the corresponding e.m.f. is determined. In the diagram pa_H-e.m.f. as shown in Fig. 22, this reference point is plotted and the theoretical straight line is drawn, the slope of which is calculated by equation (2-24). A solution (B) to be investigated is then

TABLE 26

Buffers 10^{-2} M	pH water	pa_H in 10% EGOH–60% MeOH–30% H_2O					
	+20°C	+20°C	0°C	−10°C	−20°C	−30°C	−40°C
Chloracetate	2.2	3.15	3.15	3.15	3.15	3.20	3.20
	2.6	3.80	3.80	3.80	3.80	3.85	3.85
	3.0	4.35	4.35	4.35	4.35	4.40	4.40
	3.4	4.70	4.70	4.70	4.70	4.75	4.75
Acetate	3.8	5.15	5.15	5.20	5.20	5.30	5.30
	4.3	5.70	5.70	5.75	5.75	5.85	5.85
	4.8	6.20	6.20	6.25	6.25	6.35	6.35
	5.3	6.70	6.70	6.75	6.75	6.85	6.85
Cacodylate	5.5	6.65	6.65	6.65	6.65	6.75	6.75
	6.0	7.15	7.15	7.15	7.15	7.25	7.25
	6.5	7.60	7.60	7.60	7.60	7.70	7.70
	7.0	8.10	8.10	8.10	8.10	8.20	8.20
Phosphate	6.5	8.25	8.30	8.30	8.35	8.40	8.40
	7.0	8.75	8.80	8.80	8.85	8.90	8.90
	7.5	9.25	9.30	9.30	9.35	9.40	9.40
	8.0	9.60	9.65	9.65	9.70	9.75	9.75
Tris	8.0	7.90	8.50	8.80	9.20	9.60	10.05
	8.5	8.35	8.95	9.25	9.65	10.05	10.50
	9.0	8.85	9.45	9.75	10.15	10.55	11.00
	9.5	9.30	9.90	10.20	10.60	11.00	11.45
Glycine	9.5	9.45	10.00	10.30	10.75	11.25	12.05
	10.0	10.15	10.70	11.00	11.45	11.95	12.75
	10.5	10.55	11.10	11.40	11.85	12.35	13.15
	11.0	11.05	11.60	11.90	12.35	12.85	13.65

placed into the measuring cell and the e.m.f. determined. Using the diagram pa_H-e.m.f., pa_H(B) can be graphically determined.

(*b*) pH *meter calibration*. The pH meter is here classically used and calibrated according to the normal procedure using two standard solutions. The only difference from the usual conditions is that these two solutions are a solution (A), 10^{-2} M HCl in the hydro–organic solvent whose pa_H has been calculated, and a solution (B) the pa_H of which was graphically determined. In these conditions the pH meter is calibrated for pa_H measurements in the hydro–organic solvent considered.

(*c*) *Results*. Indeed, the development of the potentiometric technique in these conditions allows the pa_H determinations to be very quickly and easily made; and therefore allows for measurements to be made of any buffer

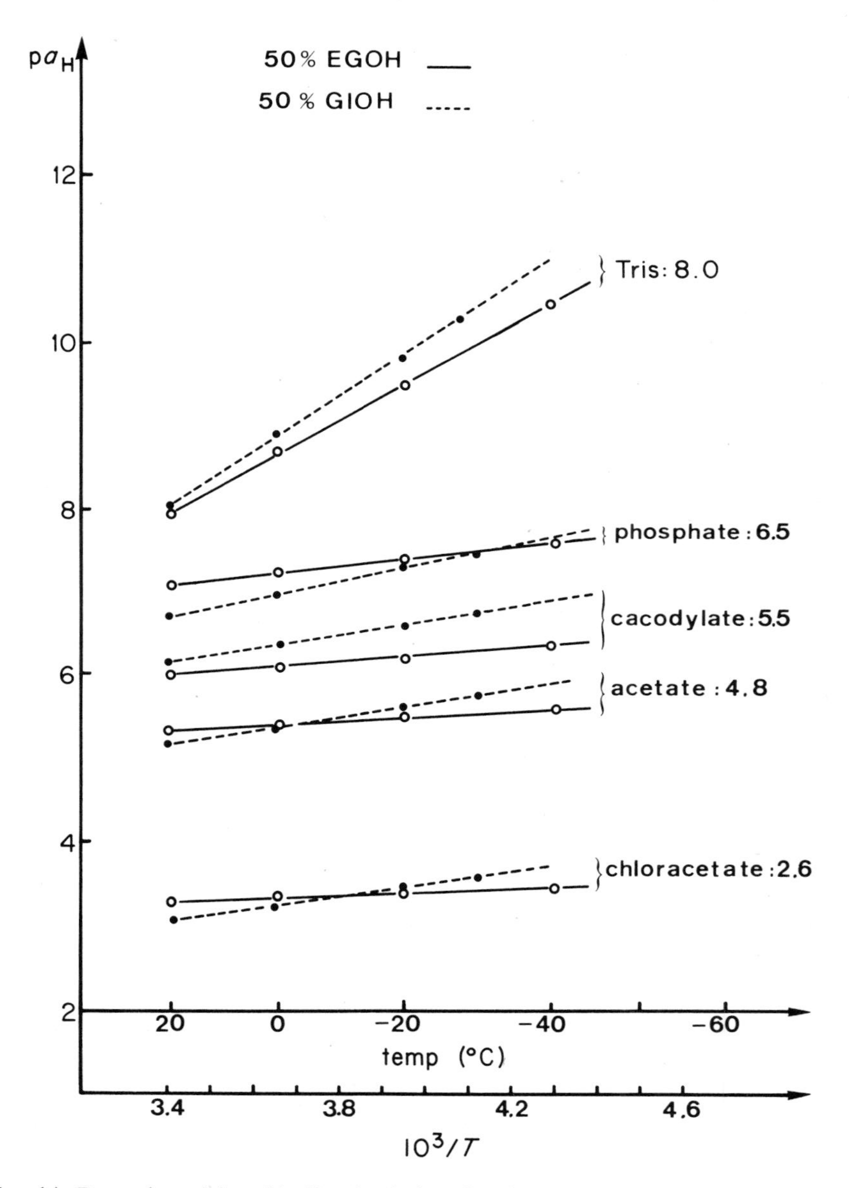

FIG. 14. Protonic activity of buffered solutions in ethylene glycol–water and glycerol–water as a function of temperature.

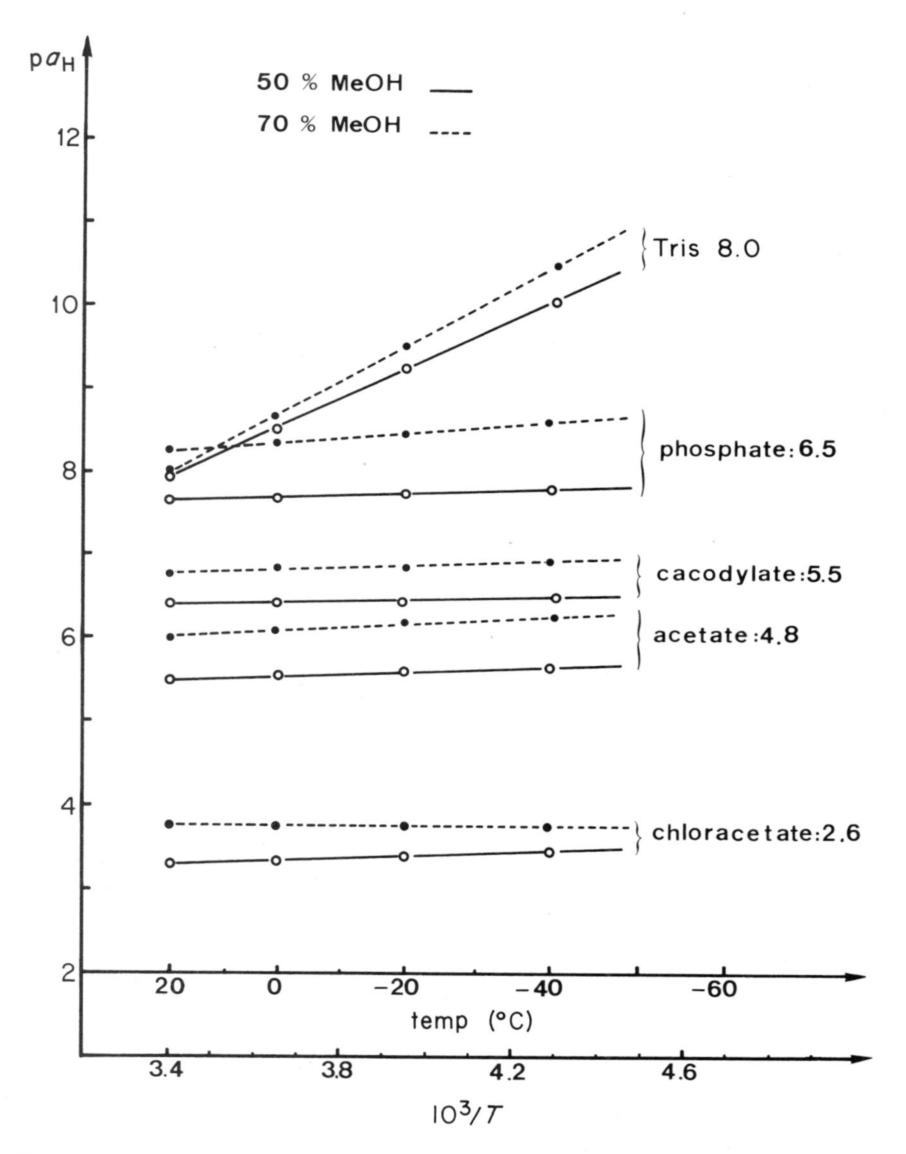

FIG. 15. Protonic activity of buffered solutions in methanol–water as a function of temperature.

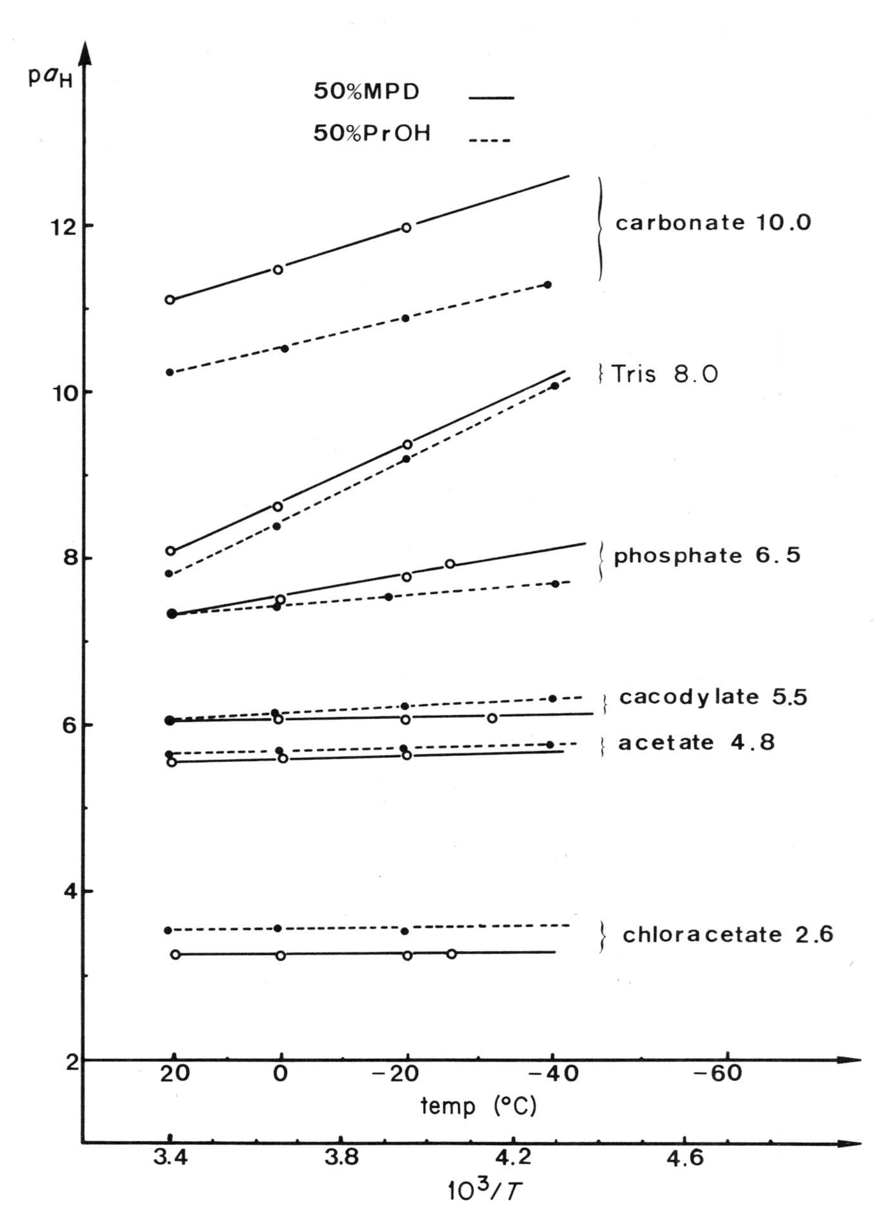

FIG. 16. Protonic activity of buffered solutions in 2-methyl-2-4–pentanediol–water and propylene glycol–water as a function of temperature.

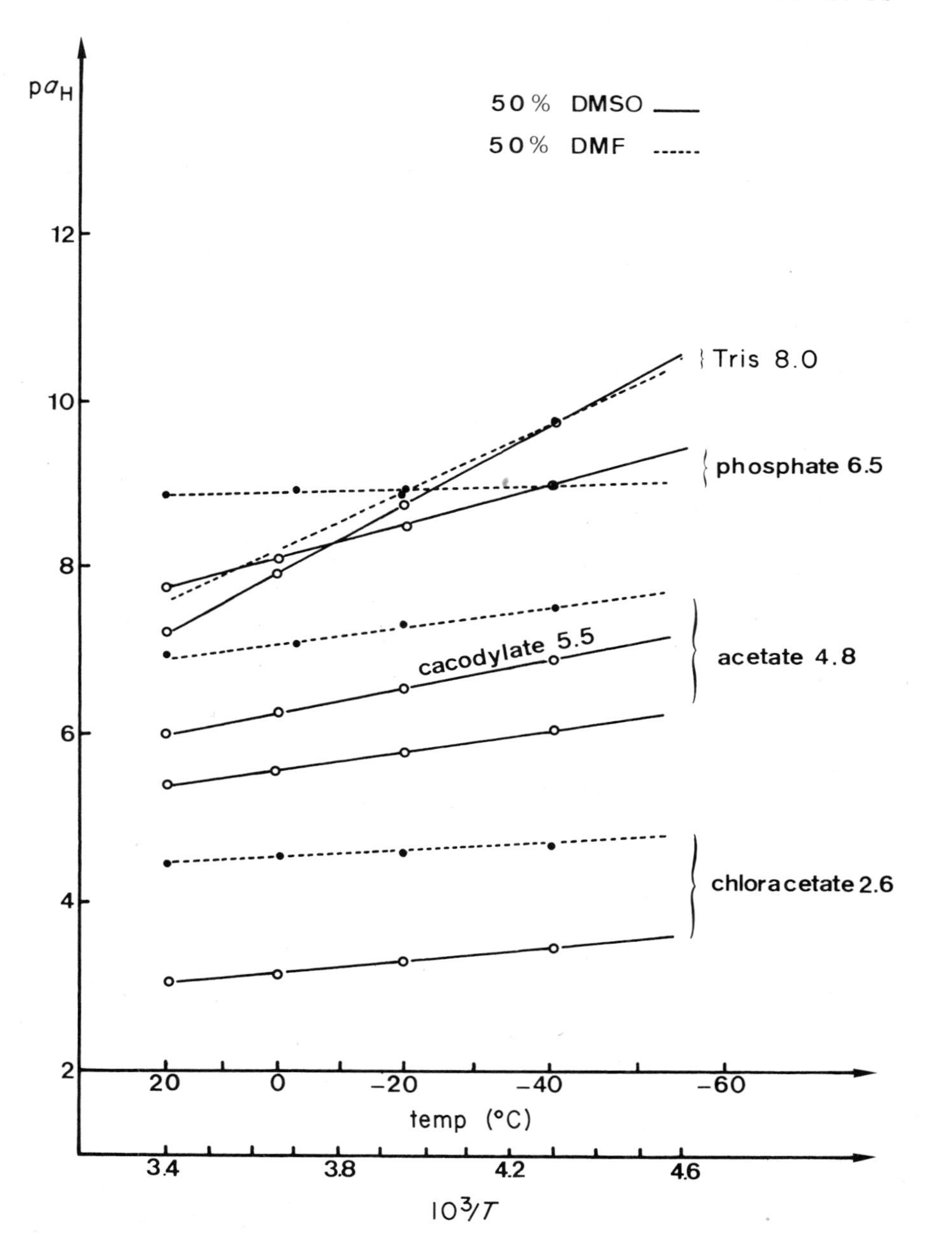

FIG. 17. Protonic activity of buffered solutions in dimethylsulfoxide–water and dimethylformamide–water as a function of temperature.

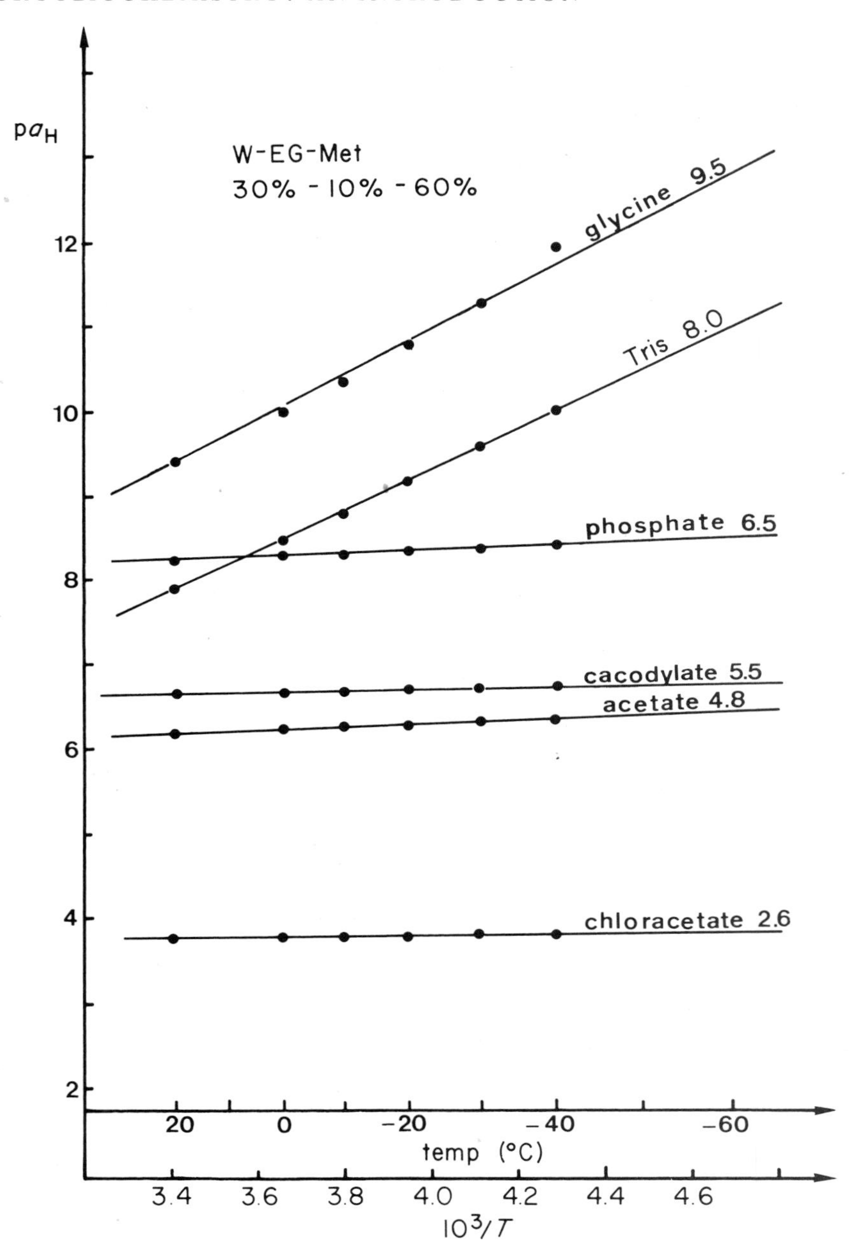

FIG. 18. Protonic activity of buffered solutions in water 30%–ethylene glycol 10%–methanol 60% as a function of temperature.

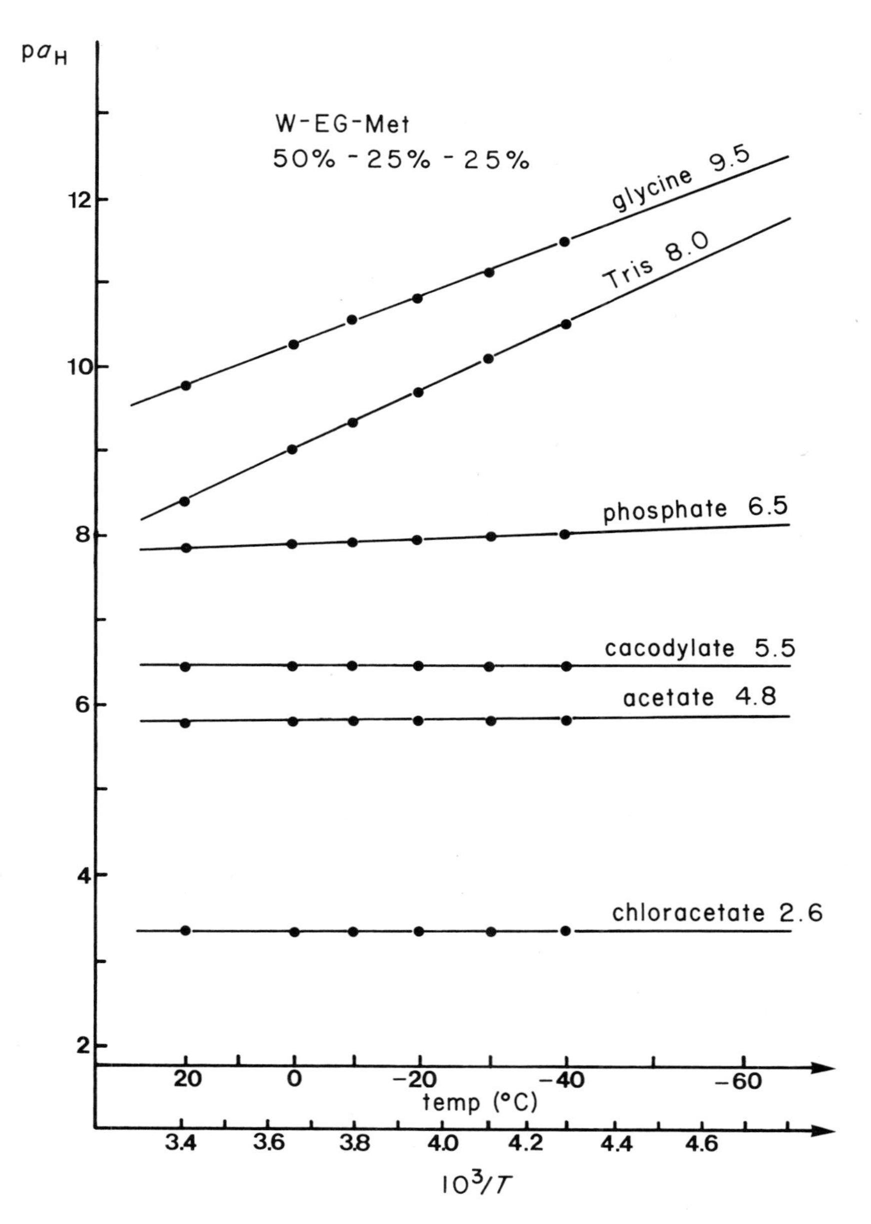

FIG. 19. Protonic activity of buffered solutions in water 50%–ethylene glycol 25%–methanol 25% as a function of temperature.

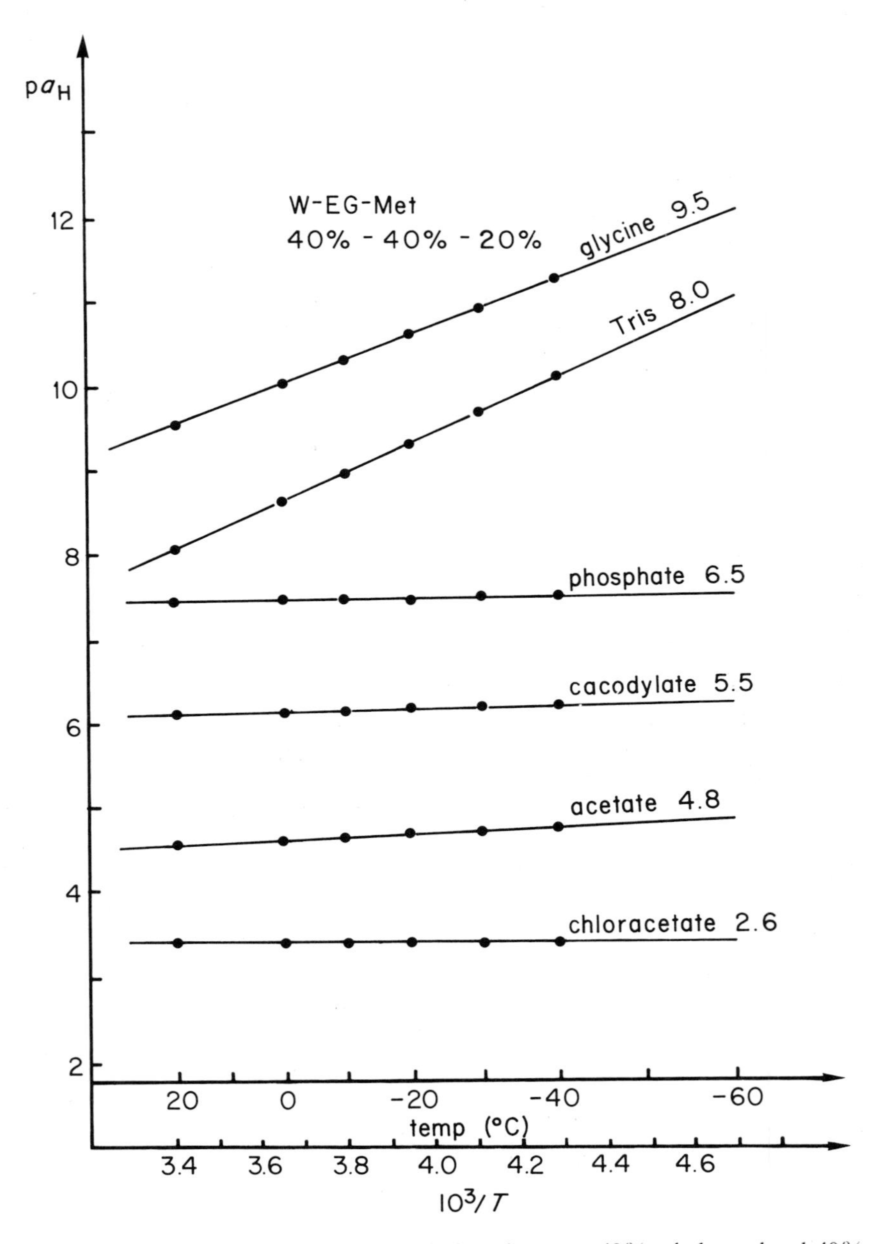

FIG. 20. Protonic activity of buffered solutions in water 40%–ethylene glycol 40%–methanol 20% as a function of temperature.

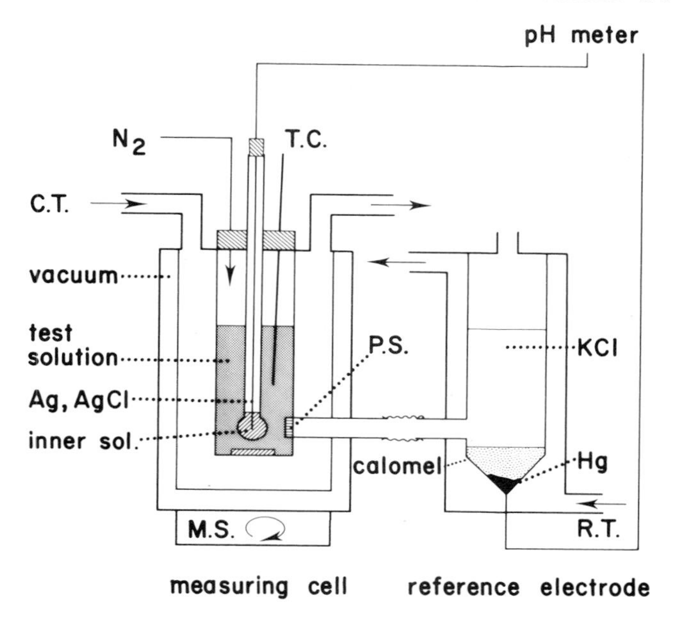

junctions:

Ag, AgCl HCl 10^{-2} 50/50	Glass	test solution	KCl 10^{-1} 50/50 calomel Hg

FIG. 21. General description of the electrodes and measuring cell. C.T., cooling thermostat; T.C., thermocouple; P.S., porous stone; R.T., room temperature thermostat; M.S., magnetic stirring.

system, in any mixed solvent. In this respect, data are summarized in Tables 27–29, concerning the pa_H of three buffer systems (cacodylate, phosphate, Tris) in three mixed solvents of water with ethylene glycol, methanol, glycerol, the volume ratios of the organic solvent being varied between 0 and 50 %, and the temperature from +20°C down to the freezing points. The pH value of each buffer, in pure water, has been arbitrarily chosen near their pK.

The pa_H values obtained here in 50 % organic solvents are in reasonable agreement with the indicator method data. The absolute error of these determinations is of the order of ± 0.1 pa_H unit, and therefore comparable to that obtained with the indicator spectrophotometry method. As we have already shown with regard to the study of an enzymic reaction in these conditions, such precision is adequate for most biochemical experiments.

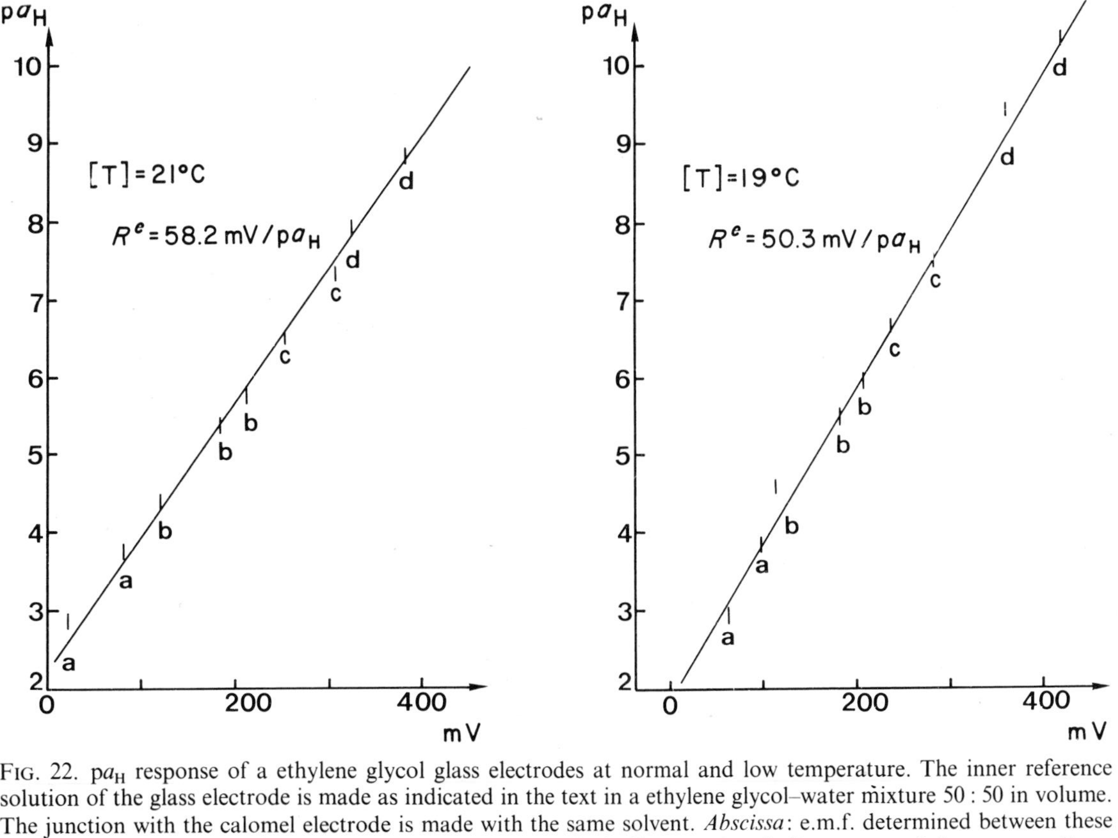

FIG. 22. pa_H response of a ethylene glycol glass electrodes at normal and low temperature. The inner reference solution of the glass electrode is made as indicated in the text in a ethylene glycol–water mixture 50 : 50 in volume. The junction with the calomel electrode is made with the same solvent. *Abscissa*: e.m.f. determined between these electrodes when immersed in ethylene glycol–water (50 : 50) buffer systems. *Ordinate*: pa_H of the same media as previously determined (20) by indicator spectrophotometry. The points are the experimental results, the straight lines represent the ideal behavior. a, chloroacetate; b, acetate; c, cacodylate; d, Tris.

TABLE 27. pa_H of buffer systems in various ethylene glycol–water mixtures between 20°C and the respective approximate freezing points. The inner reference solution of the glass electrode is made as indicated in the text in a ethylene glycol–water mixture 50/50 in volume. The junction with the calomel electrode is made with the same solvent

		Temperature (°C)					
Buffers	% EGOH	+20	0	−7	−15	−22	−35
Cacodylate	0	6.00					
	10	5.90	5.85				
	20	6.00	6.00	6.00			
	30	6.05	6.10	6.05	6.15		
	40	6.20	6.20	6.20	6.25	6.25	
	50	6.30	6.40	6.40	6.40	6.40	6.40
Phosphate	0	7.10					
	10	7.00	7.10				
	20	7.15	7.20	7.30			
	30	7.25	7.30	7.40	7.50		
	40	7.40	7.50	7.50	7.60	7.60	
	50	7.60	7.60	7.65	7.70	7.65	7.65
Tris	0	8.00					
	10	7.80	8.40				
	20	7.80	8.45	8.65			
	30	7.80	8.45	8.60	9.00		
	40	7.80	8.40	8.60	9.05	9.20	
	50	7.80	8.40	8.65	9.00	9.20	9.70

On the other hand, in this laboratory we have already applied the technique to the pH-stat method as well as to the potentiometric titrations of proteins in these conditions of solvent and temperature. Such investigations, which allow us to study the solvent and temperature effects on the net charge of the proteins as well as on their isoionic point, will be reported in chapter 3.

VALIDITY OF pa_H EVALUATIONS

Until standard "universal" pa_H scales can be established, presumably by a very sophisticated procedure, it will be necessary to check whenever possible the hydrogen ion activity referred to the standard state using reactions known as pH-dependent in aqueous solutions.

Since acidity and basicity are chemical properties, their true measure would be their effects on reactions and a large number of enzyme–catalyzed reactions, known to be markedly pH-dependent and showing a characteristic pH-activity profile, could be used to check values obtained by the

TABLE 28. pa_H of buffer systems in various methanol–water mixtures between 20° and the respective approximate freezing points. The inner reference solution and junction are made with methanol–water mixed solvent

Buffers	% MeOH	Temperature (°C)				
		+20	0	−15	−24	−32
Cacodylate	0	6.05				
	10	6.00	6.05			
	20	6.15	6.15	6.10		
	30	6.35	6.30	6.30	6.30	
	40	6.45	6.45	6.40	6.40	6.45
	50	6.65	6.60	6.70	6.60	
Phosphate	0	7.05				
	10		7.30			
	20	7.40	7.50	7.60		
	30	7.65	7.70	7.80	7.80	
	40	7.90	7.90	8.00	8.00	7.95
	50	8.10	8.10	8.20		
Tris	0	8.00				
	10	7.75	8.45			
	20	7.75	8.40	8.95		
	30	7.75	8.40	8.95	9.30	
	40	7.75	8.35	8.95	9.30	9.50
	50	7.70	8.30	8.90	9.25	9.50

indicator method. Several studies have been carried out in this laboratory on various biochemical systems and have shown that our pa_H scales in mixed solvents at any normal and subzero temperatures give satisfactory results. As an example of such studies, we would mention investigation of the tryptic hydrolysis of benzoyl-L-arginine ethyl ester (BAEE), the main results of which will be described and discussed in chapter 3.

Figure 23 shows the pa_H profiles obtained in water at 20°C and in water–glycerol at 20°, 0° and −20°C. The points shown are experimental values; the curves represent the theoretical values calculated from an expression quoted in chapter 3. The pa_H activity profiles reported in Fig. 23 allow us to draw some conclusions about the validity of the pa_H determinations and also, as we will see later (chapter 3), about the solvent and temperature effects on the ionization constant of the dissociable groups at the active site of the enzyme.

The similarity of the pH profiles obtained in various experimental conditions, as presented in Fig. 23 and particularly the good fit with the theoretical

TABLE 29. pa_H of buffer systems in various glycerol–water mixtures between 20°C and −12°C. The inner reference solution and junction are made with glycerol–water mixed solvent

Buffers	% GlOH	Temperature (°C) +20	+3	−4	−12
Cacodylate	0	6.05			
	10	6.05	6.05		
	20	6.10	6.05	6.05	
	30	6.15	6.10	6.10	6.05
	40	6.25	6.15	6.15	6.10
	50	6.40	6.25	6.20	6.15
Phosphate	0	7.10			
	10	7.15			
	20	7.15	7.15	7.15	
	30	7.20	7.20	7.15	7.15
	40	7.25	7.25	7.20	7.20
	50	7.25	7.30	7.25	7.25
Tris	0	8.05			
	10	8.00	8.40		
	20	8.00	8.40	8.55	
	30	8.05	8.35	8.50	8.80
	40	8.05	8.45	8.65	8.90
	50	8.10	8.50	8.70	8.95

curves of experimental points gathered in different buffers and for different temperatures, give reasonably convincing evidence of the validity of our pa_H scales. Their precision appears to be adequate for biochemical experiments.

However it is obvious that if the adjustment of suitable pa_H values as well as the choice of the buffer are a prerequisite to carry out enzyme activity in cooled mixed solvents, changes in pK_S of ionizing groups at the active site and corresponding shifts in the pH-activity curves must first be recorded to adjust new pa_H values to obtain an optimum enzyme-activity.

As they appear to date, the recorded changes in pK_S values of buffers and other weak electrolytes cannot be satisfactorily explained in terms of the dielectric constant variations of the media. The peculiar behavior of neutral and acidic buffers in the presence of cosolvent and as a function of temperature, the versatile effects on ionizing groups of proteins shown by different cosolvents as a function of temperature, are recorded facts which must be—and will be (see chapter 3)—evaluated to improve the understanding and use of mixed solutions.

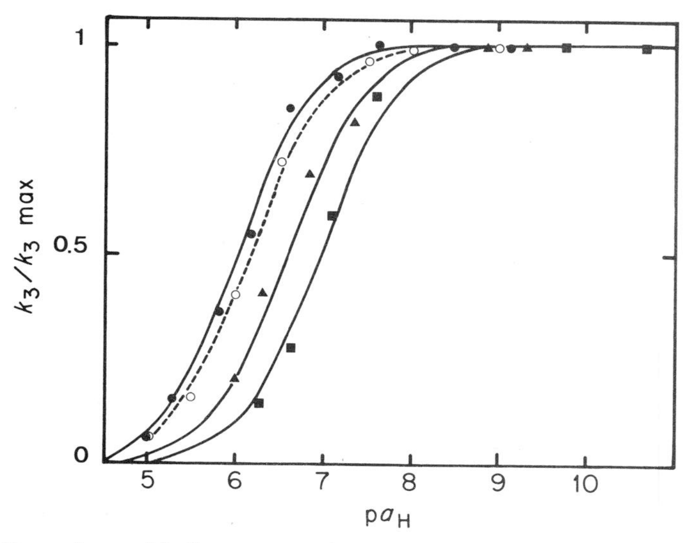

FIG. 23. Dependence of $k_3/k_{3\,max}$ on pa_H for tryptic hydrolysis of BAEE in pure water at 20°C (○) and in water–glycerol mixture (56%–44%) at 20°C (●), 0°C (▲) and −20°C (■).

Solubilities

It is known that the addition of organic solvents to aqueous salt solutions decreases the solubility of the salt and that for high salt concentrations the aqueous–organic mixtures separate into two phases, a denser phase rich in water and salt, and poor in organic solvent, and a lighter phase, rich in organic solvent with relatively little water and salt.

Quantitatively the effect of adding a protein to mixed solvents is just like that of adding a salt, and these different phenomena are termed "salting-out" effects even if such a term is technically a misnomer in describing the latter since proteins are dipolar ions and not salts.

It is also known that temperature relates directly to solubility in mixed solvents and that the solubility of salts and of most proteins decreases as the temperature is decreased.

Thus, keeping the solubility of salts, buffers and proteins in mixed solvents constant as a function of $1/T$ is a problem, and we will see that it is an art rather than a science, even if we take advantage of the background experiments already reported in the field of protein fractionation.

The underlying mechanism of solubility in water and in mixed polar solvents may be largely electrostatic, depending on the electrical energy of a system in the presence of the intense electrostatic fields around the ions (simple anions and cations, substrates and intermediates of metabolism), the dipolar ions (amino acids) or multipolar ions (proteins) which give rise to forces of great magnitude and depend on the dielectric constant of the medium. Changing the dielectric constant upon addition of organic solvent and cooling mixtures can explain changes in solubility.

Other factors will intervene, and among them the pH of the medium since it determines the net electrical charge of proteins and of many metabolites.

We will therefore examine the influence of dielectric constant and pH on solubility but we will see that the different microscopic dielectric constant around the solutes, and also a selective solvation of these solutes, causes deviations between theory and experiment. A mixed solvent is not a continuous, structureless medium, and the molecules of the more polar solvent tend to cluster around the ions according to electrostatics, forcing away the molecules of the less polar solvent.

As early as 1927, Debye (17) gave a theoretical treatment of the salting-out effect fairly close to the molecular reality and based on the assumption that the introduction of ions in an aqueous–organic mixture causes a redistribution of the solvent molecules, the water clustering preferentially around the ions and forcing away the molecules of the organic solvent. Quantitatively, this preferential (or selective) solvation may be formulated in terms of equations which we shall not try to develop, to avoid the introduction of certain mathematical concepts. The theory has been found to give reasonable results but there is a need for a more accurate theory of the salting-out effect, framed in terms of the dimensions, dipole moments and electrical polarizabilities of the individual molecules since the composition of a mixed solvent in the immediate neighborhood of the ions does not correspond to that in the bulk of the solution, but varies from point to point as one moves away from the surface of the ions.

This does not qualitatively alter the general electrostatic picture but it does require a more sophisticated calculation. Frequently in the following pages we will omit the proof of some propositions or the development of some mathematical expressions, since these are given in practically all texts on biophysical chemistry or even on electricity and magnetism.

SALTING-OUT AND SEPARATION OF PHASES

Theory

The simplest model system was considered by Debye and McAulay, 1924 (26), Scatchard, 1927 (27) and compiled by Edsall and Wyman (21) to explain the salting-out upon addition of organic solvent to salt aqueous solutions and the separation of mixed solvents in two phases upon addition of high concentrations of salt.

One dissociated salt molecule gives v ions treated as small charged sphere, v_+ cations with Z_+ charges and of radius b_c, and v_-, Z_+, b_a for the anions. When n_s salt molecules dissociate in a solvent to give v ions ($v = v_+ + v_-$) the electrostatic free energy correspondent will be

$$G_e = \left(\frac{N\varepsilon^2 v Z_+ Z_-}{2b}\right)\frac{n_s}{D} \tag{2-25}$$

where N is Avogadro's number, ε the electron charge, D the dielectric constant of solvent and b is a mean radius defined by:

$$\frac{1}{b} = \frac{2}{Z_+ Z_-}\left(\frac{Z_+}{b_c} + \frac{Z_-}{b_a}\right)$$

In fact, expression (2-25) can be simplified as

$$G_e = A\,\frac{n_s}{D}$$

where A is a constant characteristic of the salt.

Addition of organic solvent to a salt-aqueous solution, the dielectric constant of which is D_0, determines the transfer of ions to the mixture of dielectric constant D, and involves an electrostatic free energy change:

$$\Delta G_e = A\left(\frac{1}{D} - \frac{1}{D_0}\right)n_s$$

Thermodynamic calculations (29) give

$$\Delta G_e = RT \ln \frac{(N_s^\circ)\mathrm{sat}}{(N_s)\mathrm{sat}} = A\left(\frac{1}{D} - \frac{1}{D_0}\right)n_s \tag{2-26}$$

N_s is the molar fraction of salt

$$N_s = \frac{v \times n_s}{n_1 + n_2 + n_s}$$

where n_1, n_2, n_s are respectively the mole numbers of water, organic solvent and salt and v the total number of ions given by one molecule of salt. (N_s°) is the mole fraction of salt at saturation in water and (N_s)sat the same in the hydro–organic mixture.

This simplified theory assumed that the electrostatic term is the only one that causes a deviation from ideal behavior, which is an approximation. It can be seen from (2-26) formula that if $D < D_0$, $(N_s)\mathrm{sat} < (N_s^\circ)\mathrm{sat}$, i.e. addition of alcohol which decreases D also decreases the solubility of salt.

It can be calculated that if sodium chloride ($b \simeq 2$ Å) is transferred from an aqueous solution ($D_0 = 80$) into ethanol ($D = 25$) at $T = 300$°K $\log (N_s^\circ)\text{sat}/(N_s)\text{sat} = 3.3$ that is NaCl is found somewhat than 2000 times as soluble in water as in ethanol which is the experimental value.

Since the addition of alcohol to water raises the activity coefficient of salt and therefore decreases its solubility, it is interesting to check that, conversely, the addition of salt to an alcohol–water mixture can act on the solubility of alcohol. In fact, this mutual relation between these two effects is a particular case of Gibbs' theorem:

The variation in Gibbs free energy (G) of a system of m components at constant pressure and temperature, expressed as a function of the masses $n_1, n_2, \ldots, n_m$, is given by

$$dG = \mu_1 \, dn_1 + \mu_2 \, dn_2 + \cdots + \mu_m \, dn_m$$

The chemical potential μ_i of ith component is

$$\mu_i = \left(\frac{\delta G}{\delta n_i}\right) PTn_j$$

where n_j denotes that masses of all components except the ith are constant. If we consider the variation of i by the amount dn_i, and of j by the amount dn_j, the two variations will be related by the equation

$$\frac{\delta \mu_i}{\delta n_j} = \frac{\delta^2 G}{\delta n_i \, \delta n_j} = \frac{\delta \mu_i}{\delta n_i}$$

(since G is an exact differential).

Since $\delta \mu_i = RT \ln a_i$, the above can be expressed in terms of activities by

$$\frac{\delta \ln a_i}{\delta n_j} = \frac{\delta \ln a_j}{\delta n_i} \tag{2-27}$$

We see that since adding a component j increases the activity of component i, it necessarily follows that adding i to the system must increase the activity of component j.

In the three-component system considered here (where $m = 3$, water being 1, alcohol 2, and salt 3) we can infer from the above equations that addition of alcohol decreases the solubility of the salt (salting-out effect), and that addition of salt decreases the solubility of the alcohol, leading eventually to the separation in two phases.

Experimental data at least qualitatively demonstrate the above predictions: the salting-out effect of miscible organic solvents, which lower the dielectric constant of water, is observed in presence of various neutral salts. Conversely, substances such as cyanhydric acid which raise the dielectric constant produce a salting-in effect in the presence of these salts. Nevertheless, the agreement of the experimental data with the various relations is not by any means exact, the equations being only empirically useful; but the observed phenomena confirm that the forces underlying the salting-in and the salting-out effects are primarily electrostatic.

Experimental aspects

Consequently, temperature variations of mixed solvents which bring about marked changes in dielectric constant should be important in the solubility of salts and the miscibility of water and the organic moiety.

It has been observed in this laboratory that most of aqueous–organic mixtures with a high content in organic moiety ($\geq 50\%$) often give a precipitation of salt at low temperature ($< -30°C$) for concentrations higher than 0.2–0.3 M, and sometimes a separation of the two phases, the aqueous one giving an iceberg containing most of the electrolytes and other eventual solutes.

On the other hand, we will see later that low salt concentrations favor the solubility of proteins in mixed solvents, at normal and, to a lesser extent, subzero temperatures and that the main problem for enzyme–catalyzed reactions will be to find a balance between the precipitating action of organic solvents and the solvent effect of salts.

At the beginning of this chapter, when we dealt with the structure of liquid water and of its change upon addition of a cosolvent, we recalled that authors such as Eley (3) and Ben Naïm (2), assume that addition of a non-polar or partially non-polar solute drives the equilibrium of the distinct species of water towards the species of highest density.

In Eley's model, weakly protic solvents would be expected to restrict the entrance of some other partially non-polar solutes into the water structure and thus would decrease solubility as long as the integrity of water structure is retained in mixed solutions.

The experiments of Ben Naïm (2) on the solubility of argon—a non-polar gas—in water and in dilute aqueous solutions of primary alcohols and polyols provide the possibility that different cosolvents may shift the equilibria in either direction to increase or decrease the solubility of a solute with non-polar parts. But obviously, for a high concentration in cosolvent, the decrease of solubility would be the predominent one.

IONIC STRENGTH, DIELECTRIC CONSTANT AND SOLUBILITY OF PROTEINS

A protein can be considered as a zwitterion, the solubility of which is given as N/N_0 where N is the solubility at a given finite ionic strength and N_0 is the solubility at zero ionic strength in the same solvent.

For the simplest case of a spherical zwitterion of radius b with a point dipole of moment μ located at its center, Kirkwood (29) proposed the following expression for log N/N_0 as a function of low ionic strength $\Gamma/2$

$$\log \frac{N}{N_0} = \left(k \frac{\mu^2}{a(DT)^2} - \frac{b^3}{a^2} \frac{f(\rho)}{DT} \right) \frac{\Gamma}{2} \tag{2-28}$$

a is the "collision diameter" equal to b plus the mean radius of the ions in the solution, $f(\rho)$ is a function of $\rho = b/a$, k is a function of physical constants and D is the dielectric constant of the medium. The effect of a salt on the logarithm of the solubility is proportional to the first power of ionic strength.

In equation (2-28), the first term in parentheses reflects the salting-in effect due to electrostatic forces, and is proportional to the square of the dipole moment (μ) of the molecule and is inversely proportional to the square of the dielectric constant of the solvent. The second term reflects the salting-out effect which increases essentially in proportion to the volume of the molecule, and varies as the reciprocal of the first power of the dielectric constant and therefore becomes relatively negligible, as compared to the first term, when the dielectric constant goes to small values.

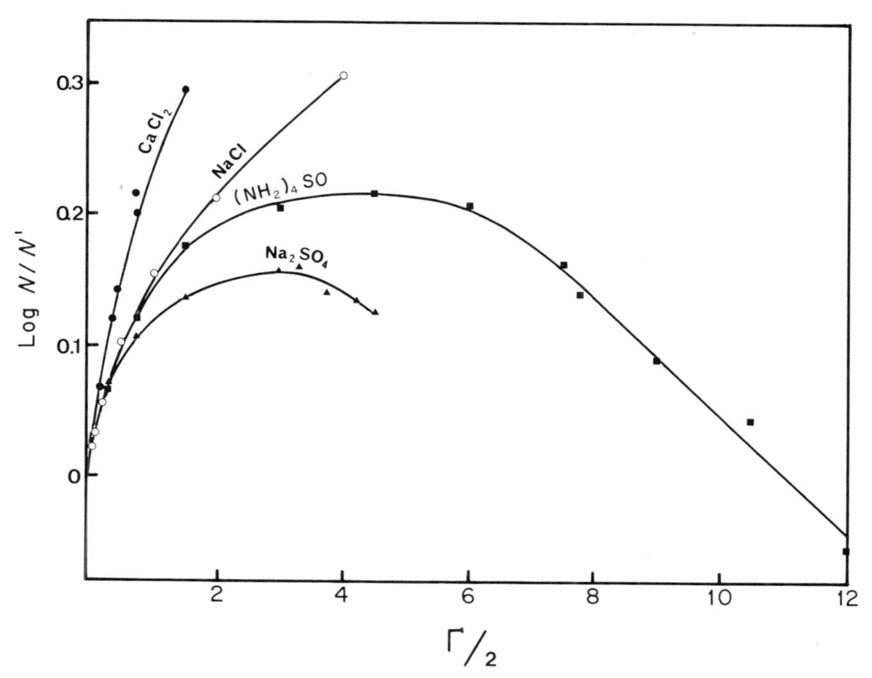

FIG. 24. Solubility of cystine in aqueous salt solutions.

Kirkwood (18) also made calculations for an ellipsoidal model of a dipolar ion, with positive and negative charges located at the foci of the ellipsoid, and the salting-in coefficient was nearly proportional to the first power of the dipole moment instead of its square, as in the above case of a sphere.

Thus the Kirkwood relation is derived from calculations using very crude models in which non-electrostatic factors are ignored. It is nevertheless true that these calculations give a remarkably good representation of the behavior of amino acids and proteins in solutions of varying ionic strength.

For example, the solubility of cystine (dipolar ion) was tested by McMeekin (30) in various aqueous salt solutions, and the solvent effect of different salts was shown to be proportional to the ionic strength in dilute solution, and that it increased in the order Na_2SO_4, $(NH_4)_2SO_4$, NaCl, $CaCl_2$, as given in Fig. 24. Adding of solvent, which decreases the dielectric constant, leads to a large decrease in the solubility of the proteins in the absence of salt and also sometimes to their denaturation. If salt is now added, we reduce the possibility of denaturation and the solubility of proteins increases. Indeed the decrease in dielectric constant enhances the electrostatic interactions between the ions of the salt and the charged groups of the protein, and thereby causes the salting-in effect to be much larger than in water.

As an example of such a process, the solubility of glycine in water, in ethanol–water mixtures and salt solutions at 25°C was reported (28) as a function of the ionic strength and of the dielectric constant D of each mixture (Fig. 25). Two major effects were then observed:

1. At zero ionic strength, the solubility of glycine decreases greatly with decreasing D.
2. The addition of solvent increases the solubility of glycine, and the rate of increase for a given increment in ionic strength is greater, the lower the dielectric constant.

According to the above results, it can be seen that cooling of mixed solvents at zero ionic strength might increase the solubility of zwitterions (amino acids, peptides, proteins) as D increases (salting-in effect), and that cooling of these mixtures in presence of salt might, in turn, decrease the solubility of the dipolar ions since D increases (salting-out). In fact, as shown by Travers in this laboratory, glycine is soluble in methanol–water (50 : 50, v/v) at concentrations up to 0.5 M at 20°C, at zero ionic strength; glycine precipitates between -10 and -50°C but remains soluble in this temperature range at 0.4 M.

Further systematic investigation is urgently needed to cope with these experimental aspects of solubility in mixed solvent as a function of temperature in the range of subzero temperatures, for we will see below that the methods applied to the fractionation of plasma components, which mostly

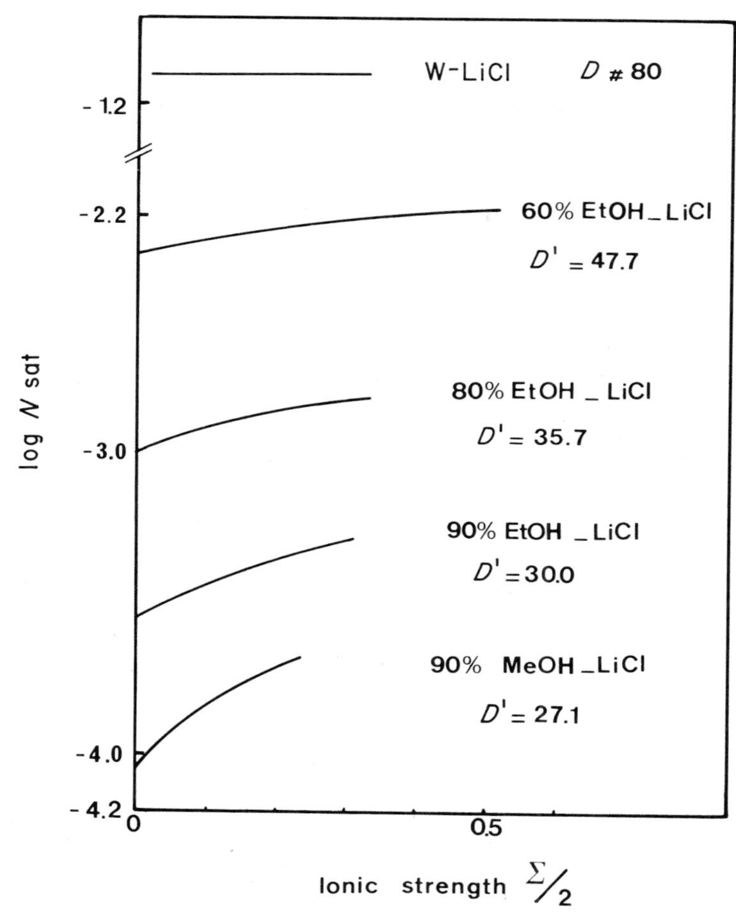

FIG. 25. Solubility of glycine in water, ethanol–water mixture and salt solutions at 25°C. D is the dielectric constant in the absence of glycine.

represent the only systematic attempts at using cooled mixed solvents, are carried out with very low mole fractions of cosolvent and at rather high temperatures (−10° to 0°C) compared to those used in the present procedure.

pH AND SOLUBILITY OF PROTEINS

The negatively charged carboxyl groups of free glutamic and aspartic acid residues and the positively charged ammonium groups of lysine, the guanidinium group of arginine, and the imidazolium group of histidine of proteins exert strong attractive forces on the water dipoles, the latter adjusting their orientation so as to attain a minimum of potential energy. Therefore these groups enhance the solubility of proteins in water. They decrease it in aqueous–organic mixtures of lower dielectric constant, in which the electrical free energy G_e (and hence the activity) of the molecules is greatly increased.

It is known that the number of charged groups in protein molecules varies with pH. At strongly acid pH values, the arginine, lysine, histidine and α-ammonium residues are all positively charged, while the carboxyl, sulfhydryl and phenolic hydroxyl groups are uncharged. At strongly alkaline pH values, the converse relation holds and the protein carries its maximum net negative charge while the groups which are positively charged in acid medium become uncharged.

$$\underset{\text{acid}}{R{-}CH(NH_3^+)(COOH)} \xrightarrow{+H^+} \underset{\text{neutral}}{R{-}CH(NH_3^+)(COO^-)} \xrightarrow{-H^+} \underset{\text{alkaline}}{R{-}CH(NH_2)(COO^-)}$$

At pH values between 4 and 8 at which most enzyme–catalyzed reactions are carried out, the net charge of a protein molecule is much smaller than in the strong acidic and alkaline range, but the total number of positive and negative charges reaches a maximum somewhere in this range. Commonly, the solubility reaches a minimum at or near the isoionic point of the protein, and increases with change of pH to either side of this point.

Thus there is one particular pH at which the protein molecule carries no net effective charge. This is not to say that the charge is zero, but that the numbers of positive and negative charges are equal, and the actual charge is $\pm$. This corresponds to the isoelectric pH of a protein (pI).

The observed solubility of the protein represents in fact the sum total of the concentration of a great number of microscopic anions, cations, zwitterions which are all present in a mobile equilibrium is solution. As the pH is changed, the relative concentration of such forms is altered. The concentrations of the cations increase as the pH is shifted to the acidic side of the isoelectric point, and those of the anions as the pH is shifted to the alkaline side

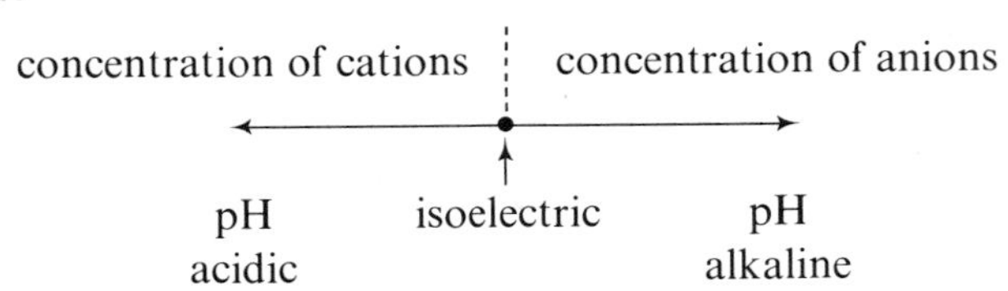

If the logarithm of the solubility is plotted against pH at constant ionic strength, the slope of the curve at any point should be proportional to the net charge of the protein at that pH.

Since the solubility of amino acids, peptides and proteins also depends on the ionic strength, the minimum solubility at the isoelectric point increases rapidly as the ionic strength is increased.

It is known that solubility curves for proteins as a function of pH fall into two general classes: (*a*) those proteins in which solubility rises on either side of the isoelectric point (the pH of minimum solubility changing only very slightly with change of ionic strength; an example is given by β-lactoglobulin in Fig. 26 (29)) and (*b*) those proteins in which increase in ionic strength decreases solubility on the acidic side of the isoelectric point while increasing

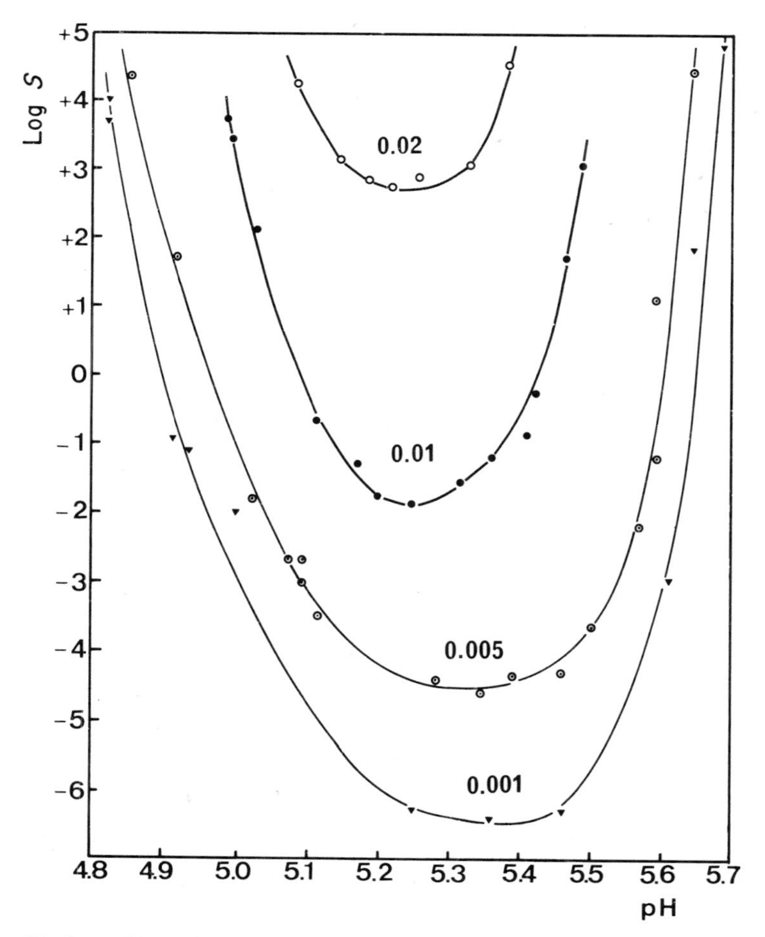

FIG. 26. Logarithm of solubility of β-lactoglobulin as a function of pH in water at four different ionic strengths indicated on each curve.

it on the alkaline side. Thus any variation of the pH due to the solvent addition and cooling of mixtures might determine variations of the solubility of proteins and therefore be used—when controlled—to assist the solubility or the precipitation of these proteins.

SOLUBILITY AND PRECIPITATION OF PROTEINS IN MULTI-VARIABLE SYSTEMS

Fractionation of protein mixtures in cooled mixed solvents

We have seen previously that proteins are most sensitive to the solubilizing action of electrolytes at low ionic strengths and in mixed solvents of low dielectric constant. The precipitation of proteins involves five essentially independent variables, namely protein concentration, salt concentration, organic solvent concentration, pH and temperature. In such a complex system, there is a wide range of possibility conditions which can be sought out for the solubility or the precipitation of any protein.

Large scale fractionation of plasma proteins has been carried out at low temperature and at low ionic strength in the presence of water–ethanol precipitants (31). This organic solvent, as well as acetone, were then used instead of high concentrations of salts employed in the salting-out procedure performed in aqueous solutions.

Since ethanol and acetone are denaturing agents, irreversible changes in the labile protein molecules can generally be minimized if the temperature is maintained sufficiently low—below 0°C and down to −10°C.

The system under consideration contains five independent variables, and if three of them are maintained constant and the other two are chosen such that one will increase and the other decrease solubility, conditions will be found and curves constructed defining the constant solubility of any protein. Some of the theoretical aspects of this subject have been reviewed (29,32) and we will summarize some technical aspects.

Certain separations of proteins can be carried out in solutions containing either an organic solvent with dipolar ions such as glycine which increase the dielectric constant of the solution. It is known that small variations of the dielectric constant of the medium can induce great changes in solubility, because of the dipole, quadrupole and higher electric moments arising from the distribution of positive and negative charges within the protein molecule and over its surface, and therefore the addition of amino acid will introduce new solubility conditions.

In contrast to the action of the dielectric constant in aqueous–organic mixtures, the salting-out effect in highly concentrated salt solutions is relatively insensitive to the specific chemical characteristics of the protein, depending mostly on unspecific factors such as size and shape.

The methods of fractionation by the low temperature—low salt—ethanol procedure have been applied since 1940 to the systematic fractionation of plasma. There has been a gradual evolution and decisive advance both in respect of the number and specificity of the fractions separated and the preservation of components in undenatured form.

In general, the separations have been achieved in the following conditions:

Concentrations of protein as low as possible to minimize protein–protein interactions, but as high as possible in order to render practical large scale processing.

Concentration of the organic solvent—mostly ethanol—and the temperature are maintained as low as possible in order to minimize protein denaturation. The mole fractions of the mixture do not generally exceed 0.163 and the temperature is maintained between 0°C and the freezing point.

Concentration of salt is as low as possible to take advantage of the diversity of protein–electrolyte interactions and to avoid the necessity of dialysis. The salt concentration is generally <0.2 M, and buffer concentrations between 0.01 and 0.05 M.

It can be seen that these conditions are widely different from ours which might be tested in order to try to improve the fractionation methods which are still far from successful.

Conditions of solubility in cooled mixtures

As stated by Cohn and co-workers in their classic paper on the fractionation of the protein and lipoprotein components of biological tissues and fluids (31), "by balancing precisely the solvent action of an electrolyte with the precipitating action of the organic liquid, widely different conditions can be defined, such that the solubility of the proteins under consideration remains constant". For each protein component, this balance at constant solvent and salt concentration will be different for each pH and temperature. In the fractionation procedure considered above, where, as we said, concentration of the organic solvent is as low as possible (for ethanol is then a denaturing agent), and experiments are carried out at only moderately low temperature such as $-5°$ or $-10°$C to avoid freezing, because this leads to higher than desired ethanol concentration in the liquid phase, temperature is a limited variable.

In our own experimental conditions, that is according to the low freezing point of most of our mixed solutions, temperature becomes a very powerful variable:

The solubility of most proteins will markedly decrease as the temperature is lowered and their separation might eventually be carried out by a stepwise lowering of the temperature. To keep the solubility of a given protein at a constant level, a precise concentration has to be found, as low as possible to avoid its precipitation at low temperatures and as high as possible to allow their investigation. Such a goal is in most of cases easily reached in the case of enzymes, which can be used in relatively low concentrations.

On the other hand, the addition of organic solvent to water and the cooling of corresponding mixtures change both the pK of amino acids, peptides, proteins, and the pH of buffers.

In these conditions, it is essential to record numerical values of pK and pH for any given volume ratio of solvent and water, and also as a function of $1/T$, for the balance of solvent and precipitating actions should be shifted by these variations upon cooling, and lead to solubilization or precipitation according to circumstances.

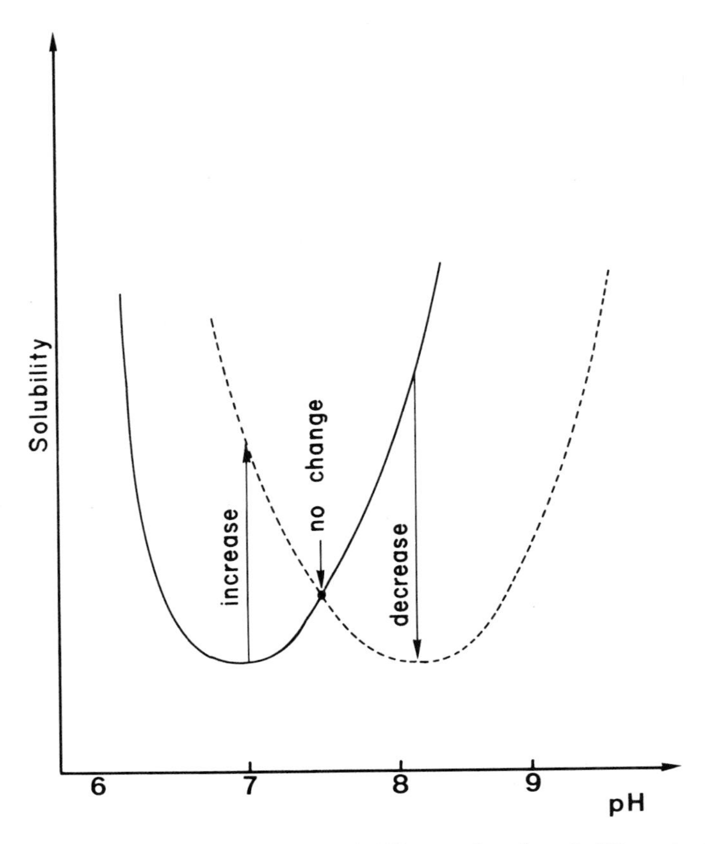

FIG. 27. Model representation of protein solubility as a function of pH in water (———) and in hydro-alcoholic solvent (-----). Arrows indicate solubility change from water to hydro–alcoholic solution.

Let us consider the eventual pK variations and their influence on solubility: If, as expected, the pK of proteins increases upon addition of organic solvent and next again during cooling of mixtures, there will be a shift of solubility curves as illustrated below in Fig. 27. According to the pH value of the medium, 3 cases can be considered:

1. The pH corresponds to the "isobestic" point between the two curves. No change, and solubility minimum in both cases.
2. The pH is on the left-hand side of the isoelectric point. An increase in solubility is to be expected.
3. The pH is on the right-hand side. A decrease in solubility can be expected. The condition to be fulfilled to obtain 2 and 3 is that solubility rises on either side of the isoelectric point.

Thus in order to control the solubility, one must know the "new" pK of proteins, as well as the pa_H of buffers. We have seen that this is now possible.

Conclusion

It is the author's hope that data gathered over several years and reported in this chapter will allow a large number of colleagues to use mixed solvents and temperature variations with the possibility of knowing the exact variations of some important parameters and to adjust the values of some of them, such as dielectric constant and pa_H.

The possibility of preparing mixed solutions at selected temperatures which are isodielectric with aqueous solutions and contain any suitable ionic content (proton activity, ionic strength) is a prerequisite in the investigation of "direct" solvent and temperature effects on protein structure and activity, both in order to carry out the study of enzyme catalyzed reactions in mixed solvents at subzero temperatures and to use cosolvents as tools in molecular biochemistry.

References

1 H. S. Franck and M. N. Evans. *J. Chem. Phys.* **13**, 507 (1945).
2 A. Ben Naïm. *J. Phys. Chem.* **62**, 1922, 3240 (1965).
3 D. D. Eley and R. B. Leslie. *Adv. Chem. Phys.* **7**, 238 (1964).
4 I. M. Klotz. *Horizons in Biochemistry*, B. Pullman and M. Kasha, Eds. Academic Press, New York (1962).
5 M. F. Perutz and H. Lehmann. *Nature* **219**, 92 (1968).
6 W. Bolton, J. M. Cox and M. F. Perutz. *J. Mol. Biol.* **33**, 283 (1968).
7 K. N. V. von Casimir, F. Keilmann, A. Mayer and H. Vogel. *Biopolymers* **6**, 1705 (1968).
8 W. Lipscomb, J. Hartstuck, G. Reeke Jr., F. Quinocho, P. Berthge, M. Ludwig, T. Steitz, H. Muirhead and J. Coppola. Brookhaven Symp. *In Biol.* **21**, 24 (1968).

9 R. Lumry and S. Ragender. *Biopolymers* **9**, 1125 (1970).
10 G. Nemethy and H. A. Sheraga. *J. Chem. Phys.* **41**, 680 (1964).
11 G. Nemethy and H. A. Sheraga. *J. Phys. Chem.* **66**, 1773 (1962).
12 J. A. Riddick and W. B. Bunger. *Organic Solvents. Techniques of Chemistry*, A. Weissberger, Ed., Vol. 2. Wiley-Interscience (1970).
13 F. Travers, P. Douzou, T. Pederson and I. C. Gunsalus. *Biochimie* **57**, 43 (1975).
14 J. Wyman. *J. Am. Chem. Soc.* **58**, 1842 (1936); *Chem. Revs.* **19**, 213 (1936).
15 J. Timmermans, A. M. Piette and R. Philippe. *Bull. Soc. Chim. Belge* **64**, 5 (1955).
16 G. Akerlof. *J. Am. Chem. Soc.* **54**, 4125 (1932).
17 P. Debye. In *Collected Papers of P. Debye*, p. 335. Interscience Publishers, New York (1954).
18 J. G. Kirkwood. *J. Chem. Phys.* **7**, 911 (1939).
19 F. Travers and P. Douzou. *J. Phys. Chem.* **74**, 2243 (1970).
20 F. Travers and P. Douzou. *Biochimie* **56**, 509 (1974).
21 J. T. Edsall and J. Wyman. *Biophysical Chemistry*, Vol. I, 472. Academic Press, New York (1958).
22 R. Shedlovsky and R. L. Kay. *J. Phys. Chem.* **60**, 157 (1956).
23 P. Maurel, G. Hui Bon Hoa and P. Douzou. *J. Biol. Chem.* **250**, 1376 (1975).
25 C. Larroque, P. Maurel, C. Balny and P. Douzou. *Anal. Biochem.* **73**, 9 (1976).
26 P. Debye and J. McAulay. *Physik. Zs.* **26**, 22 (1924).
27 G. Scatchard. *Chem. Revs.* **3**, 383 (1927).
28 J. T. Edsall and J. Wyman. *Biophysical Chemistry*, Vol. I, p. 266. Academic Press, New York (1958).
29 J. T. Edsall. *Adv. Prot. Chem.* **3**, 384 (1947).
30 T. L. McMeekin and R. C. Warner. *J. Am. Chem. Soc.* **64**, 2393 (1942).
31 E. J. Cohn, L. E. Strong, W. L. Hughes, D. L. Mulford, S. N. Ashworth, M. Melin and H. L. Taylor. *J. Am. Chem. Soc.* **68**, 459 (1946).
32 S. J. Singer. *Adv. Prot. Chem.* **17**, 1 (1962).

3

Solvent and Temperature Effects on Enzyme Activity

SOLVENT EFFECT

Many isolated reports in the literature show that any mixed solvent can be, according to the selected enzyme system, an inhibitor as well as an activator of enzyme specific activity. Such an effect is characterized by a definite degree, depending on the cosolvent concentration, and is reached fairly rapidly and thereafter is independent of time, provided that the solvent is chemically stable. It must be totally reversible by infinite dilution.

We discovered such characteristics when we tried enzymes belonging to the groups of dehydrogenases, flavoproteins, hemoproteins, luciferases, hydrolytic enzymes, the effects varying with the type of solvent and, for any solvent, with the type of enzyme experienced. In most cases, enzymes which consist of subunits were dissolved in polyol–water mixtures without losing their potentiality when they were diluted in aqueous solutions. In a number of cases, these proteins were successfully investigated in methanol–water mixtures prepared by the sampling procedure described in chapter 2.

The making of mixed solutions isodielectric with water and containing a suitable ionic content certainly enhances the stability of proteins, but is unlikely to ensure an optimal enzyme activity, i.e. the activity observed in aqueous solution. No satisfactory explanation or correction of such an "intrinsic" cosolvent effect on enzyme activity can be proposed for any particular system as long as the effect of the predictable influencing factors on protein structure and activity has not been established with suitable precision. The physico-chemical data reported in chapter 2 offer us the opportunity to adjust selected values of dielectric constant and pH* such that it becomes possible to test the actual influence of electrostatic parameters on reaction rates as compared to the intrinsic cosolvent effect mediated through the conformational change of the protein.

Such a possibility has been applied to some enzyme systems and the main results will be summarized and discussed. They show that, in the presence of high concentrations in cosolvent ($\geq 50\%$ by volume), the effect of the bulk dielectric constant value is weak compared to the cosolvent effect. This effect can be the sum of changes of the pK of ionizing groups at the active site, changes which can be "corrected" by adjusting new pH* values, and of alterations mediated through sites on the protein molecule other than the active site and which could induce a new protein conformation with new kinetic characteristics.

Since only very few systems have so far been investigated, we will not draw general rules for the use of mixed solvents and for the effects they will show towards enzyme structure and activity, but we will rather propose a methodology to investigate any particular system to be submitted to the low temperature procedure, in order both to minimize its effects on enzyme activity and to check whether changes in reaction rates occur without changes in mechanism.

Electrostatic effects on enzyme reactions

EFFECT OF THE DIELECTRIC CONSTANT

Electrostatic effects have been most often held to be primarily responsible for the cosolvent effect on reaction rates, and a number of theoretical expressions have been made to correlate this effect to the influence of the dielectric constant of the aqueous–organic mixture, particularly in the case of interactions of dipolar and ionic substrates with the enzyme, these electrostatic interactions being changed when the dielectric constant is reduced by the presence of the cosolvent.

However, deviations between theoretical treatments and experiments have been found in most cases, even in the simplest, involving ionic polyatomic reactants.

The prevailing opinion is that, in general, the electrostatic approach to the theoretical interpretation of chemical reactions is inadequate, due to the complexity of parameters influencing these reactions.

However it can be postulated that the effect of dielectric constant or electrostatic forces among reacting molecules alters the readiness with which reactants approach each other and change the rate.

The Coulombic forces between enzymes and substrates might be significant since these reactants consist largely of highly polar molecules and of ions and dipole ions carrying a net electric charge. These electrically charged structures give rise to electrostatic forces of great magnitude and their interactions can be properly understood by considering their operation.

Various equations have been established and their validity tested plotting the logarithm of the reactions rate constant ($\log k$) vs the reciprocal of the dielectric constant ($1/D$). No exhaustive listing of applications of the relation $\log k = 1/D$ will be presented here, but rather some examples where this expression has been successfully and unsuccessfully applied, to make the reader aware of the various causes of deviation from theory.

Using Coulombic energy consideration, Amis (1) derived the dielectric constant dependence of k applicable to various electrically charged and electrically unsymmetrical combinations of reactants and eventually to enzyme-catalyzed reactions.

The potential energy between an ion and a dipole being given by

$$E_c = \frac{-Z\mu_0 \cos\theta}{Dr^2}$$

the difference in Coulombic energy, ΔE_c, of the ion-dipole at two different dielectric constants D_1 and D_2 is

$$\Delta E_c = -\frac{Z\varepsilon\mu_0}{r^2}\left(\frac{1}{D_2} - \frac{1}{D_1}\right)$$

The energy of activation E_{D_1} for the same reaction at dielectric constant D_1 will result in

$$E_{D_1} = -\frac{Z\varepsilon\mu_0}{r^2 D_1}$$

$$E_{D_2} = E_{D_1} + \Delta E_c$$

and the specific velocity constant k_{D_2} at dielectric constant D_2 will be given by

$$k_{D_2} = A \cdot \exp\left(-\frac{E_{D_2}}{RT}\right)$$

with

$$\ln k_{D_2} = \ln A - \frac{E_{D_2}}{RT} = \ln A - \frac{E_{D_1}}{RT} - \frac{\Delta E_c}{RT}$$

but since

$$\ln k_{D_1} = \ln A - \frac{E_{D_1}}{RT}$$

then

$$\ln k_{D_2} = \ln k_{D_1} + \frac{Z\varepsilon\mu_0}{RTr^2}\left(\frac{1}{D_2} - \frac{1}{D_1}\right) \tag{3-1}$$

If D_2 is taken as infinite in magnitude so that all electrostatic effects between reactants vanish, and if D_1 is taken as any dielectric constant D, it follows from (3-1) that

$$\ln k_{\infty} = \ln k_D - \frac{Z\varepsilon\mu_0}{RTr^2D}$$

or

$$\ln k_D = \ln k_{\infty} + \frac{Z\varepsilon\mu_0}{RTr^2D} \tag{3-2}$$

According to the last equation, when the charge sign of the ion is taken into account, a plot of ln k_D vs $1/D$ would be a straight line of positive slope if $Z\varepsilon$ is positive, and of negative slope if $Z\varepsilon$ is negative. These predictions are in agreement with those of the Amis–Jaffé (2) equations for the rates of reactions between ions and dipolar molecules. All these equations contain parameters such as μ_0, r, D whose values are ordinarily not recorded in the literature, and reasonable values can be assumed to correlate theory with experiment. Sometimes, data have been found to be in real conflict with theory (3). The application of the Amis–Jaffé equation to both the acid and alkaline hydrolyses of esters in various mixed solvents at selected temperatures has given in most cases good correlation with theory (4).

The effect of the dielectric constant on the velocity constant k expressed by equation (3-2) might affect the Arrhenius frequency factor A in the equation for k, namely $k = A \cdot \exp(-E_A/RT)$ and/or the activation energy E_A. According to Moelwyn-Hughes (5), when the rates in solution of certain second-order ion-dipolar molecule reactions differ greatly, depending on the ion and dipole reactants selected and on the solvent used, the Arrhenius frequency factor may remain practically constant and the differences in rates arise mainly from variations in activation energy. It has been shown (6) that the rates of catalytic chlorination of ethers of the type ROC_6H_4X, depending on the nature of R and X, varied at 20°C by a factor of 3300, and these differences were accounted for by a variation in E_A of 4.8 kcal/mol while A remained constant.

There are cases in which there would be no effect of dielectric constant on reaction rates for, as pointed out by Benson (7), the Coulombic term of the simple electrostatic models for ion-dipolar molecule reactions is of the same order of magnitude as the difference in free energies of hydration of ion A^{Z_A} and the transition complex X^{Z_A}.

This author suggests that a fairer model for the formation of X^{Z_A} might result from considering it as resulting from the displacement in the solvent sheath of the ion of a solvent molecule of dipole moment μ_S by a reactant molecule of dipole moment μ_B.

Neglecting dipole–dipole interactions, the free energy of formation of the transition state would be given by:

$$G_\mu = \frac{Z_A \varepsilon \mu_B \cos\theta}{D r_B^2}\left(1 - \frac{\mu_S\, r_B^2}{\mu_B\, r_S^2}\right)$$

It can be seen that the effect of the dielectric constant D is tied up with μ- and r- the dipole moment and the distance of closest approach of a solvent molecule to the ion.

For

$$\frac{\mu_S}{r_S^2} = \frac{\mu_A}{r_B^2}$$

G_μ would be zero and there would be no effect of D.

These are some examples of semi-empirical calculations which try to predict or explain the dielectric constant effect on reaction rates. The situation might be much more complicated in the case of enzyme-catalyzed reactions which we will now examine.

Using reasonable values of the parameters involved, Hiromi (8,9) studying the α-amylase-catalyzed hydrolysis of amylose found straight lines with negative slopes when log k_2 was plotted against the reciprocals of D of the media, for the mechanistic equation

$$\mathrm{E + S} \underset{k_{-1}}{\overset{k_1}{\rightleftharpoons}} \mathrm{E - S} \xrightarrow{k_2} \mathrm{E + P}$$

Barnard and Laidler (11), considering some proteolytic enzyme systems, set up an equation which can be applied to the interaction between enzyme and substrate, and also to the subsequent breakdown of the complex.

Methanol–water mixtures with up to 25% of methanol have been used, but although this method of varying the dielectric constant has led to some interesting conclusions, it must be admitted that it is not entirely satisfactory. The results of this method must be regarded as only semi-quantitative, and once again it appears that a considerable source of uncertainty arises from the simplification of considering the solvent as a continuous dielectric and without interaction with part of the enzyme molecule.

SIMULTANEOUS OPERATION OF A DIELECTRIC AND A COMPETITIVE EFFECT

Clement and Bender (12) tried to explain the effects of dipolar aprotic solvents (dioxane, acetone and acetonitrile) on an α-chymotrypsin-catalyzed reactions in terms of simultaneous operation of a dielectric and a competitive effect and succeeded in accounting quantitatively for the results obtained. Since these effects in themselves have proved to be inadequate to explain the action of organic solvents, their combination might produce a new interpretation.

These authors considered the simplest competitive inhibition, that is

$$\mathrm{E} + \mathrm{S} \underset{}{\overset{K_m}{\rightleftharpoons}} \mathrm{ES} \longrightarrow \mathrm{E} + \mathrm{P}$$

$$\mathrm{E} + \mathrm{I} \underset{}{\overset{K_\mathrm{I}}{\rightleftharpoons}} \mathrm{EI}$$

and showed that simple competitive inhibition of such a type leads to

$$\frac{K_{m(\mathrm{obs})}}{K_{m(\mathrm{org})}} = 1 + \frac{I}{K_{\mathrm{I(org)}}} \tag{3-3}$$

where $K_{m(\mathrm{obs})}$ is the observed K_m in the mixture, $K_{m(\mathrm{org})}$ is the real K_m in that mixture, $K_{\mathrm{I(org)}}$ is the K_I in that mixture, and I is the concentration of inhibitor. It is then convenient to use the exponential form of the equation proposed by Laidler (11) from absolute reaction rate theory:

$$\ln k - \ln(k)_0 = \frac{A}{DT}$$

where k is either a rate constant or an equilibrium constant of the medium, A is a constant, D is the dielectric constant and T is the absolute temperature.

For the effect of D of the medium on K_m one can show that

$$\frac{K_{m(\mathrm{org})}}{K_{m(\mathrm{H_2O})}} = \exp(AX) \tag{3-4}$$

and for the effect of D on K_I

$$\frac{K_{\mathrm{I(org)}}}{K_{\mathrm{I(H_2O)}}} = \exp(BX) \tag{3-5}$$

The constants A and B include the temperature dependence of the dielectric constant and X is the difference of the reciprocals of the dielectric constants of the mixture and of pure water.

A combination of equation (3-3), (3-4) and (3-5) leads directly to

$$\frac{K_{m(\mathrm{obs})}}{K_{m(\mathrm{H_2O})}} = \mathrm{e}^{AX}\left[1 + \frac{I}{K_{\mathrm{I(H_2O)}}}\mathrm{e}^{BX}\right] \tag{3-6}$$

A similar equation can be derived for the solvent effect on the observed second-order rate constant k_2/K_m:

$$\frac{\dfrac{k_2}{K_{m(\mathrm{H_2O})}}}{\dfrac{k_2}{K_{m(\mathrm{obs})}}} = \mathrm{e}^{AX}\left[1 + \frac{I}{K_{\mathrm{I(H_2O)}}}\mathrm{e}^{BX}\right] \tag{3-7}$$

A and B are measures of the electrostatic interaction of the enzyme with the substrate or inhibitor.

A decrease in the dielectric constant increases the repulsion between E and S (or I) and therefore decreases the binding.

The solutions of equations (3-6) and (3-7) necessitate the use of an indirect method; considering data in a solvent mixture for two different substrates S_1 and S_2, it can be shown that the following equations hold, for a combination of mixed second order and turnover reactions and for a combination of two second-order reactions, respectively:

$$\log[k^{S_1}_{(obs)} K^{S_2}_{m(obs)}] = \frac{(A_{S_2} - A_{S_1})X}{2.303} + \log[k^{S_1}_{(H_2O)} K^{S_2}_{m(H_2O)}]$$

$$\log\left[\frac{k^{S_1}_{(obs)}}{k^{S_2}_{(obs)}}\right] = \frac{(A_{S_2} - A_{S_1})X}{2.303} + \log\left[\frac{k^{S_1}_{(H_2O)}}{k^{S_2}_{(H_2O)}}\right]$$

α-chymotrypsin-catalyzed hydrolyses can indeed be described by a combination of competitive inhibition and dielectric constant effects, and a test can be applied using the above equations.

A plot of the left-hand portions of these equations, which are experimental quantities, vs X gives straight lines with slopes of $(A_{S_2} - A_{S_1})/2.303$ and and intercept which is characteristic of the second-order rate constants and/or Michaelis constants in water.

Clement and Bender have shown that it is also possible to solve equations (3-6) and (3-7) directly, assuming that B is small in comparison to A, which means that the organic solvent–enzyme interaction is much less susceptible to a dielectric constant effect than is the enzyme–substrate interaction. Then, equations (3-6) and (3-7) take the form

$$\frac{K_{m(obs)}}{K_{m(H_2O)}} = e^{AX}\left(1 + \frac{I}{K_I}\right) \tag{3-8}$$

$$\frac{\dfrac{k_2}{K_{m(H_2O)}}}{\dfrac{k_2}{K_{m(obs)}}} = e^{AX}\left(1 + \frac{I}{K_I}\right) \tag{3-9}$$

By using data for different values of X, it is possible to solve these equations for both the unknown A, which is a function of the substrate, and K_I, which is a function of the solvent. Knowing A values of various substrates, it is possible to use equation (3-8) and (3-9) to determine K_I values of the various solvents. The inhibition constants obtained by Clement and Bender in acetonitrile, acetone and dioxane are in the same range as those determined by the hypothesis based solely on competitive inhibition. These authors have stated that

> the hypothesis for the effect of dipolar aprotic solvents on K_m or k_2/K_m embodying the two ideas of competitive inhibition by the organic solvent and an electrostatic effect of the organic solvent on the enzyme-substrate interaction is based on a reasonable model of the enzymatic process and leads to the best empirical treatment of such data that has yet been devised.

This treatment appears to be superior to that for simple competitive inhibition in its ability to correlate data at high organic solvent concentrations. For the same reason it is superior to a treatment involving only a dielectric constant effect. And the conclusion was that this hypothesis is general in its capability of treating all pertinent data in the literature and predicting new data.

More recently Mares-Guia and Figueiredo (13) applied the method to the study of the effects of organic compounds on the interaction of trypsin and competitive inhibitor, and succeeded in fitting the experimental data to the equations describing the simultaneous operation of D and competitive effects by the solvent.

However, deviations between theoretical treatments and experiment are often observed, due to the involvement of various other effects which may be difficult or impossible to elucidate. The deviations reported in a large number of chemical reactions, at lower dielectric constant due to the presence of a cosolvent, are in the direction of results obtained in the higher dielectric component (water) when used as a pure solvent, and are interpreted by the fact that water preferentially solvates the solutes, by the so-called selective or preferential solvation; and finally, as indicated below, the dielectric constant effects can be overshadowed by the "intrinsic" solvent perturbation and also, as we will see below, by the influence of other electrostatic effects.

INTRINSIC SOLVENT EFFECT

In this laboratory, Travers (10) studied several enzyme-catalyzed reactions in ethylene glycol-buffer mixtures and eventually in aqueous solutions at room temperature but in the presence of selected concentrations of glycine which is known to increase the bulk dielectric constant as we have seen in chapter 2.

The first reaction investigated in such conditions was the catalytic decomposition of hydrogen peroxide by catalase in aqueous solution with phosphate buffer pH 6.5, and next in the mixture ethylene–glycol–water (50 : 50 v/v), at pa_H^*† 6.5.

† The proton activity is termed pa_H^* as well as pH*.

TABLE 1

Solvent	[Glycine]	$D_{+20°C}$	$k(M^{-1}\ s^{-1})$
H_2O	0	80	$7\ 10^{-3}$
H_2O	1 M	102.5	$7\ 10^{-3}$
Ethylene glycol–H_2O	0	64.5	$6.5\ 10^{-4}$
Ethylene glycol–H_2O	1 M	87.1	$6.5\ 10^{-4}$

Rate constants of the first-order reaction are reported in Table 1.

It can be seen that addition of 1 M glycine, which increases D from 80 to 102.5, does not modify the rate constant value in aqueous solution, as well as in the mixture in which D increases from 64.5 to 87.1.

On the other hand, the addition of ethylene glycol (50 : 50, v/v) reversibly inhibits the catalase reaction.

The oxidation of gaïacol by hydrogen peroxide catalyzed by horseradish peroxidase has been next studied in similar conditions and main results are reported in Table 2.

TABLE 2

Solvent	[Glycine]	$D_{+20°C}$	$k(M^{-1}\ s^{-1})$
H_2O	0	80	$3.3\ 10^{-6}$
H_2O	0.5 M	91	$2.9\ 10^{-6}$
H_2O	1.0 M	102.5	$2.6\ 10^{-6}$
Ethylene glycol–H_2O (50 : 50, v/v)	0	64.5	$0.58\ 10^{-6}$
Ethylene glycol–H_2O (50 : 50, v/v)	0.5 M	76	$0.50\ 10^{-6}$
Ethylene glycol–H_2O (50 : 50, v/v)	1.0 M	87.1	$0.46\ 10^{-6}$

Concentration in peroxidase was 10^{-9} M, 2.10^{-2} M in H_2O_2 and $2.5\ 10^{-2}$ M in gaïacol.

Addition of 1 M glycine causes a decrease in activity of about 10–12% in aqueous solution, as well as in ethylene glycol–water mixture, whereas the addition of the cosolvent itself causes a decrease in activity of about 80%. On the other hand, one can see what variations in D do not influence the rate constants.

In conclusion, it can be seen that in aqueous solutions, increase of dielectric constant has little effect on the reaction rates of the two enzyme systems investigated, and that even when isodielectric with water by addition of glycine or by cooling, ethylene glycol-buffer mixtures maintain their inhibition on enzyme specific activity. At least in the chosen systems, electrostatic effects are overshadowed by the solvent perturbation on enzyme molecules. However, we used a polyol which generally protects enzyme activities and does not interact strongly with the exposed tyrosyl residues of proteins, and thus cannot seriously weaken their structure by possible hydrogen bond formation, or exchange or breakage of existing hydrogen-bonds. It now appears from many conformational studies that most reversible alterations in enzyme structure, and therefore activity, are mediated through sites on the protein molecule other than the enzymic site, enzyme molecules translating variations in *milieu* into a change in catalytic activity. The present results show that the presence of an organic solvent interacting with exposed protein side chain groups is overwhelmingly more effective than any macroscopic value of the dielectric constant.

UNCONTROLLED pH* CHANGES

The modification of the interaction energy between the substrate and the enzyme due to a variation of the dielectric constant is not the only case to be considered and various theoretical electrostatic effects could influence the rate at which a substrate binds to an enzyme. For instance the concentration of water may play a role in addition to the change in the dielectric constant if there is a decreased hydration of both the enzyme and the substrate, or even in the case of enzyme reactions in which water is one of the components of the reaction, but it is difficult to investigate the actual influence of such conjectural parameters, except by adjusting suitable concentrations of substrate, salts and cofactors so that one could recreate aqueous conditions.

Meanwhile the influence of any cosolvent on the dissociation constant of weak electrolytes, buffers and ionizing groups at the active site of the enzyme might play a key role in the solvent effect on reaction rates.

When additions of a cosolvent are made to an aqueous solution, they should be done with concentrations in buffer suitably adjusted to correct changes in the dissociation constant. As an example of the kind of error which can be made when experiencing enzyme catalyzed reactions without a sufficient knowledge about the cosolvent effect on buffers, let us mention the case of the NADH-cytochrome b_5 reductase activity in the mixture ethylene glycol–water (50 : 50, v/v).

This flavoprotein serves as a catalyst in electron transfer from pyridine nucleotides to a hemoprotein, cytochrome b_5, and is located in liver microsomes. It can be extracted and purified, and used in the test tube to carry out the catalysis of the following reactions

$$\mathrm{NADH} + 2\,\mathrm{cyt}\ \mathrm{b_5(Fe^{3+})} \xrightarrow{\text{enzyme}} \mathrm{NAD^+} + \mathrm{H^+} + 2\,\mathrm{cyt}\ \mathrm{b_5(Fe^{2+})}$$

or

$$\mathrm{NADH} + 2\,\text{ferricyanide}\,(\mathrm{Fe^{3+}}) \xrightarrow{\text{enzyme}} \mathrm{NAD^+} + \mathrm{H^+} + 2\,\text{ferrocyanide}\,(\mathrm{Fe^{2+}}) \quad (3\text{-}10)$$

Such reactions are quite pH-dependent, as shown by Strittmatter and colleagues (14) and the pH-activity profile of reaction (3-10) is shown in Fig. 1 where it can be seen that the maximum of activity is obtained at pH 5.6. It is obvious that any addition of a high concentration of cosolvent to aqueous solutions necessitates an adjustment for an accurate pa_{H}^* value, and also the definition of the solvent effect on the pH-activity curve. It can be seen in Fig. 1 that there is a shift of this curve in ethyleneglycol–water mixtures (50 : 50, v/v), as well as an inhibition of the enzyme activity, and that the activity maximum is now at pH* 6.6. Let us mention that, whereas it was impossible to test the enzyme activity in water around pH 5.5 because of the instability of the enzyme molecule at that pH value (tests were usually carried out at pH 7.5 that is on the shoulder of the pH-activity curve), it becomes possible to test the activity at the new value of its optimum (pH* 6.6), the enzyme being stable at that pH.

SOLVENT EFFECTS ON THE ACID DISSOCIATION CONSTANTS OF GROUPS AT THE ACTIVE SITE OF AN ENZYME: THE CASE OF THE TRYPTIC HYDROLYSIS OF BAEE

Since proton activity values of most common buffers in a large number of mixtures are now available or measurable, as we saw in Chapter 2, this factor which can influence the enzyme specific activity can be eliminated, but changes of the acid dissociation constants of ionizing groups at the active site can be expected, recorded, and new pH* conditions can be set up to eliminate this effect.

pH-*activity curves in mixed solvents*

It has been assumed (15) that if conditions can be found where changes in the pK values of the groups at the active site of an enzyme are the dominant factor in the effects of solvents on enzyme activity, cosolvents could be used to determine the charge types of these groups. Accordingly the types of charges at the active site can, theoretically, be determined by measuring the pH-activity curve in water and in mixed solvents.

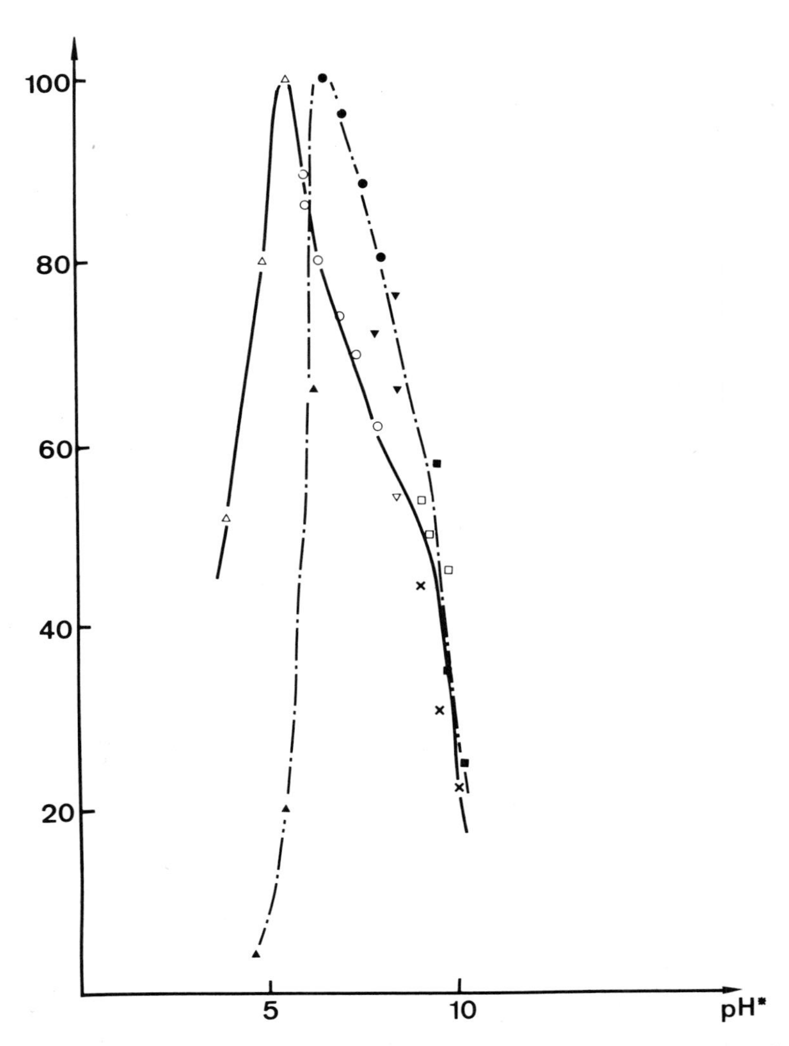

FIG. 1. pH-activity profiles of the NADH-cytochrome b_5 reductase in aqueous solution (———) and in the mixture ethylene glycol (EG)–water in the volume ratio 50 : 50 at +20°C (–·–·–).

Bell-shaped pH-activity curves imply a requirement for at least two ionizing groups for activity, one in the base form and the other in the acid form. The ionization constants K_1 and K_2 of these groups are given by:

$$K_1 = \frac{[AH^-]\cdot[H_3O^+]}{[AH_2]} \quad \text{for } AH_2 + H_2O \rightleftharpoons AH^- + H_3O^+$$

and

$$K_2 = \frac{[A^{--}]\cdot[H_3O^+]}{[AH^-]} = \frac{[A^{--}]\cdot[H_3O^+]^2}{[AH_2]\cdot K_1}$$

$$\text{for } AH^- + H_2O \rightleftharpoons A^{--} + H_3O^+$$

Concentration in each form is given by a Michaelis function:

$$\begin{aligned}
[A] &= [AH_2] + [AH^-] + [A^{--}] \\
&= [AH_2]\cdot\left(1 + \frac{K_1}{[H_3O^+]} + \frac{K_1K_2}{[H_3O^+]^2}\right) = [AH_2]f \\
&= [AH^-]\left(\frac{[H_3O^+]}{K_1} + 1 + \frac{K_2}{[H_3O^+]}\right) = [AH^-]f^- \\
&= [A^{--}]\left(\frac{[H_3O^+]^2}{K_1K_2} + \frac{[H_3O^+]}{K_2} + 1\right) = [A^{--}]f^{--}
\end{aligned}$$

Two examples of the functions $1/f$, $1/f^-$ and $1/f^{--}$ are represented in Fig. 2. It can be seen that the shape and the magnitude of the maximum depend on the difference between the values of pK_1 and pK_2. Let us mention that these constants are characteristic of the ionization of the molecules regardless of the ionization constants of some groups.

The ionizing groups responsible for catalysis can be respectively neutral acids of the type

$$H_2O + AH \rightleftharpoons H_3O^+ + A^-$$

and cationic acids of the type

$$H_2O + BH^+ \rightleftharpoons H_3O^+ + B$$

It is well established that on addition of organic solvents to aqueous solutions the pK of neutral acids increases whereas the pK of cationic acids decreases. The corresponding modification of the Michaelis function $1/f^-$ is shown in Fig. 3. The ionizing groups are respectively cation-neutral (*a*), neutral-cation (*b*) and neutral-neutral (*c*).

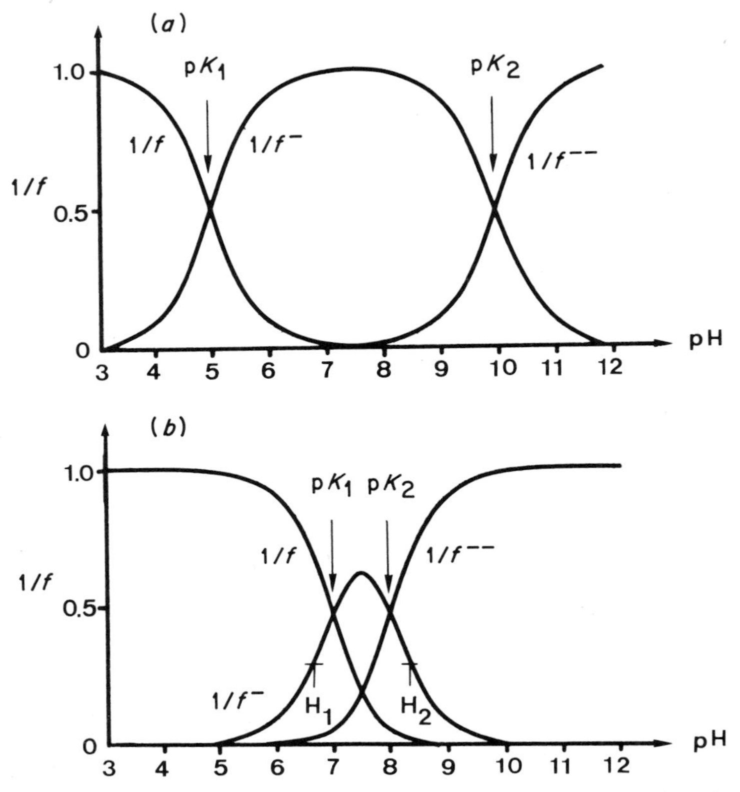

FIG. 2. Mode of variation of the pH function with pH. (*a*) Reciprocals of pH functions for a system with two ionizing groups with p*K*s of 5 and 10 respectively. (*b*) Reciprocals of pH functions for a symmetrical system with p*K*s of 7 and 8.

The addition of cosolvent can cause a shift of the pH-optimum and increase or decrease the maximum activity. The types of charges involved at the active site could be determined according to the sign of the change in the p*K* values.

We studied in this laboratory the pH-dependence of the tryptic hydrolysis of benzoyl-L-arginine ethylester (BAEE) both in water and in different mixtures at normal and subzero temperatures (29,30). This enzyme system has been chosen for it apparently involves only one ionizing group in the active site. The V_m and K_m(app) of the reaction have been determined in each mixed solvent between pa_H^* 4 and 9–10, and at 20, 0 and −20°C. The pK_{ES} of the ionizing group involved in the active site of the enzyme substrate complex, was determined from the dependence of V_m on pa_H^* by the method of Dixon and Webb (19).

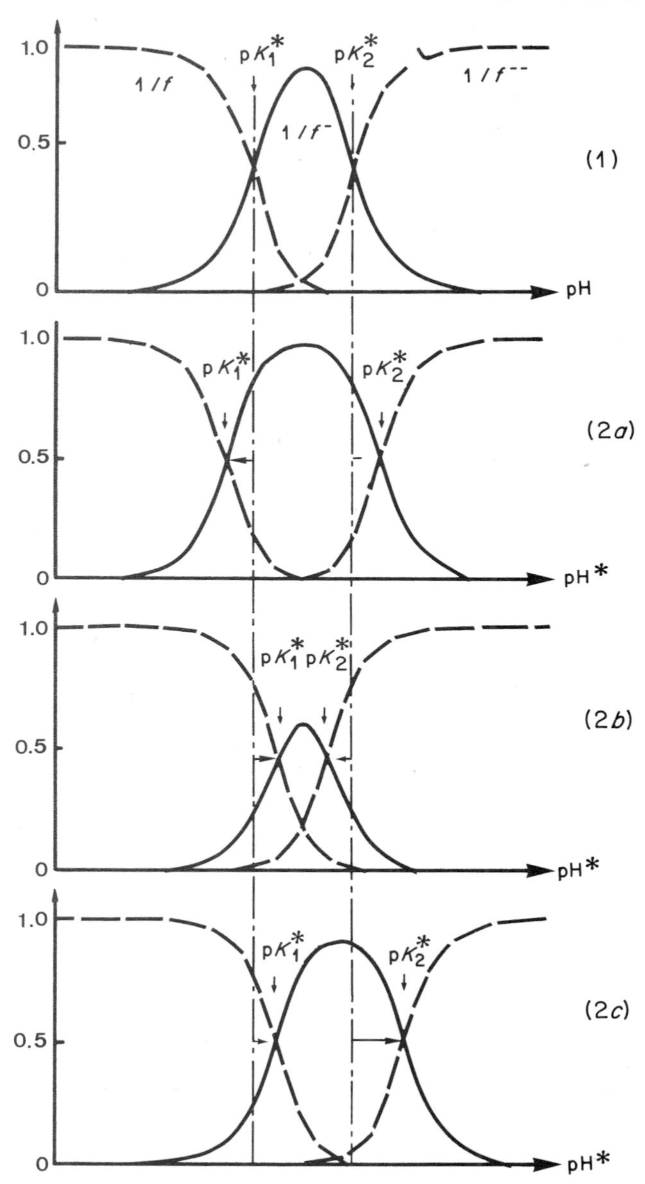

FIG. 3. Organic solvent effect on the dissociation constant of neutral and cationic ionising groups, and its implications on the reciprocals of the pH functions. (1) reciprocals of pH function for a system with two ionizing groups 1 and 2; (2*a*) solvent effect in the case 1 cationic, 2 neutral; (2*b*) solvent effect in the case 1 neutral, 2 cationic; (2*c*) solvent effect in the case 1 and 2 neutral.

TABLE 3

Mixed solvents		Temperature (°C) 20	0	−20	E(act) ΔH_i
Water	k_3	8.7 ± 0.2			11.0 ± 0.5
	K_m(app)	$\begin{cases} <10^{-5} \\ 4.10^{-6} \end{cases}$			11.0 (20)
	pK_{ES}	$\begin{cases} 6.2 \\ 6.25\ (20) \end{cases}$			7.0 (20)
Water–methanol (50 : 50)	k_3	6.8 ± 0.2	1.40 ± 0.05	0.11 ± 0.01	15.2 ± 2
	K_m(app)	$1.1\ 10^{-4}$	3.10^{-5}	$2.2\ 10^{-5}$	
	pK_{ES}	5.7 ± 0.1	5.7 ± 0.1	5.7 ± 0.1	0.0
Water–ethylene glycol (50 : 50)	k_3	11.0 ± 0.2		0.22 ± 0.02	14.4 ± 2
	K_m(app)	$8.4\ 10^{-5}$		$3\ 10^{-5}$	
	pK_{ES}	6.1 ± 0.1		6.5_5 ± 0.1	3.8 ± 0.3
Water–1-2 propane-diol (50 : 50)	k_3	11.0 ± 0.2	2.2 ± 0.05	0.19 ± 0.02	15 ± 2
	K_m(app)	$7.0\ 10^{-5}$	$5\ 10^{-5}$	$4.5\ 10^{-5}$	
	pK_{ES}	5.8_5 ± 0.1	6.05 ± 0.1	6.3 ± 0.1	3.8 ± 0.3
Water–glycerol (56 : 44)	k_3	10.7 ± 0.2	2.4 ± 0.05	0.25 ± 0.02	15 ± 2
	K_m(app)	$3.5\ 10^{-5}$	$<10^{-5}$	$<10^{-5}$	
	pK_{ES}	6.0_5 ± 0.1	6.6 ± 0.1	7.0 ± 0.1	8.1 ± 0.5
Water–dimethylsulf-oxide (50 : 50)	k_3	14.3 ± 0.3		0.18 ± 0.02	16.1 ± 2
	K_m(app)	$1.8\ 10^{-4}$		$1.3\ 10^{-5}$	
	pK_{ES}	5.45 ± 0.1		6.3_5 ± 0.1	7.6 ± 0.5

Kinetic parameters at pa_H^* 9.0

The rate constant $k_{3\,max}$ (deduced from V_m), as well as K_m(app) attained in this work at pa_H^* 9.0, on the alkaline plateau of the pa_H^* profiles, are given in Table 3 for the various solvents and temperatures investigated.

At 20°C, $k_{3\,max}$ appears to be slightly higher in the mixed solvents than in pure water except for methanol where the converse is observed. On the other hand, K_m(app) is in all cases increased by the presence of the organic solvent.

When the temperature is decreased from 20 to −20°C, $k_{3\,max}$ is greatly decreased according to the Arrhenius relationship:

$$\ln k = -\frac{E(\text{act})}{RT} + \text{constant}$$

The activation energy of the limiting step of this reaction (k_3), determined under various conditions, is also given in Table 3. Although these data are rather imprecise (only two or three experimental points) E(act) appears to be slightly increased in the presence of the organic solvents. K_m(app) is, on the other hand, not greatly affected by the decrease in temperature.

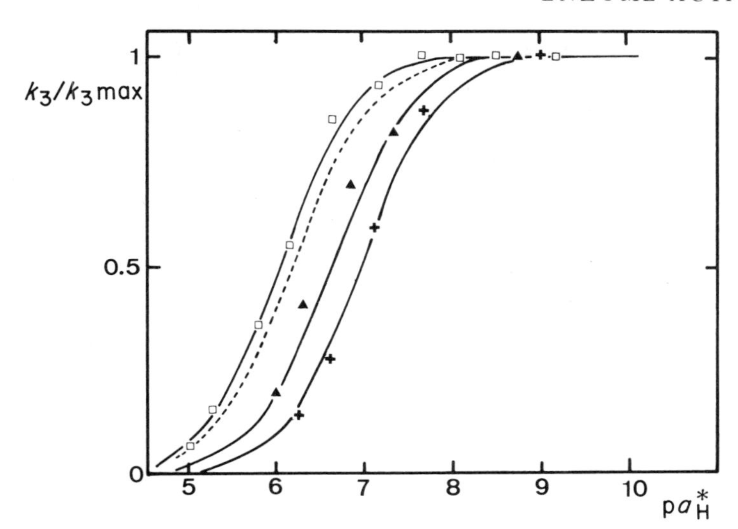

FIG. 4. Dependence of $k_3/k_{3\,max}$ on pa_H for tryptic hydrolysis of BAEE in pure water at 20°C and in water–glycerol mixture (56 : 44) at 20°C (□), 0°C (▲) and −20°C (+), ----, water; ——, glycerol–water.

pa_H^* profiles in mixed solvents at room temperature: 20°C

The variations of k_3 (normalized to the maximum values, $k_{3\,max}$ on the alkaline plateau of the pa_H^* curves) with pa_H^* are shown in Fig. 4 in the case of water and water–glycerol mixture. The values of pa_H^* (protonic activity in the presence of the organic solvent) are taken from tables given in chapter 2. Similar profiles have been obtained for the other mixed solvents. The points shown are experimental values; the solid curve represents the theoretical values calculated from the formula:

$$\frac{k_3}{k_{3\,max}} = \frac{1}{1 + \dfrac{a(H^+)}{K_{ES}}}$$

where $a(H^+)$ is the protonic activity in the mixed solvent and K_{ES} the ionisation constant of the ionizing group involved in the enzyme–substrate complex, as determined from the Dixon and Webb plots of log V_m against pa_H^*.

The values of pK_{ES} determined in pure water and in the various solvents are given in Table 3. The value of 6.20 obtained in water is in good agreement with that quoted by Gutfreund (20), i.e. 6.25. On the other hand, pK_{ES} is decreased in the presence of the various organic solvents, the extent varying considerably from solvent to solvent.

Shifts in the pa_H^* dependence of $k_3/k_{3\,max}$, observed in passing from water to mixed solutions at room temperature, must be considered in terms of the effects of the solvent on pK_{ES}.

Two major aspects of the effect of the solvent are the electrostatic effect and the solvation effect (21). While the latter is difficult to interpret in the absence of a satisfactory molecular theory, the electrostatic effect can be

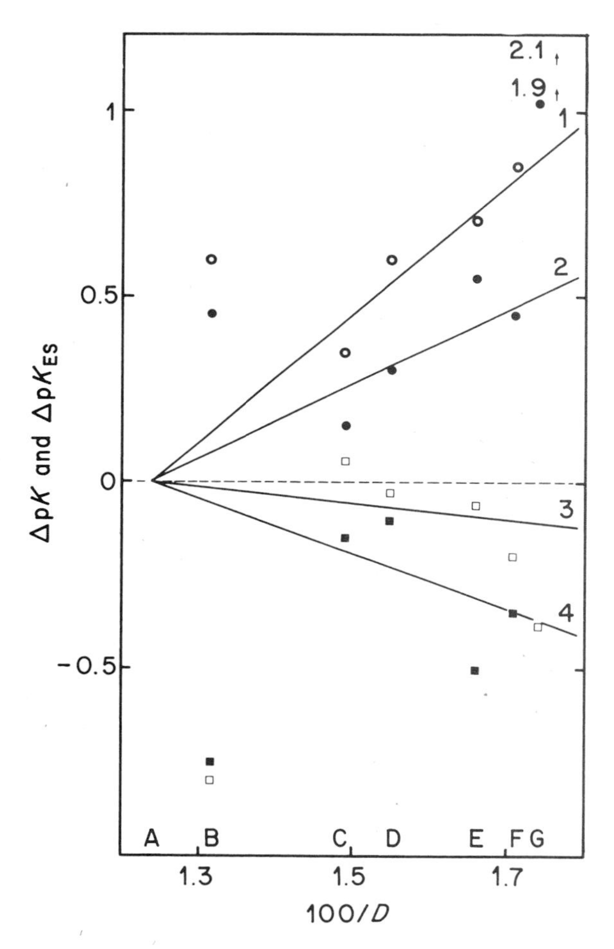

FIG. 5. Plots of ΔpK of several buffer molecules and ΔpK_{ES} against $1/D$ at 20°C in several mixed solvents. $\Delta pK = pK$ (mixed solvent) $-$ pK (water), D dielectric constant of the mixtures. A, pure water; B, water–dimethylsulfoxide (50 : 50); C, water–glycerol (56 : 44); D, water–ethylene glycol (50 : 50); E, water–methanol (50 : 50); F, water–1-2-propanediol (50 : 50); G, water–dimethylformamide (30 : 70). 1 (○), acetate; 2 (●), cacodylate; 3 (□), Tris; 4 (■) trypsin.

qualitatively accounted for by the Born equation (22) which, in view of the approximation involved in its formulation, agrees surprisingly well with experimental data (23). Figure 5 shows the relation between ΔpK (difference between pK values obtained in water and in mixed solvents) and $1/D$ (D, dielectric constant) for various buffer molecules. A reasonably linear correlation is observed for water, alcohols and polyols (amphiprotic solvents); dimethylsulfoxide and dimethylformamide (aprotic solvents) are different. On the other hand, the solvent effect appears to be dependent on the nature of the ionizing group, as already observed (24), and this is clearly shown in Fig. 5, by the differences between the slope of the plots obtained with neutral (acetic, cacodylic) and cationic acids (Tris).

The results concerning the effect of solvent on the pK_{ES} of trypsin can be analyzed in the light of these conclusions. This was the basis of the method of Findlay *et al.* (15), whose precision can be greatly enhanced by the use of our pa_H^* scales. In Fig. 5 we have also plotted ΔpK_{ES} as a function of $1/D$ and a reasonably linear correlation is obtained for water, alcohols and polyols; as in the case of buffer molecules dimethylsulfoxide differs. The ionizing group at the active center of trypsin should be a cationic one as indicated by the slope of this plot. In fact it is well known that it is an imidazolium of a histidine residue (20,25,26). In passing from water to water–dioxane 50/50, Inagami and Sturtevant (27) found a pK_{ES} shift from 6.06 to 5.8 in reasonable agreement with our results. Finally all the available evidence suggest that the mechanism of tryptic hydrolysis is unaffected by the presence of organic solvents, although a solvent effect may be observed on enzyme activity (27,28,13).

Temperature effects on the acid dissociation constants of weak electrolytes. pa_H^* profiles in mixed solvents at subzero temperatures

Figure 4 shows the pa_H^* profiles obtained in water–glycerol at 0° and −20°C. pK_{ES} values, obtained under these conditions from the Dixon–Webb plots, are reported for the various solvents in Table 3. Except in the case of methanol, large variations of pK_{ES} are observed with decreasing temperature for a given mixed solvent. This variation is in the opposite direction to that caused by the solvent. Such a temperature effect on pK_{ES} can be analyzed in terms of the enthalpy of ionization ΔH_i of the ionizing group concerned. The Van't Hoff relationship

$$\frac{d(\ln K)}{dT} = \frac{\Delta H_i}{RT^2}$$

allowed us to determine ΔH_i in the various mixed solvents. These data are reported in Table 3. The accuracy of ΔH_i is obviously not very good (data from only two or three temperatures); nevertheless ΔH_i appears to be very dependent on the organic solvent. It is, however, independent of temperature between 20° and −50°C, in all our experiments including buffer molecules and trypsin. It depends on the nature of the ionizing group: somewhat low for the neutral acids (between 0 and 6 kcal mole^{-1}) and in general higher for the cationic acids (7–15 kcal mole^{-1}) as already observed in pure water. Moreover the enthalpy of ionization appears to be strongly dependent on the organic solvent used, and a general tendency can be established for buffer molecules, from the tables listed in chapter 2, which is: ΔH_i (methanol) $<$ ΔH_i (ethylene glycol) $\approx$ ΔH_i (1-2-propanediol) $<$ ΔH_i (glycerol) $\approx$ ΔH_i (dimethylsulfoxide). The enthalpy of ionization of the ionizing group at the active site of trypsin (Table 3) measured in glycerol and dimethylsulfoxide is in good agreement with the value of 7.0 kcal mole^{-1} quoted by Gutfreund (20). It is however lower in the other solvents and appears to be zero in methanol. This rather surprising discovery may reflect the solvent effect already described in the case of buffers, for the same tendency is observed here.

We show in Fig. 6, the pH profiles obtained in water–methanol (50 : 50) and in water–1-2-propanediol (50:50) at 20, 0 and −20°C, in order to illustrate the large scattering of the enthalpy of ionization of the histidine residue 46 of trypsin obtained in different solvents. Such a scattering of data,

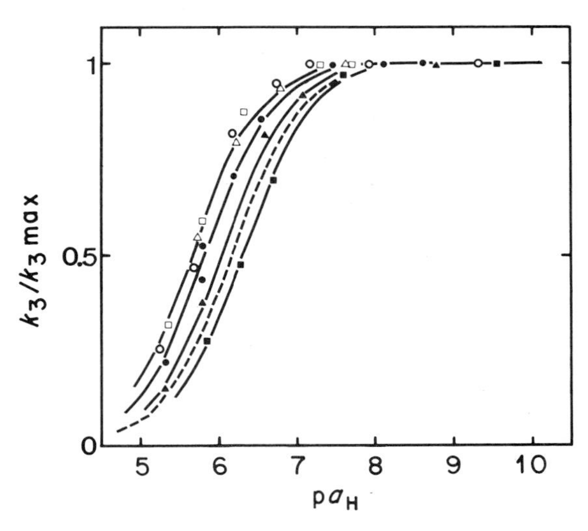

FIG. 6. Dependence of $k_3/k_{3\,max}$ on pa_H for tryptic hydrolysis of BAEE, in pure water (– – – –), in water–methanol 50 : 50, v/v at 20°C (○), 0°C (△), −20°C (□), and in water–1-2-propanediol 50 : 50 v/v at 20°C (●), 0°C (▲), and −20°C (■).

TABLE 4. Ionization thermodynamic functions of various dissociable species in different solvents (M, water–methanol 50 : 50 in vol; E, water–ethylene glycol 50 : 50; P, water–1-2-propanediol 50 : 50; G, water–glycol 56 : 44) at 20°C, and compensation temperature. ΔG° and ΔH° are given in kcal mole^{-1}; ΔS° in cal mole^{-1} deg^{-1}. The absolute uncertainties are: ± 0.15 on ΔG°, ± 0.3 on ΔH°, ± 2 on ΔS°, and ± 10°k on T_c

		H_2O	M	E	P	G	T_c(°K)
Histidine 46	ΔG°	8.37	7.63	8.17	7.84	8.10	
of trypsin	ΔH°	7.0	0.0	3.8	3.8	8.10	315
	ΔS°	−4.67	−2.60	−14.9	−13.8	0.0	
Chloroacetic acid	ΔG°	3.90	4.82	4.70	5.02	4.42	
	ΔH°	−1.15	1.05	1.05	0.50	2.90	235
	ΔS°	−17	−12.9	−12.5	−15.4	−5.2	
Acetic acid	ΔG°	6.36	7.30	7.17	7.5	6.84	
	ΔH°	−0.09	0.80	2.10	0.80	3.90	250
	ΔS°	−21.4	−22.2	−17.3	−22.8	−10.0	
Cacodylic acid	ΔG°	8.17	8.90	8.57	8.78	8.37	
	ΔH°	—	0.80	2.30	1.30	3.90	250
	ΔS°	—	−27.6	−21.4	−25.5	−15.3	
Phosphoric acid (II)	ΔG°	9.82	11.12	10.45	10.66	9.85	
	ΔH°	0.99	0.90	2.6	2.1	4.9	225
	ΔS°	−29.6	−34.9	−26.8	−29.2	−16.9	
Tris(hydroxymethyl	ΔG°	11.02	10.72	10.72	10.6	10.92	
amino-methane)	ΔH°	10.9	10.9	13.3	12.0	14.8	310
	ΔS°	−0.3	0.61	8.8	4.78	13.2	

already shown in Table 3, can be further seen by comparison of Fig. 6 with the pH profiles obtained in water–glycerol presented in Fig. 4.

The thermodynamic functions, standard free energy, enthalpy and entropy, ΔG°, ΔH°, ΔS° relative to the ionization of the various ionizable species in four mixed solvents, are presented in Table 4. K ionization constant and ΔH° are direct experimental data, whereas ΔG° and ΔS° have been calculated from the classical relationships:

$$\ln K(T) = -\frac{\Delta G^\circ(T)}{RT}$$

$$\Delta G^\circ(T) = \Delta H^\circ - T\Delta S^\circ$$

In all cases, ΔH° is found constant over the whole temperature range explored.

Since enthalpy and entropy of ionization of histidine residue 46 of trypsin and of the various weak electrolytes have been determined for each mixed solvent, the pairs $\Delta H^\circ - \Delta S^\circ$, relative to the various solvents can be plotted on a graph and a straight line is generated for each dissociable species. Such a result is indicative of the existence of an enthalpy—entropy "compensation process" involved in the ionization of weak acids in mixed solvents.

There is compensation when in equilibrium or rate processes, the variation of a parameter i (pH, ionic strength, substituted compounds, solvent composition, etc.) generates a linear relationship between ΔH_i° and ΔS_i°, the temperature remaining constant. Usually, ΔH_i° and ΔS_i° are the total standard enthalpy and entropy of the process, involving both chemical and solvation part processes.

Such a relationship can be expressed as follows:

$$\Delta H_i^\circ = \alpha + T_c \Delta S_i^\circ$$

T_c, is the slope of the plot ΔH_i° vs ΔS_i° is the compensation temperature. α is the intercept at $\Delta S_i^\circ = 0$ and represents the standard free energy of the chemical part process.

As said above, one way to generate the pairs $\Delta H - \Delta S$ is to vary the medium composition by changing the nature of the cosolvent or increasing amounts of a same cosolvent, and then to record equilibrium constants or kinetic parameters at selected temperatures.

Temperature compensation values (T_c) are deduced from linear plots ΔH_i° vs ΔS_i° and lie in a relatively narrow range, from about 250°K–315°K.

The linear enthalpy–entropy plots are shown in Fig. 7, where the results concerning the histidine residue 46 of trypsin are presented with those relative to several weak electrolytes. The similar behavior of these different ionizable species, whatever their carrier entity (small molecules or proteins) has to be noted. The linearity is remarkable for the four mixed solvents, whereas the points relative to pure water are not generally located on their respective compensation lines. The compensation temperature, T_c, listed in Table 4 is deduced from the slope of each plot, taking into account only those points relative to mixed solvents. Alternative compensation plots (not shown), ΔG° vs ΔH°, have been drawn and give the same compensation temperature within ± 10°K. Under our solvent conditions, T_c, markedly different for neutral acids (240 $\pm$ 10°K) and for cationic acids (310 $\pm$ 5°K), is nearly identical for all the members of each series. The above results and experimental conditions have been published, see references (29) and (30).

We have shown above that the use of pa_H^* scales, established in mixed solvents, allow the precise determination of the solvent effect on the ionization constant of dissociable groups at the active site of enzymes; and according to the method of Findlay *et al.* (15), their identification. At this stage it

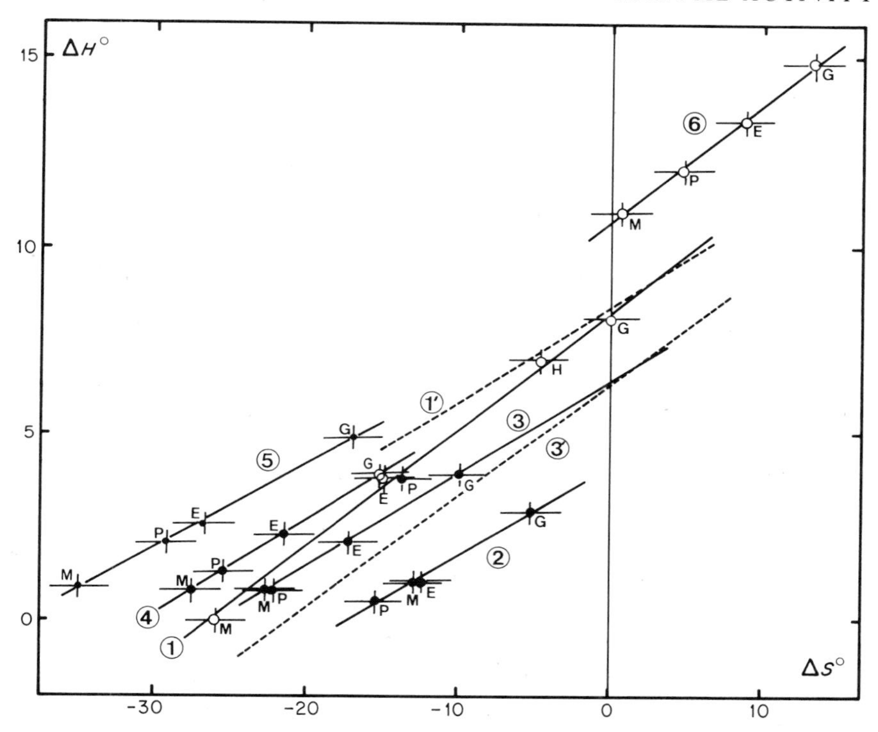

FIG. 7. Enthalpy–entropy compensation plots for the ionization of various dissociable species, 1, histidine (imidazolium) 46 of trypsin; 1′, histidine (imidazolium) of Angiotensin II from ref. (33); 2, chloroacetic acid; 3, acetic acid; 3′, acetic acid from ref. (32); 4, cacodylic acid; 5, phosphoric acid; 6, Tris (hydroxymethyl aminomethane), in different solvents. M, water–methanol 50 : 50, v/v; E, water–ethylene glycol 50 : 50, v/v; P, water–1-2-propanediol 50 : 50, v/v; G, water–glycerol 56 : 44, v/v.)

seemed interesting to use as a further characteristic parameter of the groups studied, their enthalpy of ionization. However, it can be seen from Fig. 6 that when measured in various mixed solvents, the enthalpy of ionization varies strongly from solvent to solvent, and accordingly cannot be used in the identification of dissociable groups.

The results presented in Fig. 7 and Table 4 show that this apparent scattering of the enthalpy of ionization in the different mixed solvents, is the consequence of an enthalpy–entropy compensation process involved in the ionization of any weak electrolyte in aqueous solution. Although such a result has already been found under various conditions (31,32,33), the molecular significance of this compensation process is not yet fully understood, nor is it our purpose to discuss this point. Rather, as this phenomenon seems to be general in our solvent conditions, it might be used as a tool for the investigation of the active site of enzymes. This approach, according to the above results reveals two important findings:

1. The nature of the environment of the active site (hydrophilic or hydrophobic): as stated by Lumry *et al.* (31) and Ives *et al.* (32) the occurrence of a compensation pattern for any rate or equilibrium process, might be ascribed to the thermodynamic contribution of the solvation part process. Accordingly, as far as enzymes are concerned, any exposed ionizable group would manifest a compensation pattern. This is the case for the histidine residue 46 of trypsin, which is well known to be exposed to the solvent. Alternatively, absence of any compensation pattern, may be expected for buried groups located in a hydrophobic environment.
2. The nature of the ionizable groups at the active site (neutral or cationic): Fig. 7 and Table 4 show that under our solvent conditions, T_c is very different for neutral and cationic acids; such a distinction could therefore be used for the identification of any exposed ionizable group of an enzyme.

Compensation patterns obtained here can be compared to those generated in the ionization of the same weak acid by a different parameter. Figure 7 shows the plots of compensation lines obtained by Ives *et al.* (32) and Paiava *et al.* (33), respectively, for the ionization of substituted acetic acids and of histidine (imidazolium) residue of substituted Angiotensin II, both in pure water. While T_c values obtained by these authors are different from those we observed in mixed solvents, the two pairs of compensation lines cross exactly or very nearly at $\Delta S^\circ = 0$, as theoretically expected, ΔG° (chemical) being constant, whatever the compensation pattern generated for a given molecular species.

As already stated, the physical significance of the compensation temperature, T_c, is not clearly understood (31). However, its practical implications are simple. In the case of histidine residue 46 of trypsin for instance, a compensation pattern with $T_c = 315°K$ theoretically involves the constancy of ΔG° and accordingly of pK at 315°K, whatever the solvent used; this tendency is shown in Fig. 8, where the pK of histidine is plotted against the reciprocal of temperature. All the linear plots corresponding to the various solvents converge at the same point; $T_c = 319 \pm 5°K$ can hence be determined. Thus, the solvent effect on the ionization constant of histidine residue 46 of trypsin (and of any group presenting a compensation pattern on ionization in these conditions) vanishes at the compensation temperature. This is a result of great practical significance. On the other hand, it should be noted that the solvent effect on the ionization constant is strongly dependent on temperature, and changes its sign when passing through the compensation temperature. Thus at normal temperature (i.e. 300°K), constancy of ΔG°

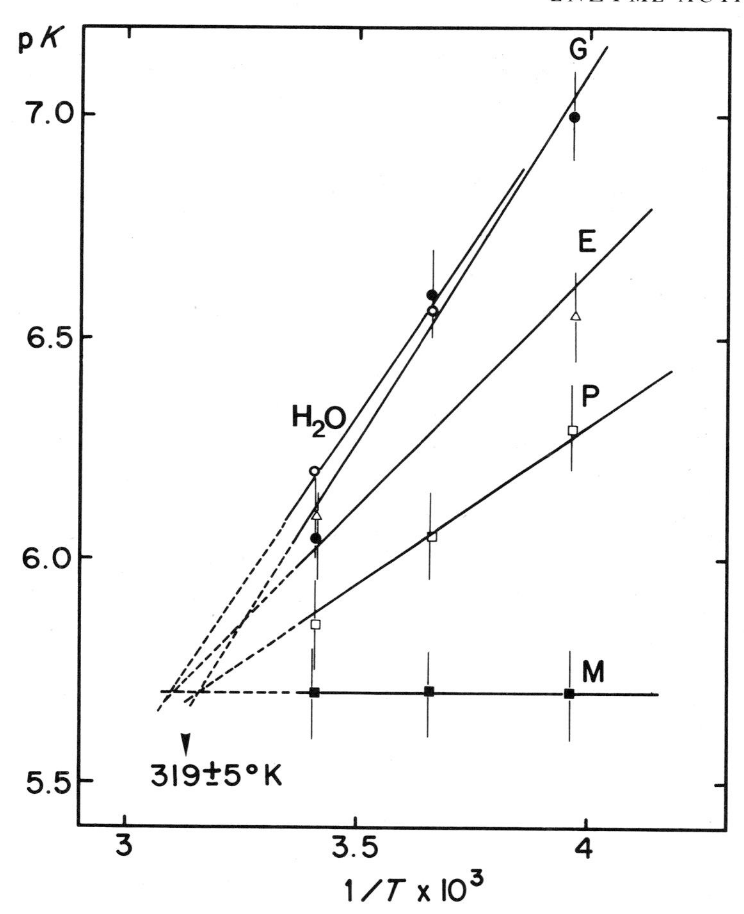

FIG. 8. pK of histidine (imidazolium) 46 of trypsin, as a function of the reciprocal of temperature, in various solvents. M, water–methanol 50 : 50, v/v; E, water–ethylene glycol 50 : 50; P, water–1-2-propanediol 50 : 50; G, water–glycerol 56 : 44.

is nearly reached for cationic groups ($T_c \approx 310°K$), whereas it is not attained for neutral groups ($T_c \approx 240°K$). Such a situation might explain the fact that at normal temperature a large positive solvent effect is observed on the ionization constant of neutral groups, while it is negative and rather weak for the cationic groups. In conclusion the results presented here suggest that any complete investigation of the active site of an enzyme could be undertaken via the "compensation" law, by the study of conjugated solvent-temperature perturbations on the ionization of dissociable groups involved in it.

Main implications of the above kinetic studies

Let us consider the various implications of the kinetic analysis of the tryptic hydrolysis of BAEE which can be considered as a typical example:

Firstly, it can be seen that recording pH*-activity profiles of enzyme-catalyzed reactions which are pH-dependent should permit the adjustment of pH* conditions insuring optimal activity in any mixed solution at any normal or subzero temperature. It is obvious that new buffer concentrations must be eventually adjusted instead of the concentration defined in normal conditions of medium and temperature.

Secondly, the comparative recording of the shape of any pH-dependent enzyme catalyzed reaction in water, in mixed solvents at selected temperatures, as well as kinetic analysis in terms of the parameters $K_m$app and V_m give evidence of whether the reaction mechanism is affected by the presence of a cosolvent and by variations of temperature. These tests are of invaluable significance since it is essential to find out whether there is a strict correlation between any reaction mechanism observed in normal and abnormal conditions of medium and temperature.

Thirdly, the above studies allow us to check the eventual validity of using cosolvents and (or) temperature variations to study the ionizing groups at the active sites of enzymes. We have seen that the method proposed by Findlay *et al.* (15) of using cosolvents to identify the sign of such groups could be misleading since cosolvents act differently on the dissociation constants of these groups. On the other hand, it has been shown that, when measured in various mixed solutions, the enthalpy of ionization of the groups varies markedly from solvent to solvent and accordingly cannot be used safely in their identification. However we have seen that the solvent effect on the dissociation constants depends on temperature and that enthalpy–entropy plots allow determination of the type of charge of the ionizing groups, whereas absence of such patterns might be expected to indicate groups located in a hydrophobic environment.

Finally we can conclude that the type of investigation described in this section was one of the most useful we conducted in mixed solvents at any temperature, from both a practical and theoretical point of view. Nevertheless, we have checked that once all requirements for ensuring an optimal activity have been fulfilled (mixed solutions isodielectric with water, suitable ionic strength and pH*), there is always a "remaining" cosolvent effect on reaction rates. It must be established that it is in any case fairly rapidly obtained upon addition of cosolvent, independent of time and fully reversible upon removal of the cosolvent.

Investigators interested in experimentation on enzyme catalyzed reactions at subzero temperatures should verify these criteria and then decide on an accurate enzyme concentration. However, further investigation of the "remaining" reversible cosolvent effect is necessary to provide additional information and possibly to improve the low temperature procedure with respect to use of cosolvents as "tools" to investigate the behavior of enzymes in mixed solvents.

Intrinsic cosolvent effect on enzyme structure and activity

MAIN OBSERVATIONS AND PROBLEMS ABOUT CHANGES IN CONFORMATION RELATED TO ACTIVITY

Minor or localized structural changes, occurring well before detectable denaturation is evident, may modify the structure of the enzyme or of its active center and, therefore, its reactivity.

One of the parameters most often held to be responsible for such changes is the direct interaction of the cosolvent with the enzyme through hydrogen bonding with appropriate groups.

The influence of hydrogen bonding on many reaction rates has been reviewed and discussed in a number of papers (see ref. 34), its occurrence has been demonstrated in a number of cases.

It has been stated that a particular solvent will cause a deceleration of the rate of a reaction if the active centers that take part in the reaction are blocked by hydrogen bonding with the solvent, and that an acceleration will be produced by a solvent which, by hydrogen bonding, promotes the electron shift necessary for the reaction. Solvents strongly interacting with groups such as imidazolium (9) and with the exposed tyrosyl residues of proteins by hydrogen bond formation, or exchange, or breaking of existing hydrogen bonds, could seriously alter the protein structure and therefore its specific activity. However, the case of the hydrogen bonding influence on enzyme reaction rates is typical of the difficulty of explaining solvent effects through any theoretically and experimentally accessible or conjectural parameter *a priori* able to modify the structure and the activity of enzymes.

It now appears that a number of reversible alterations in enzyme activity can be similarly induced by various perturbants which could operate by inducing new protein conformations, and that such alterations can be mediated through a number of sites on protein molecules, through several types of interaction between the cosolvent and the sites (hydrogen bonding, non specific interactions by dipolar and dispersion forces, etc.) and also possibly via changes in water structure at the level of hydration shells.

Thus conformation changes caused by perturbant agents including cosolvents are postulated but as yet scarcely demonstrated; the reversible effect observed in enzyme activity could be explained by the "induced fit" concept (35) which assumes that the binding of ligands causes a significant change in the average positions of the atoms in the protein, ultimately at the active site, "crippling" the enzyme molecule through physical alterations of its conformation; the modified enzyme would then still be active but function at a reduced rate. On the desorption or removal of the perturbant agent, the enzyme would return to its initial conformation and activity. Direct evidence for protein conformation changes under the influence of cosolvents could give a satisfactory explanation for the effect of cosolvents on reaction rates. In most earlier studies of cosolvent effects on protein structure, authors tried to determine changes in the degree of organization of helical content by all the available spectroscopic techniques, but unfortunately with a total lack of precise values for the physico-chemical parameters changing on addition of cosolvent which could thereby contribute to the destabilization of protein molecules.

In recent years, progress has been made through the investigation of localized or minor structural changes occurring well before any detectable "global" conformation change. The amount of such a change needed to affect the enzyme specific activity is unknown but one could assume that only a fraction of an Ångström change in orientation of catalytic groups may produce very significant changes in reaction rates.

As an example of such a modification, Lipscomb and co-workers (36) have observed a conformational change expressed in the movement of a catalytic group at the active site of carboxypeptidase upon binding of the substrate molecule. These authors have elucidated the structure of the enzyme at 6 Å resolution and have also carried out investigations at 2.8 Å resolution. Working on crystallized enzymes, where enzyme activity is reduced, and using poor substrates, they were able to detect binding at the active site. All the substrates are bound in regions close to the zinc atom in a depression of the molecule and the groups moving under their influence appear to be a tyrosine residue and an aspartate residue either or both of which could be involved in the catalytic action. The movement of these residues is respectively 15 Å and 2Å.

This kind of work is an exception and several methods are now currently used to investigate the kind of localized perturbation which, through sites on the protein surface or through unmasking of buried residues, could be responsible for changes in enzyme activity. We will briefly deal with these methods, their goals and perspectives.

Solvent perturbation method

The absorption spectra of the chromophoric groups buried in the protein bulk and exposed to the medium can be distinguished by the so-called solvent perturbation method; the spectra of groups buried in the interior of the protein molecules are not affected by small solvent perturbations, provided the conformations remain unaffected by the addition of a certain percentage of miscible organic solvent (in concentrations 10 to 20%). By contrast, the spectra of chromophoric residues in contact with the solvent are sensitive to changes introduced in the physical properties of the medium by the additive: refractive index and dielectric constant, as well as solvent–solute interactions in the immediate vicinity of the residues, cause measurable shifts in the absorption spectra. Thus, solvents such as polyols can be used as perturbants without affecting the native conformation of the protein (37,38,39).

The protein, with some residues exposed and some others buried, will show only a fraction of the total shift of an equivalent amount of free chromophoric groups, or of the totally unfolded protein chain, and the fraction of the difference spectrum will be a relative measure of the fraction of chromophoric residues exposed to the solvent.

The small spectral shifts resulting from such solvent perturbations are measured by differential spectrophotometry with the use of tandem double cells to correct for solvent differences (40,41).

Polyols (such as glycerol, ethylene glycol, propylene glycol) at concentrations up to 20% are used since they are unlikely to induce changes in the conformation of many proteins and since they cause red-shifts in the spectra of tyrosine residues. The method has been applied to the study of location and sometimes changes in location of tyrosyl residues of protein molecules (40). It could be used either with increasing concentrations of polyols to check any eventual conformational change due to solvent mixtures prepared for low temperature experiments.

It has been shown for example (42) that ribonuclease has 50% of its tyrosyl residues exposed to the solvent, so that the spectral perturbation (red-shift) of the protein molecule with all of its disulfide bridges cleaved should be twice that of the native protein. The absorbence difference between water and 20% of polyols is nearly twice for urea-denatured ribonuclease than for native ribonuclease in neutral and slightly acidic solution. Below pH 3.0, the absorbence difference of native ribonuclease shows an appreciable rise suggesting that most tyrosyl residues are progressively exposed in acid solution. In this pH region, similar changes in pH-dependent difference spectra related to the exposure of the same groups have been observed.

Some residues are almost certainly partially exposed to the solvent; partial exposure may be due to the fact that the residue is situated in a crevice and only capable of interacting with small-radius perturbants, or it may be due to the range effect of the perturbant on the chromophores situated near, but not on, the surface of the protein. In these conditions, slight perturbations might modify the initial spectra.

These different properties suggest a more general application of the method of the solvent-perturbation spectroscopy to structural studies of proteins in mixed solvents and to problems of locating the environment and specific involvement of chromophoric groups in such artificial conditions. Additional measurements of the helix content by optical rotatory dispersion, as well as titration curves, might be carried out in the same conditions to check whether other signs of conformational changes are apparent or not.

As an example of the investigation of local changes in protein structure in the presence of an organic solvent, let us mention observations made in this laboratory by Guinand *et al.* (66) with β-lactoglobulin in the presence of increasing concentrations of ethylene glycol at pH* 5.4.

Difference spectra show a non-linear variation of absorbance as a function of the concentration in ethylene glycol between 20 and 50%, suggesting a specific action on the chromophores (tryptophan and tyrosine residues). The same spectra show a linear variation in salty aqueous solutions between 0.1 and 5 M (NaCl). It was checked that the non-linear variation observed in mixed solution was not due to an eventual change in proton activity. A normal "red-shift" of the absorption bands is observed in any case, due to the variation of the refraction index of the media to which the chromophores are exposed.

The non-linearity reported above explains why one observes in the same conditions a decrease of the levorotation of the protein.

Thermal perturbation spectroscopy

This procedure is similar to the one above and consists of perturbing the temperature in a range where there is no "thermal transition", to determine the number of chromophores exposed to the solvent as well as those buried beneath the protein surface (41,42). Their absorbence must vary linearly as a function of the temperature. Non-linearity would be a proof of local changes.

Difference spectra can be recorded in such conditions. In the case of β-lactoglobulin (pH* 5.4) such spectra present 3 positive maxima at 277, 283 and 293 nm, which should correspond to the absorption of the 2 tryptophan and 6 tyrosine residues exposed to the solvent in these pH* conditions. It can be checked that the absorbence of these maxima varies linearly as a function of temperature in aqueous solution, as well as in salt solutions (0.1 to 5 M NaCl), respectively between 20° and °0C, 20° and −20°C.

In the presence of ethylene glycol, linearity is observed between +20 and −20°C, and it can be concluded that temperature does not introduce an additional perturbation of the protein molecule.

In aqueous solution at pH 5.2, the absorption spectrum of the β-lactoglobulin (which consists of two subunits and will be termed later N_2) presents two negative bands culminating at 296 and 306 nm due to two kinds of tryptophan residues in different environmental conditions since a decrease in temperature gives two different changes.

There are two tryptophan residues per protein molecule (one on each subunit) which are responsible for the absorption at 296 nm. They are exposed to the solvent (at pH 5.2). Temperature variation determines a blue shift. The tryptophan residues absorbing at 306 nm appear to be normally buried and have an anomalous behavior: at about pH 8.4 this band disappears and the band at 296 nm increases and four tryptophan residues are then titrable.

At pH 5.4, the intensity of the 306 nm band is constant both in the presence of ethylene glycol and in salt solutions; therefore the corresponding tryptophan residues are normally buried as stated above.

Finally, let us mention that proteins with prosthetic groups or coenzymes showing intense characteristic absorption spectra in the near ultraviolet and visible regions can be studied by spectrophotometry, spectral changes reflecting structural perturbations in the region of these chromophores.

Investigations by titration of anomalous groups—potentiometric titrations

Since, in most proteins, one out of every 3–4 amino acid residues contains a titrable acidic or basic group which is in direct contact with the solvent, and therefore can accept or release hydrogen ions at this location without requiring any modification of the protein conformation in its vicinity, hydrogen ion titration curves of proteins can be recorded which are characteristic of the native forms. The simplest information obtained from such curves is a count of the number of groups titrated, but the eventual differences of group counting between one set of conditions and another can indicate how a protein differs under the two conditions and will be useful to detect any possible conformational change, when some groups become suddenly in immediate contact with the solvent.

Perhaps one of the most striking examples to be recalled here is the case of ribonuclease. It has been shown long ago (43) that only 3 of the 6 phenolic groups of the protein can be titrated while it is in its native form; these 3 groups have an essentially normal intrinsic pK, and the 3 phenolic groups which are not available for titration can be titrated at 25°C near pH 13, or in 8 M aqueous urea, or ethylene glycol.

Ribonuclease has been found to be significantly soluble in several cosolvents—formamide, dimethylformamide, dimethylsulfoxide and ethylene glycol, and even 2-chloroethanol. Several different criteria (ultraviolet absorption spectrum, spectrophotometric titration behavior, and optical rotary dispersion) indicate that the structure of this protein in ethylene glycol is indeed different from that in water. On the other hand all six hydroxyl groups titrate normally and reversibly, a result suggesting that the hydrophobic regions of the molecule which occur in water are disrupted by the cosolvent. This disruption is reversible since the enzymatic activity of rubonuclease dissolved in ethylene glycol is recoverable in aqueous solution.

Even if the potentiometric study in ethylene glycol solution carried out by Sage and Singer (43) was a rudimentary one, it was a pioneering work since potential measurements in mixed solvents should be of considerable interest in protein chemistry and particularly, as we will see below, in the study of the cosolvent effect on protein structure.

A POSSIBLE APPROACH: A COMPARATIVE STUDY OF THE EFFECTS ON NET PROTON CHARGE OF PROTEINS AND ON THEIR ACTIVITY

As pointed out by Lumry (31):

> chemical mechanisms of protein function are uniquely dependent on the conformational behavior of proteins, but the kinds of rearrangements they can undergo are yet hypothetical and the conformational nature of protein function is not yet understood at the ball and stick level.

Several independent criteria can be applied simultaneously to determine the effect of mixed solvents on the conformation and activity of enzyme molecules, but they most often lack the necessary sensitivity and specificity to detect any correlation between changes in enzyme activity and changes in structure. For instance, most kinetic measurements are carried out on overall reactions and not in terms of parameters K_m (app) and k_{cat} which allow the determination of the type of competition introduced by the presence of the cosolvent. On the other hand, it should be noted that the ionization constants of exposed groups on the protein surface will be modified by the cosolvent, affecting the enzyme activity through conformation changes possibly mediated by the active site. A common assumption is that competitive inhibition means a direct interaction of the cosolvent on the active site, through hydrogen bonding or other possible types of interaction. In fact, this assumption must be checked on any system under investigation. Some normally "buried" ionizing groups might be unmasked by the cosolvent and the net proton charge $\bar{Z}$ of the protein could therefore be significantly changed inducing a new conformation and a change in activity.

These different possibilities prompted us to carry out comparative studies of reversible inhibition and of changes in the potentiometric properties of enzyme molecules. This work is at a very early stage but we will see that the first results are quite promising.

Once the full recovery of the enzymatic activity in aqueous solution from solution in mixed solvents is demonstrated, showing that the cosolvent effect is reversible, kinetic and potentiometric investigation can be carried out.

Kinetic measurements

Since we have seen that the solvent effect appears to be in most of cases a reversible inhibition of enzyme specific activity, we can treat it quantitatively by use of the Michaelis–Menten relationship defining the quantitative relationship between the enzyme reaction rate v and the substrate concentration [S] when both V_m and K_m are known.

Let's recall that in the simplest case

$$E + S \underset{k_2}{\overset{k_1}{\rightleftharpoons}} ES$$

and

$$ES \underset{k_4}{\overset{k_3}{\rightleftharpoons}} E + P$$

both reactions being considered reversible.

The derivation of the Michaelis–Menten equation, which is that of Briggs and Haldane, begins by considering the rates of formation and decomposition of ES, and this rate is given by

$$\frac{d[ES]}{dt} = k_1([E] - [ES])[S]$$

The rate of formation of ES from E + P, which is very small, is neglected. The rate of decomposition of ES is given by

$$-\frac{d[ES]}{dt} = k_2[ES] + k_3[ES]$$

In the steady state, where the ES concentration remains constant,

$$k_1([E] - [ES])[S] = k_2[ES] + k_3[ES]$$

leading to

$$\frac{[S]([E] - [ES])}{[ES]} = \frac{k_2 + k_3}{k_1} = K_m$$

where K_m is the Michaelis–Menten constant.

The steady-state concentration of the ES complex is:

$$[ES] = \frac{[E][S]}{K_m + [S]}$$

The initial rate v being proportional to the concentration of ES,

$$v = k_3[ES]$$

when the substrate concentration is so high that all the enzyme in the system is in the ES form, the maximum velocity V_m is reached and given by

$$V_m = k_3[\mathrm{E}].$$

v can be expressed:

$$v = k_3 \frac{[\mathrm{E}][\mathrm{S}]}{K_m + [\mathrm{S}]}$$

and one can write

$$\frac{v}{V_m} = \frac{k_3 \dfrac{[\mathrm{E}][\mathrm{S}]}{K_m + [\mathrm{S}]}}{k_3[\mathrm{E}]}.$$

Solving for v,

$$v = \frac{V_m[\mathrm{S}]}{K_m + [\mathrm{S}]} \tag{3-11}$$

When both V_m and K_m are known, the above equation defines the quantitative relationship between the reaction rate and the substrate concentration. In the present case, K_m is represented by

$$K_m = \frac{k_2 + k_3}{k_1}$$

When the rate limiting step in the overall reaction is

$$\mathrm{ES} \xrightarrow{k_3} \mathrm{E} + \mathrm{P}$$

Then

$$K_m = \frac{k_2}{k_1}$$

The Michaelis–Menten equation can be transformed by taking the reciprocal of both sides of equation (3-11)

$$\frac{1}{v} = \frac{1}{\dfrac{V_m[\mathrm{S}]}{K_m + [\mathrm{S}]}} = \frac{K_m + [\mathrm{S}]}{V_m[\mathrm{S}]}$$

$$\frac{1}{v} = \frac{K_m}{V_m[\mathrm{S}]} + \frac{[\mathrm{S}]}{V_m[\mathrm{S}]}$$

and

$$\frac{1}{v} = \frac{K_m}{V_m} \cdot \frac{1}{[\mathrm{S}]} + \frac{1}{V_m}$$

This is the Lineweaver–Burk equation: by plotting $1/v$ vs $1/[S]$, a straight line is obtained with a slope of K_m/V_m and with an intercept of $1/V_m$ on the $1/v$ axis. This double reciprocal plot allows us to obtain V_m with greater accuracy than from the simple plot of v vs [S].

The intercept on the abscissa of the Lineweaver–Burk plot is $-1/K_m$.

Valuable information could be obtained from the kinetic study of enzyme catalyzed reactions in mixed solvents using the Lineweaver–Burk plot.

Enzyme inhibition is broadly classified into two types, reversible and irreversible. Reversible inhibition is characterized by an equilibrium between enzyme and inhibitor, with an equilibrium constant K_i which is a measure of the affinity. The degree of inhibition depends on the inhibitor concentration which is usually obtained rapidly and thereafter is independent of time.

Reversible inhibition can be treated quantitatively in terms of the Michaelis–Menten relationship and can be competitive or non-competitive, the first one being reversed by increasing the substrate concentration, and the second not.

Competitive inhibitors act by increasing the effective K_m, the inhibitor competing with the substrate for the enzyme; substances such as those related in structure to the substrate combining with the enzyme at the same site as the substrate are competitive inhibitors.

Non competitive inhibitors have no effect on K_m but reduce V_m; these inhibitors are substances which combine at a site so far removed from the active site that they have no influence on the binding of the substrate. Organic solvents can belong to this last group, but we will see that they can act as competitive inhibitors in some reactions.

Several other possibilities of inhibition have been described, and there is also the possibility that an inhibitor may act in more than one way, giving mixed types of inhibition.

The kinetics of the various types of inhibition can be briefly reviewed and expressed in terms of the Lineweaver–Burk plots which have been widely used for distinguishing types of inhibition.

(*a*) *Fully competitive inhibition*. The inhibitor combines with the substrate-binding site and the situation can be represented by the following equations:

$E + S \rightleftharpoons ES$ equilibrium in the Michaelis case, with dissociation constant K_s

$E + I \rightleftharpoons EI$ which undergoes no further reaction and is in equilibrium with dissociation constant K_i.

$$ES \xrightarrow{k} E + P$$

Dealing with the Michaelis case, we can write the following equations, in which p is the concentration of ES and q that of EI:

$$(e - p - q)s = K_s p$$
$$(e - p - q)i = K_i p$$
$$v = kp$$

Solving these equations for v, we obtain

$$v = \frac{ke}{1 + \frac{K_s}{S}\left(1 + \frac{i}{K_i}\right)}, \quad \text{where } V_m = ke \text{ and } K_m = K_s\left(1 + \frac{i}{K_i}\right)$$

If the same system is considered without assuming that the step $E + S \rightleftharpoons ES$ reaches equilibrium and inserting rate constants into the above equations, we obtain

$$E + S \underset{k_{-1}}{\overset{k_{+1}}{\rightleftharpoons}} ES$$

$$E + I \underset{k_{-2}}{\overset{k_{+2}}{\rightleftharpoons}} EI$$

$$ES \xrightarrow{k_{+3}} E + P$$

Application of steady state kinetics gives

$$k_{+1}(e - p - q) = (k_{-1} + k_{+3})p$$
$$k_{+2}i(e - p - q) = k_{-2}q$$
$$v = k_{+3}p$$

Solving these equations for:

$$v = \frac{k_{+3}e}{1 + \frac{k_{-1} + k_{+3}}{k_{+1}s}\left(1 + \frac{k_{+2}i}{k_{-2}}\right)}$$

and expressing

$$\frac{k_{-1} + k_{+3}}{k_{+1}} \quad \text{as } K_m$$

$$\frac{k_{-2}}{k_{+2}} \quad \text{as } K_i$$

and

$$v = \frac{V_m}{1 + \frac{K_m}{s}\left(1 + \frac{i}{K_i}\right)}$$

Thus, the competitive inhibitor produces an apparent increase in K_m by the factor $1 + i/K_i$. The apparent K_m increases without limit as i is increased, and at any finite inhibitor concentration, the limiting velocity with excess of substrate is always equal to V_m, which is the maximum velocity for the uninhibited reaction.

(*b*) *Partially competitive inhibition.* Equations differ from those given above in assuming that a complex EIS may exist and are the following

$$\mathrm{E + S} \rightleftharpoons \mathrm{ES} \tag{3-12}$$

$$\mathrm{E + I} \rightleftharpoons \mathrm{EI} \tag{3-13}$$

$$\mathrm{ES + I} \rightleftharpoons \mathrm{EIS} \tag{3-14}$$

$$\mathrm{EI + S} \rightleftharpoons \mathrm{EIS} \tag{3-15}$$

$$\mathrm{ES} \xrightarrow{k} \mathrm{E + P}$$

$$\mathrm{EIS} \xrightarrow{k} \mathrm{EI + P}$$

If we assume that the inhibition is purely an effect on affinity, the complexes ES and EIS break down at the same rate, that is they have the same rate constant k and the overall velocity is the sum of these two reactions. If this were not so, S would affect both V_m and K_m.

Writing K_s, K_i, K_s' and K_i' for the equilibrium constants of reactions 3-12–3-15 respectively and p, q and p' for the concentrations of ES, EI and EIS respectively we have:

$$\begin{aligned} (e - p - p' - q)S &= K_s p \\ (e - p - p' - q)i &= K_i p \\ qs &= K_s' p' \\ pi &= K_i' p' \\ v &= k(p + p') \end{aligned}$$

Solving for v, we obtain the equation

$$v = \frac{ke}{1 + \dfrac{K_s}{s} \dfrac{1 + \dfrac{i}{K_i}}{1 + \dfrac{i}{K_i}\dfrac{K_s}{K_s'}}}$$

when i is very large this equation reduces to

$$v = \frac{ke}{1 + \dfrac{K_s'}{s}}, \quad \text{where} \quad \begin{aligned} V_m &= ke \\ K_m &= K_s' \end{aligned}$$

and we may consider that there is a new enzyme species EI with a different affinity for the substrate $1/K'_s$. This type of inhibition may be distinguished from the fully competitive type by varying the inhibitor concentration at fixed substrate concentration, but increases to a definite level when all the enzyme is combined with inhibitor, and can then increase no further.

(*c*) *Non-competitive inhibition.* In this type of inhibition the inhibitor does not influence the combination of the substrate with the enzyme, but influences only V_m. The complex EIS may break down at a different speed from ES, the velocity being the sum of the two reactions; or it may not break down at all and the velocity is purely that of the breakdown of ES, the inhibition being equivalent to a reduction in the amount of active enzyme.

Equations are the following

$$\mathrm{E + S} \rightleftharpoons \mathrm{ES} \qquad \text{equilibrium constant } K_s$$
$$\mathrm{E + I} \rightleftharpoons \mathrm{EI} \qquad \text{equilibrium constant } K_i$$
$$\mathrm{EI + S} \rightleftharpoons \mathrm{EIS} \qquad \text{equilibrium constant } K_s$$
$$\mathrm{ES + I} \rightleftharpoons \mathrm{EIS} \qquad \text{equilibrium constant } K_i$$
$$\mathrm{ES} \xrightarrow{k} \mathrm{E + P}$$

and

$$(e - p - p' - q)s = K_s p$$
$$(e - p - p' - q)i = K_i p$$
$$qs = K_s p'$$
$$pi = K_i p'$$
$$v = kp \qquad \text{when EIS may not break down}$$

leading to

$$v = \frac{\dfrac{ke}{1 + \dfrac{i}{K_i}}}{1 + \dfrac{K_s}{s}}, \quad \text{where } V_m = ke \times \frac{1}{1 - \dfrac{i}{K_i}}$$

$$K_m = K_s$$

When EIS breaks down at a different velocity from ES,

$$v = kp + k'p'$$

and

$$v = \frac{\dfrac{ke + k'\dfrac{i}{K_i}e}{1 + \dfrac{i}{K_i}}}{1 + \dfrac{K_s}{s}}$$

This equation can be written

$$v = \frac{ke}{\left(1 + \dfrac{K_s}{s}\right)\dfrac{1 + \dfrac{i}{K_i}}{1 + \dfrac{i}{K_i}\dfrac{k'}{k}}}, \quad \text{where } V_m = ke \times \frac{1 + \dfrac{i}{K_i}\cdot\dfrac{k'}{k}}{1 - \dfrac{i}{K_i}}$$

Therefore in both cases, $K_m = K_s$ is unaffected, and V_m is modified by the inhibitor.

(*d*) *Mixed-type inhibition.* An inhibitor may act on both V_m and K_m, giving a mixture of competitive and non-competitive inhibitions and are classified as mixed-type inhibitors. Various mixed-type inhibitions can arise in a number of different ways and equations for mixed inhibitions have been derived and published in the literature (44).

(*e*) *Graphical presentation of inhibition by the Lineweaver–Burk plot.* There are five ways of plotting the effect of substrate concentration on enzyme reaction velocities and therefore five ways of identifying the type of inhibition. The Lineweaver–Burk method of plotting is the most popular for distinguishing types of inhibition and its characteristics are shown in Fig. 9.

It can be seen that with fully competitive inhibitors, the lines intersect on the vertical axis, whereas with fully non-competitive inhibitors they intersect on the base line, and between the two axes with mixed-type inhibitors. Dixon (44) has given a simple graphical method to obtain inhibitor constant K_i directly without calculation.

Titration curves as a method of detecting the occurrence of conformation changes

We have seen that a combination of physical and chemical measurements is required to specify precisely the nature of a conformation change and to correlate it with the solvent effect on enzyme specific activity. Titration curves provide a simple method of detecting the occurrence of conformation

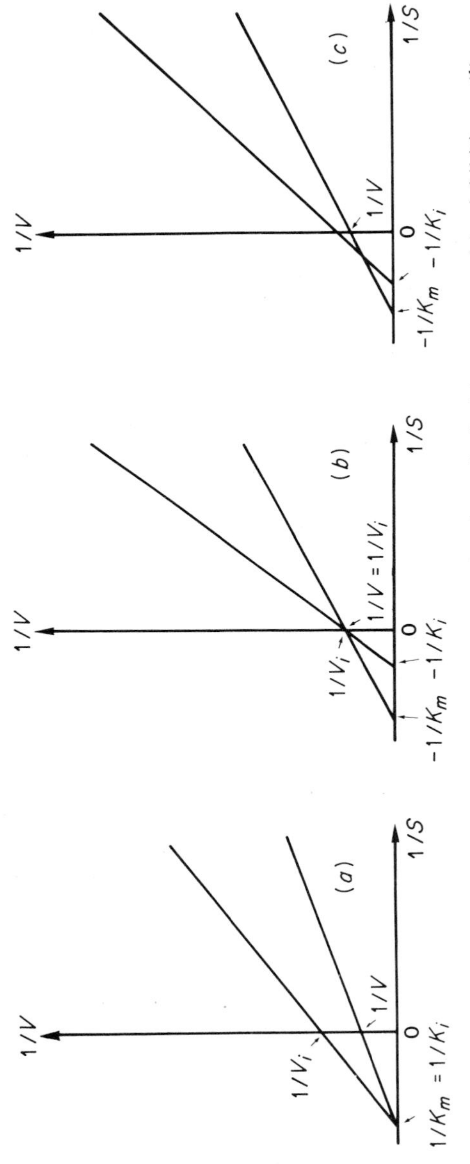

FIG. 9. Graphical determination of inhibition constant (Dixon's method). (*a*) non-competitive inhibitions; (*b*) competitive inhibitions; (*c*) mixed-type inhibitions.

changes, and since the simplest information gained from such curves is a count of the number of groups titrated, the differences observed in counting groups between one set of conditions and another could tell us directly how a protein differs under the two conditions and eventually why one observes differences in activity.

Since most titration curves consist of three S-shaped sections, it is possible to count separately the groups titrated respectively in the acid, neutral, and alkaline regions, and to observe differences by examining the protein in different solvents, water and mixtures with different volume ratio. We have seen in chapter 2 that one can relate the pH to an arbitrary defined "activity" of hydrogen ions simply by setting $\mathrm{pH} = -\log a_{\mathrm{H}}^{*}$ and that the dissociation constants of model compounds can be determined in terms of such an arbitrary scale, relative to the e.m.f. of a suitable cell.

In these conditions, titration curve differences can be recorded as a function of the concentration of cosolvent and may be of these types. (*a*) the total count of groups remains the same, but their division into the regions enumerated above has altered. (*b*) The total number of groups has altered, i.e. new titrable groups have appeared (unmasking of previously buried groups). In both cases, the S-shaped titration curves differ in the way shown in Fig. 10. (*c*) The numbers counted in each of the regions remain the same, but the shape of the curve is altered, and the difference appears as in Fig. 11.

It is often difficult to find out which type of difference is involved because of the rapid denaturation occurring at extreme pH values. Titration curve differences may then be measured directly by means of a pH-stat. This apparatus is designed to add acid or base to keep the pH unchanged, and the total amount required between the beginning and end of the reaction is a measure of the difference between the corresponding titration curves at that pH. Measurements made at a series of pH values will then produce difference titration curves such as shown in Figs. 10 and 11, and used in conjunction with the complete curves.

Difference counting has already been used in a large number of investigations, revealing the presence of anomalous groups in denatured proteins, showing deviations from expected pH values characteristic of the state of the protein, providing evidence concerning the physical or chemical modifications of proteins. It might be used to investigate the behavior of enzyme proteins in mixed solvents, and we will give examples of such investigations.

Examples of joint investigations

Examples of the study of enzyme inhibition by organic solvents in terms of Lineweaver–Burk plots are rather scarce. Let us mention again the work by Clement and Bender (12) who tried to explain the effects of dipolar aprotic

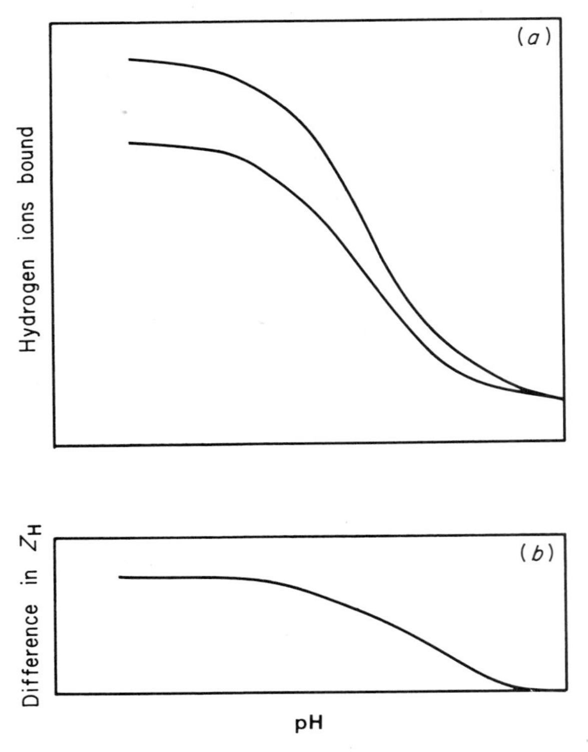

FIG. 10. Differences between titration curves for a given type of group in two different states of a given proteins. The difference is in the number of titrable groups. (*a*) actual titration curves; (*b*) plot of the difference between them.

solvents (dioxane, acetone and acetonitrile) on α-chymotrypsin catalyzed reactions in terms of simultaneous operation of a dielectric and a competitive effect and succeeded in quantitatively accounting for the results obtained. Since these effects in themselves have proved to be inadequate to explain the action of organic solvents, their combination might produce a new interpretation already considered above.

In this laboratory, Maurel and Travers (45) investigated the oxidation of gaïacol by hydrogen peroxide catalyzed by horseradish peroxidase, in ethylene glycol–water and in methanol–water mixtures in different volume ratios. These authors observed a mixed-type inhibition in certain conditions and in the presence of ethylene glycol, whereas they recorded kinetic data reflecting competitive inhibition in methanol. These different observations cannot be satisfactorily interpreted but they show the critical role of the cosolvent on the reaction and therefore on the enzyme, and the need to try

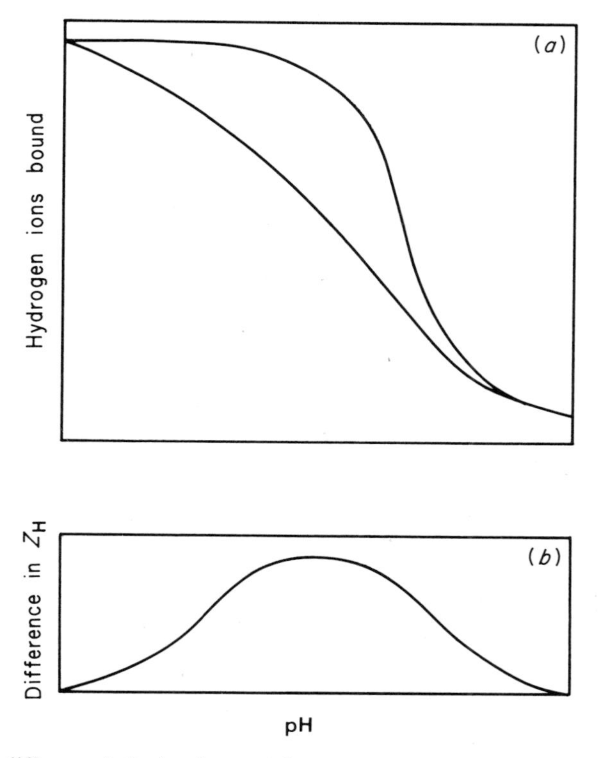

FIG. 11. The difference is in the shape of the curve, the number of titrable groups being the same. (*a*) actual titration curves; (*b*) plot of the difference between them.

and correlate the solvent effect on any enzyme catalyzed reaction with its effect on protein structure. Such a correlation was attempted in the following two cases.

(*a*) *Solvent effect on the formation of compound I of the horseradish peroxidase and on the titration curve of this enzyme.* Anticipating the applications of the low temperature procedure (chapter 5) let us say that the oxidation of a number of substrates by hydrogen peroxide, catalyzed by the horseradish peroxidase (termed HRP)

$$H_2O_2 + 2AH \xrightarrow{HRP} 2H_2O + 2A_{oxidized}$$

involves in fact several steps and can be "thermally controlled" at low temperatures.

TABLE 5

Methanol (% in vol.)	0	10	30	50	70
k_1 : mole^{-1} s^{-1} 10^{-6}	4.2	3.9	3.3	2.8	2.2

In these conditions, discussed further in chapter 5, it is possible to resolve the following steps:

1. $$\text{E(HRP)} + H_2O_2 \underset{k_{-1}}{\overset{k_1}{\rightleftharpoons}} \text{compound I}$$

2. $$\text{compound I} + \text{AH} \longrightarrow \text{compound II} + A_{ox.}$$

3. $$\text{compound II} + \text{AH} \longrightarrow \text{E(HRP)} + 2H_2O + A_{ox.}$$

In these conditions and in the absence of the substrate AH, it is possible to study step (1) and accumulate compound I. At subzero temperatures, the reverse reaction $\text{I} \xrightarrow{k_{-1}} \text{E} + H_2O_2$ can be inhibited and the direct kinetic study of the reaction $\text{E} + H_2O_2 \xrightarrow{k_1} \text{I}$ is achieved by stopped flow. The kinetics of this reaction were studied in the mixtures methanol–water in various volume ratios ranging from 10–90 to 70–30, v/v, ethylene glycol–water 50–50, and dimethylformamide–water 50–50, between 25 and −40°C. Kinetic recordings of the formation of compound I are reported in Fig. 12. In these conditions, the rate constant k_1 was measured, for example, at normal temperature as a function of methanol concentration and its decrease is shown in Table 5.

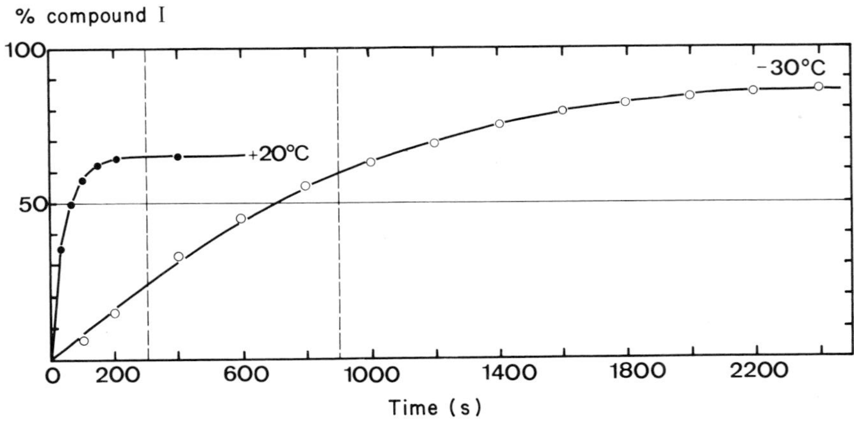

FIG. 12. Kinetics of the first step of the peroxidase reaction leading to compound I. Formation of compound I as a function of time at different temperatures.

It can be seen that log k_1 as a function of the percentage in methanol is a straight line with a negative slope, see Fig. 13. Thus it appears that the addition of increasing amounts of methanol affects the binding of hydrogen peroxide to the HRP. This effect is fully reversible by infinite dilution. It might be due to a conformation change of the enzyme, and we checked that it cannot be explained by changes in viscosity or dielectric constant of the medium.

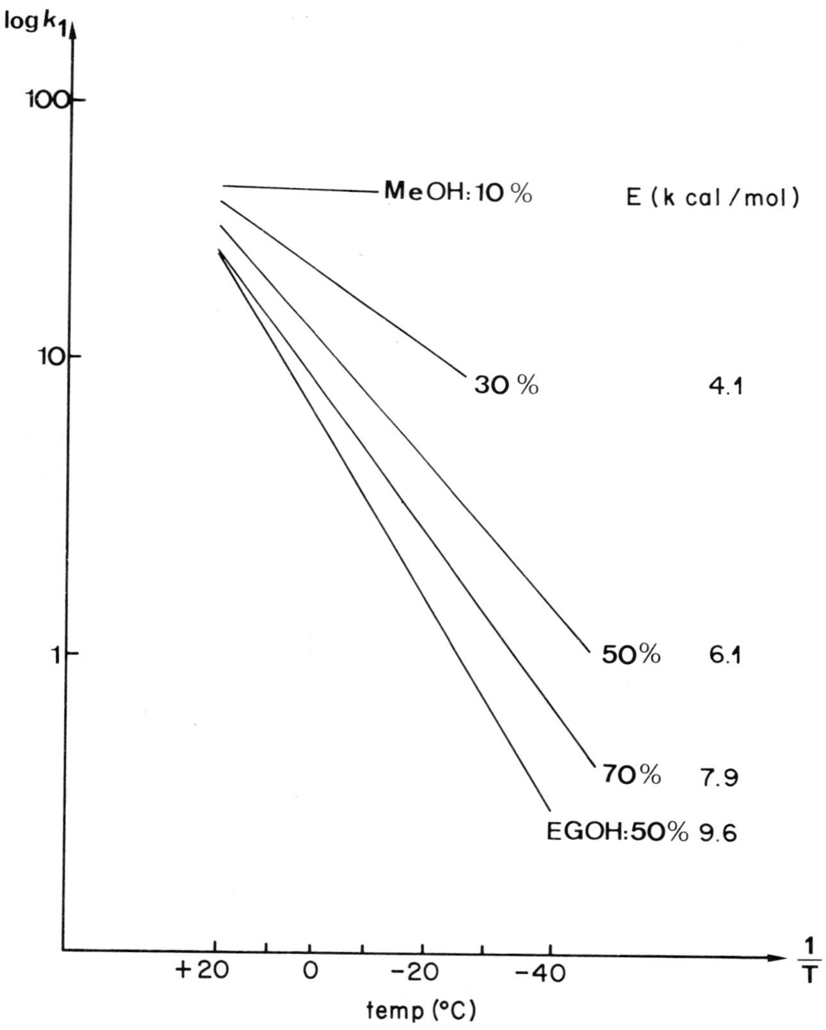

FIG. 13. Evolution of the rate constant k_1 of the formation of compound I in various mixtures as a function of temperature.

Then we have checked whether the cosolvent effect can be due to a modification of the "net charge" of the protein, influencing its affinity for the substrate.

We recalled previously that one out of every 3–4 amino acid residues contains a titrable acidic or basic group which is in contact with the solvent. The modification of the pK of such residues outside the active site could induce reversible conformation changes, and it is known that only slight modifications in the orientation of a catalytic group may determine very large modifications in enzyme specific affinity. It is obvious that both cosolvents and temperature variations bring about pK changes. Such changes might be evidenced by hydrogen ion titration curves of proteins, in water and in mixed solvents.

Using the procedure already described, we recorded the titration curves of HRP in water and in mixtures of methanol–water. Such curves are shown in the Fig. 14 where it can be seen that 50% by volume of methanol has markedly changed the "normal" titration curve of the enzyme. Such an effect could explain the observed changes in k_1 values in proportion as methanol is added to aqueous solutions of HRP.

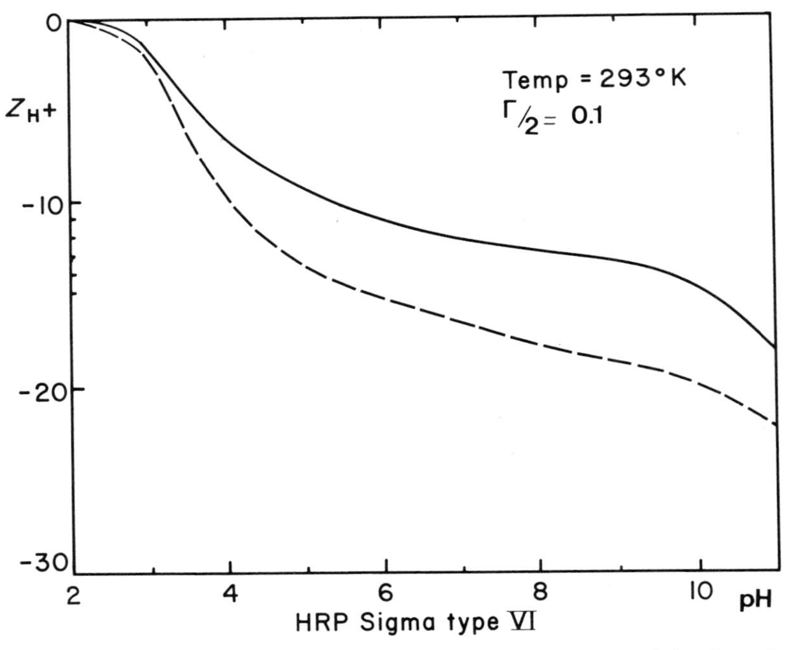

FIG. 14. Titration curves of horseradish peroxidase in water and in the mixture methanol–water (50 : 50, v/v) at 20°C. ———, H_2O; ----, 50% MeOH.

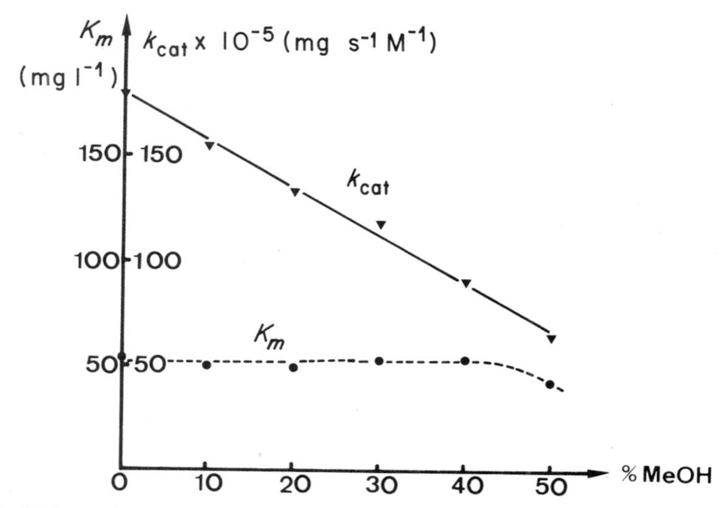

FIG. 15. Values of K_m and k_{cat} of the hydrolysis of *Micrococcus lysodeikticus* by lysozyme as a function of the concentration in methanol.

(*b*) *Lytic activity of lysozyme*. Bacterial cell walls (such as *Micrococcus lysodeikticus*) are "solubilized" by lysozyme (46,47) and we have recorded both the K_m and the V_m of this reaction in aqueous solution and in methanol–water mixtures. Assuming the following reaction:

$$E + S \xrightleftharpoons{K_m} E - S \longrightarrow \cdots \xrightarrow{k_{cat}} E + P$$

where E is lysozyme and S the bacterial cell walls, it is possible to measure K_m and V_m values as a function of the concentration of methanol, and to check that the K_m value remains practically constant up to 50% by volume of methanol (Fig. 15). Meanwhile it can be seen in Fig. 16 that the titration curve of lysozyme is practically unchanged in ethylene glycol–water mixture. A similar result if obtained in presence of methanol might explain the constancy in the affinity of E for S. This result reminds us that the native conformation of lysozyme is more stable towards pH changes at room temperature than that of most other proteins.

The variations of V_m observed in the presence of methanol (Fig. 15) should be due to a solvent effect on the rate-limiting step (k_{cat}), through hypothetical changes in solvation of intermediates and (or) changes occurring at the level of cell walls. It is necessary to extend these types of investigations to many other enzyme systems to check whether there is always a correlation between cosolvent effects on the affinity of E for S and changes in titration curves of the protein. Such observations might help to understand the "intrinsic"

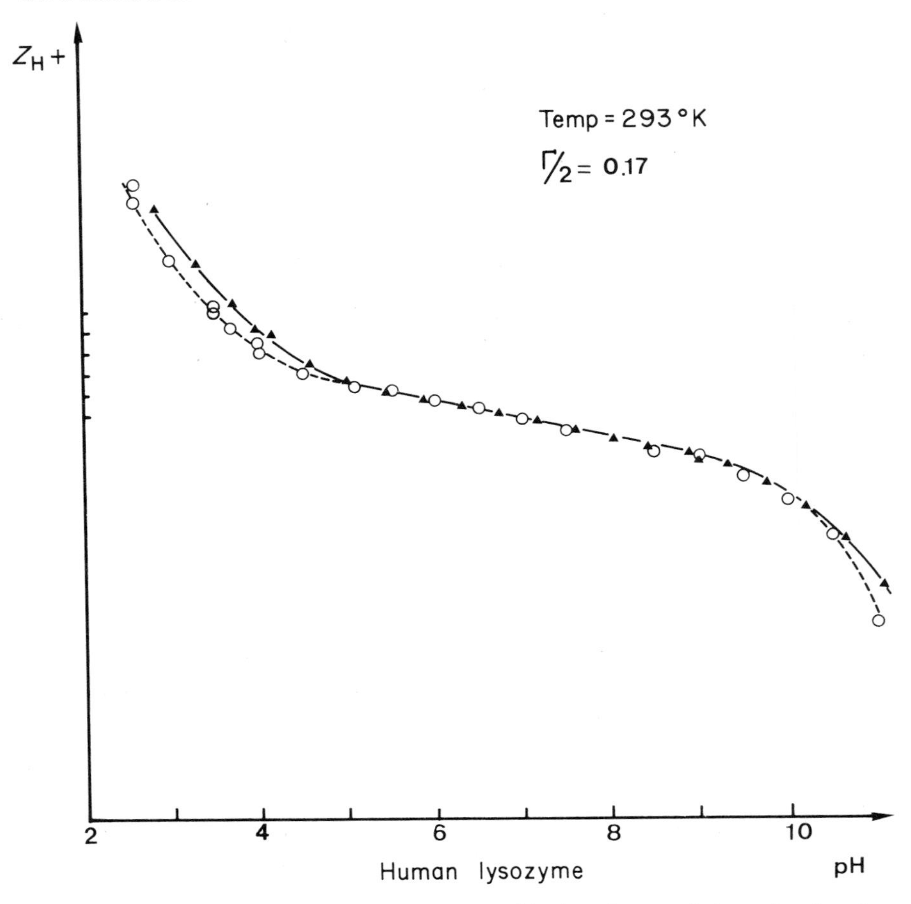

FIG. 16. Titration curves of lysozyme in water and in the mixture ethylene glycol–water (50 : 50, v/v) at 20°C. —▲—, 50% EGOH; ---O---, H_2O.

solvent effect on a number of enzyme reactions and then to try to discover new conditions for improving enzyme activity in mixed solvents.

Solvent effects on quaternary structure

Up to now we have only considered the case of the single chain enzyme. New problems arise when considering enzymes which consist of two or more subunits and it is interesting to consider, even briefly, the possible cosolvent effects on such types of systems.

It can be assumed that many types of bonding are responsible for the quaternary structure of these proteins; these are electrostatic forces between charged groups forming salt bridges, hydrogen-bonding between hydrogen donors and acceptors, hydrophobic surfaces between juxtaposed subunits, dispersion forces stabilizing secondarily specific association initiated by the above forces. Thus subunits can associate and, therefore, dissociate in more than one way, but a single type of binding force is often predominant: electrostatic forces (salt and hydrogen bridges) can be considered to be predominant from the action of electrolytes on the stability of many oligomeric proteins; proteins which are not sensitive to electrolyte concentration, but are easily dissociated to some extent by detergents, would have their subunits mainly held together by hydrophobic forces. Such possibilities require definite experimental proof and the establishing of the relative contribution of each kind of bond, which would vary for a given protein from one medium to another; and, for a particular solvent, as a function of various physico-chemical parameters changing upon cooling; as we have seen, this would be the case for the pK values of dissociable groups.

If hydrophobic interactions were significantly involved in non-covalent binding between subunits, the addition of moderate concentrations of weakly protic solvents to aqueous solutions should produce dissociation without significant conformational changes in the subunits. Dissociation should be eased by cooling since hydrophobic bonds become significantly weaker as the temperature is lowered (48,49).

These predictions could be checked on many proteins which are not sensitive to electrolyte concentration, but are known to be easily dissociated by detergents interfering, through their hydrocarbon portion, with hydrophobic bonds. The order of effectiveness of the weakly protic solvents as destabilizers on surfaces of contact between subunits, leading to increased subunit dissociation should be: propylene glycol > ethylene glycol > glycerol, since this is the order of their decreasing capacity to dissolve hydrocarbons. On the other hand, these solvents should be harmless with regard to intrapeptide hydrogen bonding since the non-covalent binding between subunits might be expected to be weaker than the non-covalent binding maintaining the native conformation within the subunits themselves.

It has been shown (50) that non-polar interactions at the contact surfaces of α- and β-hemoglobin subunits can be destabilized, leading to increased subunit dissociation with increasing hydrocarbon content of ureas and amides. In addition, two polar contacts (salt bridges) between identical chains also seem to be split as a result of subunits dissociation.

Intensive investigations should be carried out on a number of oligomeric enzymes showing an abnormal behavior when their catalyzed reactions are studied in mixed solvents. The methods of investigation described above, which can give information about the actual size of protein molecules, should be used.

The effects on protein structure will vary with the type of solvent used. In a large number of cases, oligomer proteins can be dissolved in polyol–water mixtures without losing their potentiality once they are diluted in aqueous solutions.

Moreover, we will see in the next section that β-lactoglobulin, a dimer α-β, exhibits an unchanged pK value of dissociation into subunits in ethylene glycol–water mixtures at room temperature, whereas some local perturbations at the level of each subunit are observed in these conditions.

Primary alcohols added to aqueous solutions at room temperature seem to be more harmful towards oligomeric proteins. We already have mentioned the case of hemoglobin in methanol–water mixtures (30 : 70; 50 : 50, v/v); we could indicate further examples of oligomeric proteins which are altered by this volume ratio and which must be investigated at low temperatures in solutions prepared by the sampling procedure already described. It is not yet known whether corrections of dielectric constant and/or adjustment of new pH* values could prevent such alterations, but, after all, we are primarily interested in carrying out investigations at low temperatures and we have a satisfactory sampling procedure to do so in primary alcohols and on a large number of enzyme systems.

Any solvent effect will vary with the type of protein used. Strong interactions of the solvent with a protein will make the solvent a denaturant, and weaker interactions will cause conformational changes, possibly causing dissociation into subunits. Preferential binding of solvent or water to particular chromophoric and non-chromophoric protein side chain groups could lead to perturbations to a greater or lesser extent than that expected from the composition of the solvent mixture. Partially exposed side chain groups will have a particular behavior according to the nature and concentration of the organic solvent. Therefore, any protein to be submitted to low temperature experiments must be tested carefully using all available methods, to determine the best solvent, the best volume-ratio, and the extent of the perturbation.

As previously stated, although a solvent may have little effect on conformation at a pH at which a protein has a great stability, under other conditions of pH the same solvent can destabilize the native form of the protein and bring it to the verge of collapse. This is also to be expected when the dielectric constant of the medium is markedly lowered by addition of a high percentage of solvent, such as methanol and methylpentanediol.

Comparative effects of cosolvents and temperature variations on reaction rates

We have seen that cosolvents to be used as "antifreeze" must affect enzyme specific activity reversibly; reduction in reaction rates can be compensated

TABLE 6

Reduction of reaction rates	by solvent effect	by temperature (+20° → −20°C)
Lysozyme	2(40% methanol)	200
Horseradish peroxidase	5(50% ethylene glycol)	400
NADPH-cytochrome P_{450} reductase	3(50% ethylene glycol)	100

by increasing the enzyme concentration. However, it is necessary to check whether the effect of changing (lowering) temperature is more pronounced than the cosolvent effect on reaction rates, and some examples of both effects are listed in Table 6. It can be seen that changing the temperature from +20° to −20°C reduces the rates much more than the reduction induced by 50% by volume in cosolvent, a result which shows how much cooling reduces any initial enzyme activity expressed in terms of reaction rates.

We will now deal with the important problem of this temperature effect, used to slow down reactions without changing their mechanism.

TEMPERATURE EFFECT

Enzyme specific activity at subzero temperatures: Linearity and non-linearity of the Arrhenius relationship

Rate constants k of chemical reactions are temperature dependent and obey the Arrhenius relationship

$$k = A \exp\left(-\frac{E}{RT}\right)$$

This expression can be written

$$\frac{\mathrm{d}(\ln k)}{\mathrm{d}T} = \frac{E}{RT^2} \qquad (3\text{-}16)$$

where $E = \Delta H^* + RT$, ΔH^* being given by

$$\Delta G^* = \Delta H^* - T\Delta S^* = -RT \ln K^*$$

The effect of temperature on enzyme specific activity is usually expressed in terms of the temperature coefficient Q_{10} which is the factor by which the velocity is increased on raising the temperature by 10°C, and from equation (3-16), it is readily shown that approximately

$$E = \frac{RT^2 \ln Q_{10}}{10}$$

When the logarithm of the velocity of a reaction is plotted against $1/T$, a single straight line is often observed, and the values for the energy of activation E of the reaction can be determined from such plots. Single straight lines obtained over a wide range of subzero temperatures could signify that the reaction mechanism is not altered, for it would be a coincidence if the mechanism was changed without any variation of activation energy. It is possible to make rough predictions about the effect of temperature on the reaction velocities when their activation energy is known, and to foresee, according to the expression $k = A\exp(-E/RT)$, the rate reduction at different selected subzero temperatures.

The rate of a reaction being arbitrarily standardized to 1 at normal temperature, the expected reductions of rate for selected activation energy values at various temperatures are shown in Table 7. Thus reactions too fast to be recorded at normal temperature would be within range at subzero temperatures. These estimations suppose that Arrhenius plots are linear.

TABLE 7. Reduction of reaction rates

Activation energy Temp (°C)	4 kcal/mol	8	12	16
+20	normalized rates to 1 for 20°C			
0	1.7	2.8	5	8
−20	3	9.5	28	90
−40	6	37	220	1300
−60	14	185	2500	31 000
−80	38	1320	48 000	1 500 000

In fact, for a number of cases reported in the literature, mostly in the usual range of temperatures, and sometimes below 0°C in mixed solvents, the graphs have a discontinuity of slope and approximate to two straight lines meeting at an angle. This could indicate that there is a change in the value of activation energy at the transition temperature and therefore eventually a change in the reaction mechanism.

Lumry and Kistiakowsky (56) have shown that breaks in the Arrhenius plots in fluid media cannot be demonstrated with the data available which give rise to smooth curves following a rather considerable straight line portion for the higher temperatures used. These authors pointed out that sharp breaks in the Arrhenius plots would require the assumption of consecutive reactions with activation energies differing by hundreds of kilocalories per mole, and that their existence is very unlikely. They proposed the interpretation that a reversible inhibition of the enzyme by one or more other constituents of the *milieu* is responsible for the non-linearity of the Arrhenius plots.

Assuming that this non-linearity is due instead to the reversible formation of a catalytically inactive enzyme, Kavanau (57) interpreted the phenomenon by the fact that the concentration of a catalytically active enzyme decreases sharply with the temperature. Considering that the high temperature denaturation of an enzyme unfolds its conformation to such an extent that the specific structure of the active center is lost, Kavanau suggested that the reverse process may be an important factor in the formation of a catalytically inactive enzyme at low temperature.

Thus, the formation of intramolecular hydrogen bonds at low temperatures might lead to a situation for the enzyme in which it is insufficiently unfolded to align its active centers properly or to have them exposed to the substrate; and at such a low temperature, reversible denaturation would not be inconsistent, according to Kavanau, with the experimental observations of the effects of pH and ion inhibitors on enzyme activity.

The reactive configuration of the enzyme would be possessed in only a relatively narrow temperature range, being lost at both high and low temperatures.

In order to determine whether the above hypothesis was consistent with the kinetics of biochemical processes, Kavanau proposed an analytical expression accurately describing the temperature dependence of the velocity constants for a wide variety of processes. The thermodynamic functions determined with its aid should be more reliable and readily accessible than those determined from measuring the slopes of individual plots. As a starting point for the construction of such an expression, an additional parameter was introduced into the Arrhenius equation; thus

$$k_r = \mathrm{d}e^{-E_0/(T - T_0)} \qquad (3\text{-}17)$$

An arbitrary temperature scale related to a zero point temperature, T_0, is introduced, and it is ascertained that a curve of this form fits the data from a wide variety of biological processes.

A further modification of equation (3-17) gave a theoretical kinetic equation which holds for values of the exponent $E_0/(T - T_0)$ in the range 1 to 10 and which reduces to the Arrhenius equation for larger values of the exponent:

$$k_r = p\left[1 - \frac{2}{\sqrt{\pi}}\Gamma_z\left(\frac{3}{2}\right)\right] \qquad (3\text{-}18)$$

where p is a parameter, $z = E/RT$, and $\Gamma_z(\frac{3}{2})$ is the incomplete gamma function of $\frac{3}{2}$ which is equal to $\int_0^z z^{1/2}e^{-z}dz$.

Making the substitution of $x = E_0/(T - T_0)$ for z in equation (3-18), one obtains

$$k_r = p\left[1 - \frac{2}{\sqrt{\pi}}\Gamma_x\left(\frac{3}{2}\right)\right] \qquad (3\text{-}19)$$

Equation (3-19) was found to fit a wide variety of experimental data with great accuracy, including data on enzyme catalyzed reactions, and this supports the contention that an equilibrium indeed exists between reactive and catalytically inactive enzymes.

Equation (3-19) expresses the condition in which the concentration of the reactive enzyme varies markedly with temperature through a transformation in the temperature scale itself, an effective zero point temperature, T_0, being defined, at which the concentration of reactive enzyme is essentially zero. According to this interpretation, the enzymatic reaction in the low temperature region is governed by the number of enzyme molecules in the reactive condition, the formation of these reactive molecules involving relatively high energies. According to Kavanau, the average number of hydrogen bridges which must be broken to form reactive molecules decreases as the temperature is raised, and the rate-determining stage assumes a greater importance depending on the number of activated complexes formed, a process involving relatively low energies of activation.

Kavanau presented evidence to show that the behavior of a number of enzyme systems—unfortunately limited to hydrolytic enzymic reactions—was in accord with the above interpretation for the temperature range above 0°C.

Later, Maïer *et al.* (53) sought out evidence in support of this interpretation based on data for another hydrolytic enzyme (alkaline phosphatase) and for an oxidizing enzyme (peroxidase) in mixed solvents in the temperature range +20° to −30°C. They found a deviation from the Arrhenius relationship due to some effect of the low temperature on the proteins, and they checked that there was good agreement between the experimental data and the theoretical curves.

The curves were based on the following interpretation: if there is an equilibrium between active and inactive enzymes (respectively E_A, E_I):

$$E_A \rightleftharpoons E_I$$

where

$$[E_0] = [E_A] + [E_I]$$

E_0 referring to the total enzyme,

$$K = \frac{[E_I]}{[E_A]}$$

The rate of formation of products (p) in the region where the reaction is zero order with respect to substrate is expressed by

$$\frac{d[p]}{dt} = k[E_A]$$

If the equilibrium is such that substantially all of the enzyme is in the E_I form:

$$\frac{d[p]}{dt} = \frac{k[E_0]}{K}$$

This condition will be met at the lower temperature provided ΔH for the equilibrium is negative. An Arrhenius plot will give a straight line with an activation energy $U - \Delta H$, and since ΔH is negative, the activation energy will be greater than U by the absolute value of ΔH.

If the enzymes were in the active form E_A, one would have

$$\frac{d[p]}{dt} = k[E_0]$$

this condition being met at the higher temperature for ΔH negative, and an Arrhenius plot would give a straight line for an activation energy U.

In the intermediate region where there is the equilibrium $E_A \rightleftharpoons E_I$,

$$\frac{d[p]}{dt} = \frac{k[E_0]}{K + 1}$$

It was verified that indeed when an Arrhenius plot for the higher temperature region is extrapolated to the lower temperature region, the difference between the extrapolated and experimental values is equal to $\log(K + 1)$.

The dependence of K, the equilibrium constant for inactivation, on temperature was obtained in the case of the alkaline phosphatase in 40% methanol at subzero temperatures and showed a good agreement with the Van't Hoff equation which was a necessary condition for the above interpretation of experimental results, and the inactivation of the enzyme was definitely attributed to an increase in intramolecular hydrogen bonding.

Nevertheless, Westhead and Malmström (54), investigating enolase activity, concluded that data supported in part the interpretation of Lumry and Kistiakowski reported above, and that the complexity of the effects involved in the reaction media pointed to the danger of making deductions from the effects of organic solvents and temperatures without a sufficiently thorough study of such effects. We should add that this also holds without a knowledge of the evolution and of numerical values of the various physical chemical parameters, and without an intense study of each given enzyme system.

Since one of the major problems of low temperature biochemistry is to correlate a mechanism in mixed solvents at subzero temperatures with the mechanism existing in normal conditions of medium and temperature, we have carried out intensive experiments on the Arrhenius plots. In many cases we found a non-linearity of these plots at low temperatures, and we investigated whether such a non-linearity was due either to artifacts or to a change in activation energy.

First, the proper significance of the values for E must be questioned if, for instance, the true value of the velocity is not obtained in proportion as the temperature is lowered. An example can be given with a peroxidative reaction catalyzed by the enzyme horseradish peroxidase (HRP) where, as seen above

$$H_2O_2 + 2AH \xrightarrow{HRP} 2H_2O + 2A_{ox}$$

(the oxidizable substrate AH was gaïacol).

Let us consider this reaction in the mixture ethylene glycol–water (volume ratio 1 : 1, pH 6.0) between +4° and eventually −40°C; the reaction kinetics at selected temperatures are plotted in Fig. 17 and it can be seen that as temperature is lowered an "induction" phase appears and lasts longer. Such curves represent the "prestationary" state of the reaction.

When the logarithm of the velocity corresponding to the segment AB of Fig. 17 is plotted against $1/T$, the resulting graph represents a continuity of slope (see Fig. 18). However, when the logarithm of the velocity corresponding to the segments of the prestationary state is plotted against $1/T$, the graph has a discontinuity of slope and approximates to two straight lines meeting at an angle (see Fig. 19).

This indicates that explanations of discontinuities at subzero temperatures should be carefully checked and that experiments to record the Arrhenius plots should be prepared by kinetic measurements at selected temperatures.

In fact, further experiments on other enzyme systems and in various mixed solutions are needed to complete our information and to find out eventually some other factors responsible for the observed fall off in some Arrhenius plots at very low temperatures. For instance, some enzyme-catalyzed reactions will be increasingly diffusion-controlled as viscosity increases (as a function of $1/T$), and changes in proton activity might also be responsible for breaks in the Arrhenius plots.

In many cases the non-linearity of Arrhenius plots might be linked to cold-inactivation of enzyme systems. At the very beginning of the observation of the non-linearity in aqueous solutions, it was supposed that curves

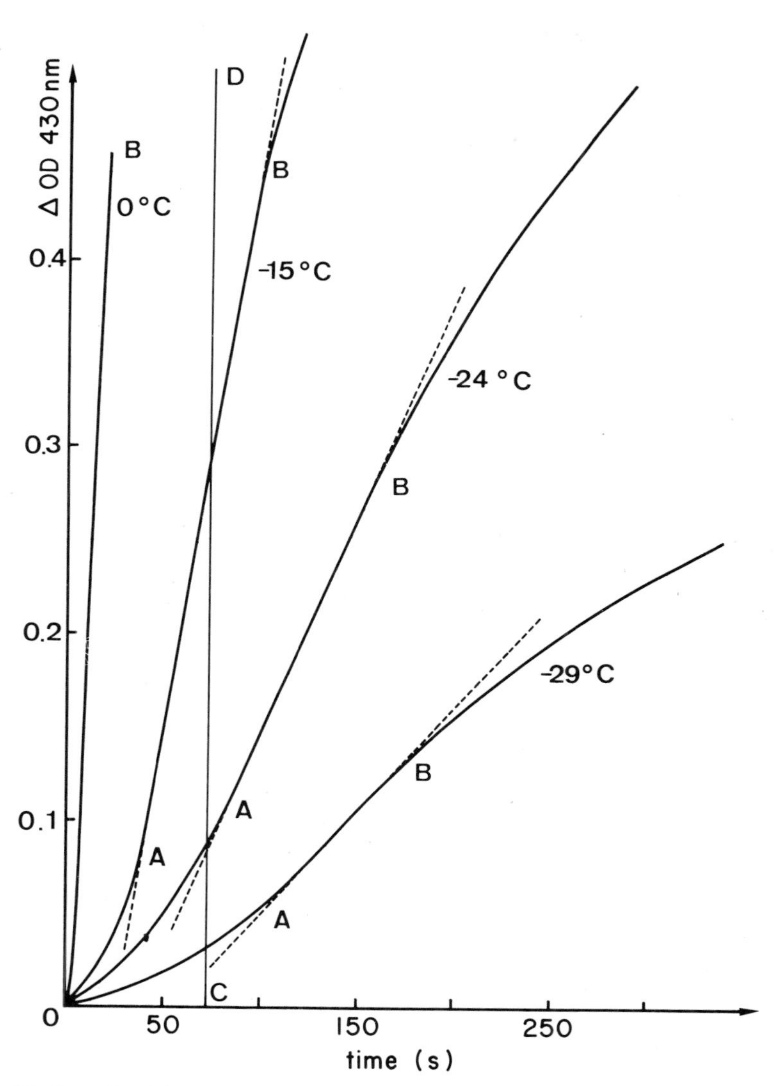

FIG. 17. Reaction kinetics of the enzyme-catalyzed oxidation of gaïacol by hydrogen peroxide in ethylene glycol–water (50 : 50, v/v) at different temperatures.

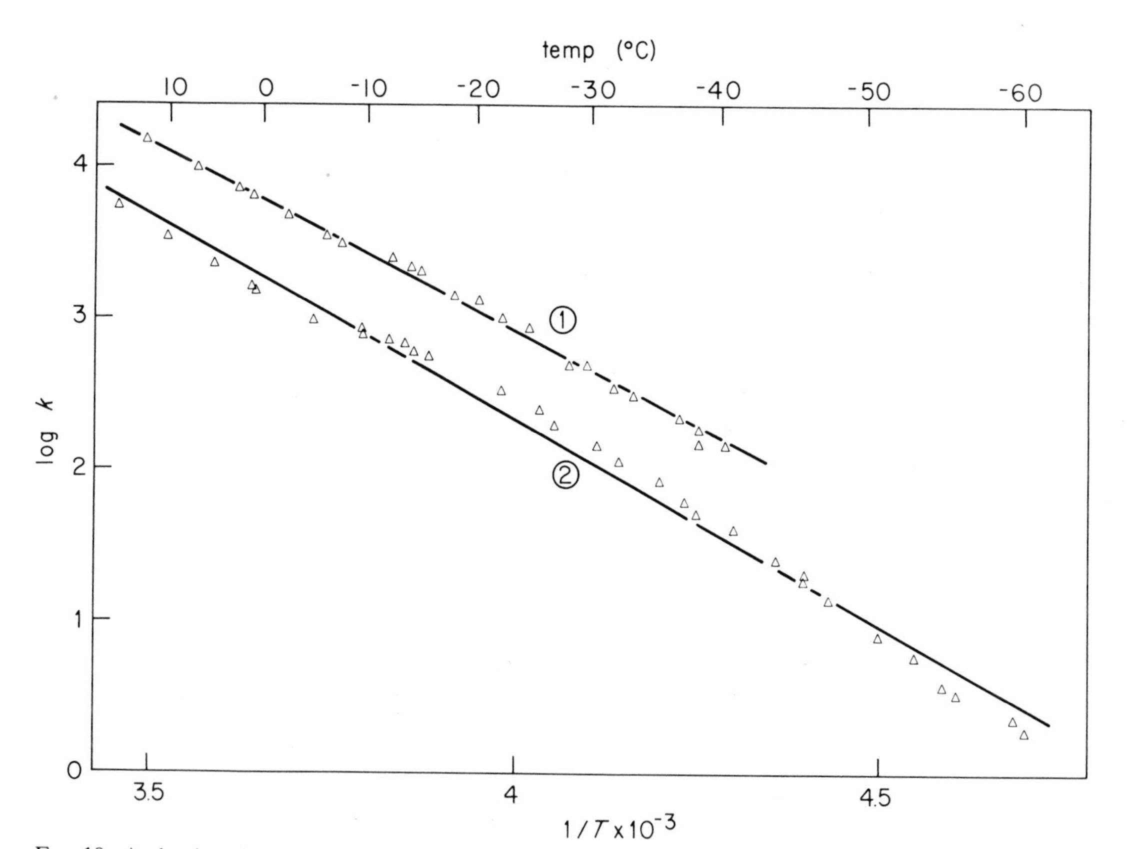

FIG. 18. Arrhenius plots obtained when the logarithm of the velocity in segments A, B of Fig. 17 are plotted against $1/T$. (1) in ethylene glycol–water (50 : 50, v/v); (2) in methanol–water (60 : 40, v/v).

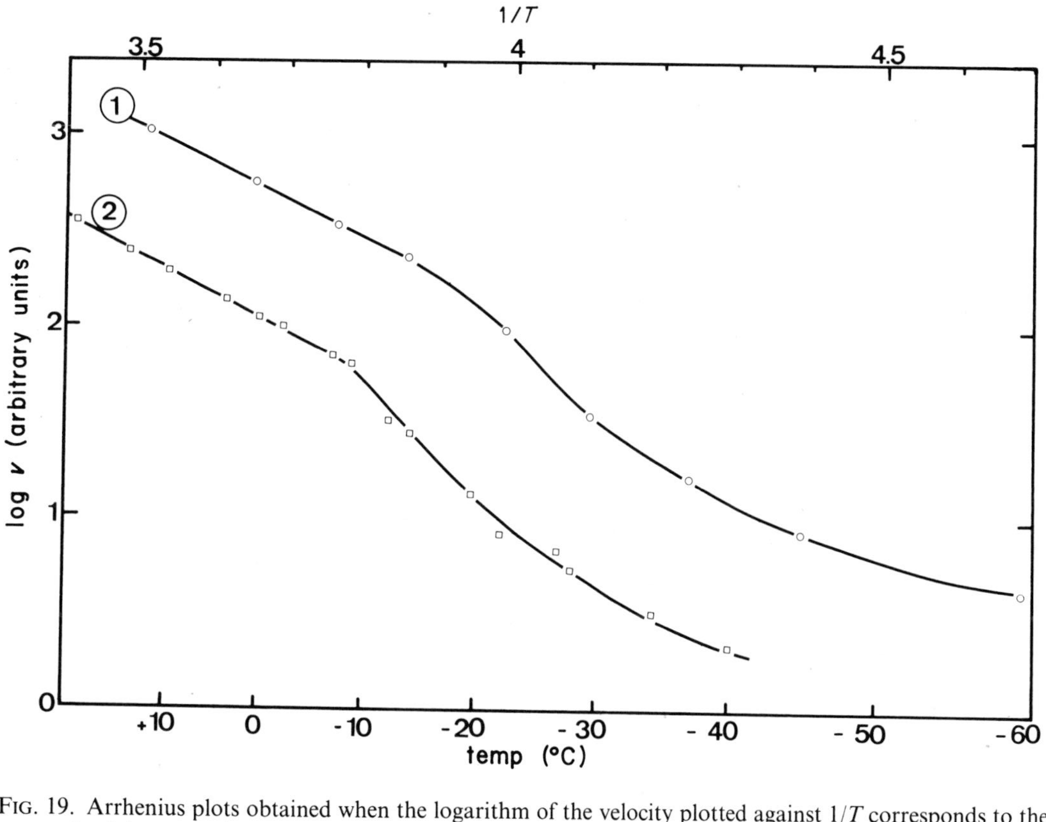

FIG. 19. Arrhenius plots obtained when the logarithm of the velocity plotted against $1/T$ corresponds to the segments of the prestationary states reported in Fig. 17. (1) methanol–water (60 : 40, v/v); (2) ethylene glycol–water (50 : 50, v/v).

obtained by such plots represented joined segments of straight lines with sharp breaks at certain "critical" temperatures, the slopes of these segments corresponding to $-E_n/R$ of a set of n consecutive reactions of activation energies E_n, which respectively would act as "pacemakers" in different temperature ranges (51). The data were generally limited to the region above 0°C for obvious reasons in aqueous solutions; nevertheless, experi-

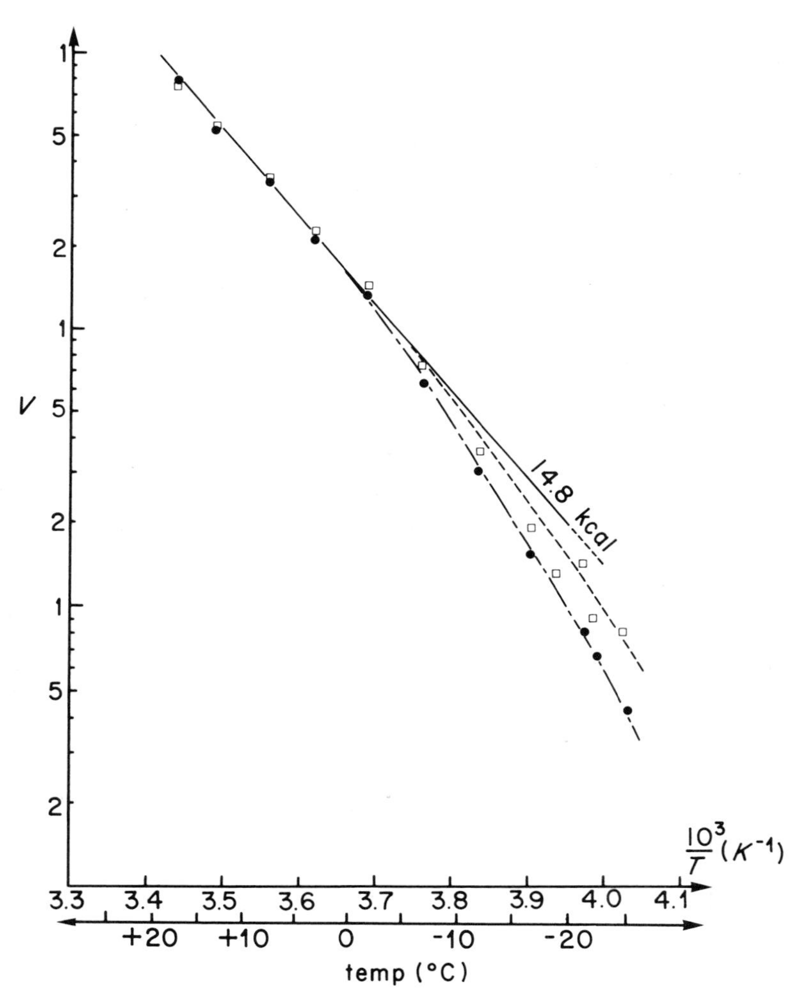

FIG. 20. Arrhenius plot of the flavoprotein catalyzed reduction of ferricyanide by NADH, in ethylene glycol–water mixture (50 : 50, v/v), in the presence of different buffers. —□—, phosphate; —●—, Tris.

ments were also carried out in frozen media (52,53,54) and Sizer (55) suggested that enzymes can exist in two configurations, above and below a certain critical temperature, but the meanings of such experiments remain obscure.

Thus a number of falls-off in the Arrhenius plots in mixed solvents could result from "trivial" causes, as described above in the case of the peroxidase-catalyzed oxidation of gaïacol by hydrogen peroxide and, as we will see, in

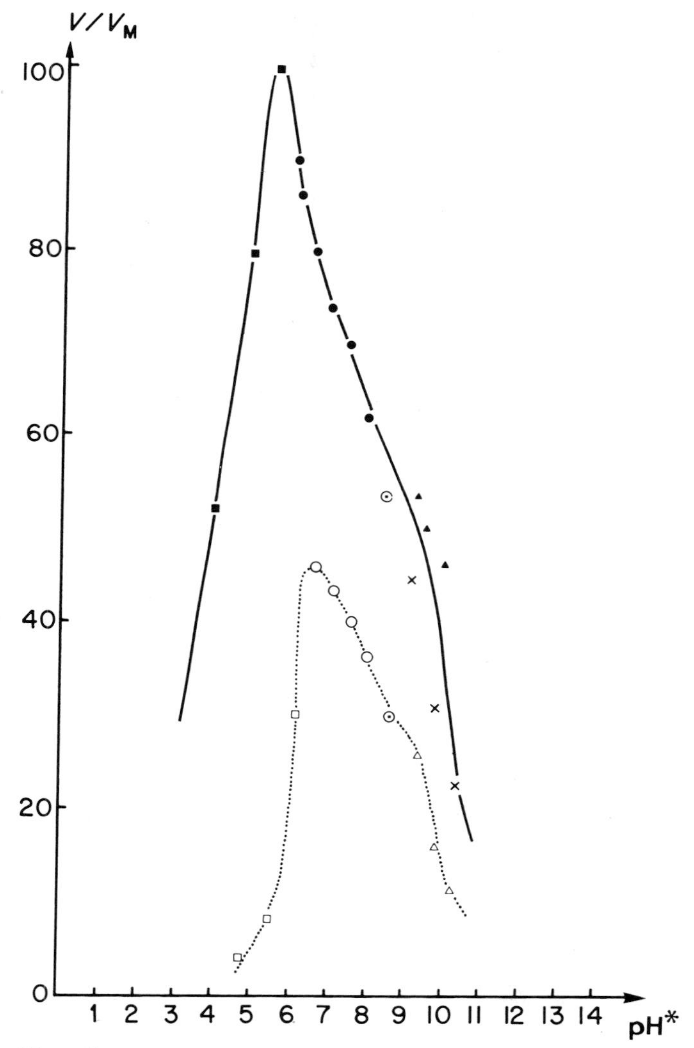

FIG. 21. pH profiles of the flavoprotein catalyzed reduction of ferricyanide by NADH in ethylene glycol–water mixture (50 : 50, v/v) (----) and in aqueous solution (———).

the case of the reduction of ferricyanide catalyzed by a flavoprotein, the NADH-cytochrome b_5 reductase. The NADH-cytochrome b_5 reductase extracted from rat liver microsomes was experienced in this laboratory by Thore in ethylene glycol buffer mixtures (50:50, v/v) between $+20°$ and $-30°C$ with ferricyanide as electron acceptor from the substrate NADH.

Investigations were carried out with both phosphate and Tris buffers at pH* 7.5 (at 20°C), i.e. in pH* conditions ensuring both a good stability of the enzyme and moderate activity.

In these conditions, this enzyme system does exhibit normal behavior in phosphate buffered mixtures and an anomalous one in tris buffered mixtures, when data are plotted on the basis of the Arrhenius equation (see Fig. 20). Such opposite behavior seems to be due simply to the fact that the pH* of phosphate buffered solutions remains practically unchanged between $+20°$ and $-30°C$, whereas the pH* of tris buffered solutions increases by about one unit. Considering the pH-activity profile of the reaction (see Fig. 21) it can be seen that increasing the pH* decreases the enzyme specific activity, and this could at least partially explain the progressive fall-off in the Arrhenius plot.

However, the widespread occurrence of non-linearity of Arrhenius plots both in aqueous and mixed solutions in the usual range of temperatures could be due both to experimental conditions and possibly to the decrease in concentration of catalytically active enzymes.

We have seen that enzyme molecules can be altered by cosolvents and then function at a reduced rate; one can also assume that enzymes which are built up of subunits could be inactivated by dissociation, temperature variations being able to shift the reversible equilibrium between oligomer proteins (P) and their subunits (S), $P \rightleftharpoons S$, and therefore between the active form P and inactive subunits S. Let us examine an example of this behavior.

Temperature effects on quaternary structure

Temperature effects below 0°C are still mostly conjectural. Theoretically, the displacement of the equilibrium native state $\rightleftharpoons$ denatured state of proteins would depend from the exothermicity or endothermicity of the corresponding transition. Since the transition native $\rightleftharpoons$ denatured form of many proteins in mixed solvents appears to be an endothermic process, a decrease in temperature would favor the native form and the above transition, which is observed in the range of above zero temperature, could be reversed at subzero temperature. In fact, we have as yet no proof of such behavior. The only observation is that the denaturation of proteins occurring in some mixed media at normal temperature can be "quenched" or at least delayed infinitely at a lower temperature. Information about the low temperature effect on proteins whose prosthetic group could be released from the apoprotein in mixed solvents at room temperature are lacking. Indeed, very little is known about the "real" effects of low temperatures on the tertiary structure of the enzymes we investigated in these conditions to obtain the temporal resolution of their catalyzed reactions. It is to be presumed that changes in pK of ionizable amino acid residues located on the surface of the proteins should react to temperature variations according to their enthalpy of ionization; such changes could influence both the conformation and the specific activity of some enzymes but we have as yet no evidence of such a temperature-dependent behavior.

The situation is quite different in the case of oligomeric proteins, i.e. proteins which consist in two or more identical or nearly identical subunits. Some information is available in the literature, which mentions the particular behavior of oligomeric proteins during the cooling of their aqueous solutions.

COLD INACTIVATION

These enzymic proteins are inactivated when cooled down to 0°C or even frozen below 0°C. Some of them have been found to be protected against this often irreversible inactivation in polyol–water mixtures. In each case examined, cold inactivation is accompanied by changes in molecular weight.

Dissociation into subunits on cooling has been reported in the case of erythrocyte glucose-6-phosphate dehydrogenase, carbamyl phosphate synthetase, adenoside triphosphatase of beef heart mitochondria, argino succinase and pyruvate carboxylase.

In contrast to these enzymes, others such as urease (58) and 17β-hydroxysteroid dehydrogenase (59) undergo a series of aggregation reactions on cooling. In these cases, conformational changes could be a necessary prerequisite for the formation of molecular aggregates, since it has been shown that reactivation is observed almost exclusively in the low molecular weight species, except in the case of urease, although the reversible cold inactivation seems to be due to associations between molecules. Soluble adenosine triphosphatase, which consists of identical subunits, loses its activity at 4°C and does not undergo a reactivation on warming. 17β-hydroxysteroid dehydrogenase undergoes a series of inactivating transitions on cooling; cooling induces either reversible dissociation of the enzyme into subunits or an irreversible formation of molecular aggregates (60).

Erythrocyte glucose-6-phosphate dehydrogenase is reversibly cold inactivated by removal of NADP and its reduced form (NADPH). Several metastable and subactive states are found at 0°C, i.e. a monomer and a dimer without bound NADP, and a dimer with bound NADP. Reactivation is achieved by warming. Such a reactivation is pH dependent between pH 6.0 and 9.0 with a pH optimum of 7.2 (60).

A similar pH dependence is observed in the case of glutamic acid decarboxylase, where the coenzyme pyridoxal phosphate dissociated at 0°C, and the apoenzyme undergoes a transition, native $\rightleftharpoons$ denatured, which can be reversed at room temperature in the presence of an excess of coenzyme. It has been shown that media of high ionic strength and of high dielectric constant have a protective effect upon enzyme activity at 0°C, suggesting that the loss of activity results from the interaction of electrostatic forces on the enzyme surface. On the other hand, the limited range of pH within which temperature has an influence upon stability of the enzyme, pH 6.0–7.5, suggests that the critical electrostatic interaction depends upon the state of ionization of a single type of dissociable group.

Other enzymes which can be cold-inactivated should be tested as a function of ionic strength and of pH; among them enzymes, which consist of coenzymes or prosthetic groups non-covalently bound, are perhaps the more interesting, and it seems that besides the dehydrogenases, many flavoproteins might exhibit interesting behavior.

We will stress, with Talalay and his colleagues (59), that the widespread and often correct belief that strict maintenance of cold (around 0°C) is an essential condition for success in enzyme isolation and storage can no longer be considered as acceptable, for many enzymes display far greater stability at room temperature than in the icebox. As we will see below, the use of polyols might avoid the cold-inactivation of most of cryosensitive enzymes around 0°C and should permit in the meantime experimentation with these enzymes in cooled solutions at subzero temperatures.

CRYOPROTECTION

Several enzymes known as cold-inactivated are protected (stabilized) by the addition of polyols (in most cases, glycerol) to their aqueous solutions, and such a cryoprotective effect again reminds us of the behavior of β-lactoglobulin.

Varying concentrations of glycerol protect the following enzymes against cold inactivation: 17β-hydroxysteroid dehydrogenase, beef heart mitochondria and yeast adenosine triphosphatases, carbamyl phosphatase synthetase, phosphorylase 3, etc., and it is probable that such stabilization would be observed with the other cold-inactivable enzymes (61).

A similar protection has been reported against inactivation and/or inhibition of hybridization of enzymes during freezing–thawing and it is interesting to notice that the protective polyols also are effective in preventing cryoinjury of certain intact cells and organelles. Lactate dehydrogenase and triosephosphate dehydrogenase (61) are among these enzymes and would presumably be protected in cooled mixed solutions.

The protective action observed in the various enzymes cited above might be due to the fact that the dissociation into subunits ($N_n \rightleftharpoons nR$) is perfectly reversible, for the polyol protects each R form against the reversible trans-conformation R → S, but also leads eventually to the transition $nR \xrightarrow{-T} N_n$, or even to the stabilization of the native form N_n in the mixed solvent.

To date, the mechanism of the stabilizing effect of polyols against cold inactivation is obscures in spite of much speculation about cooled and frozen media.

According to authors such as Talalay (59), the protective action in the liquid phase could be due to the stabilization of networks of "structured" water molecules which appear to be essential to the maintainance of a native conformation. These networks and their importance to the structure of proteins has been widely established and precisely reviewed (62). Some refined hypotheses, which account for the protective effect of polyols in terms of solvent structure and solvent protein interactions, have been advanced but their validity must await a far better knowledge of the solvent shells of proteins than is presently available (63).

The situation is of course quite similar and even much more complicated in the case of frozen solutions. It has been stated that the inactivation of aqueous samples during freezing and thawing could be due to the fact that when water around protein molecules crystallizes into hexagonal ice by freezing at temperatures above −75°C or so, the "structured" water on the protein surface which is assumed to be in the form of cubic pentagonal polyhedral structures (64,65) might be transformed into the hexagonal form of ice, disturbing first the hydrophobic interactions and then breaking intra-molecular hydrogen bonds (62). The protective effect of polyols might therefore be due to their stabilization of the structured water on the protein surfaces. Other authors suggested that protection might be a function of the colligative properties of mixed solvents, reducing the concentration of salt and pH changes in eutectic mixtures (63).

Changes in hydrophobic interactions might in turn play an important part in cold-inactivation of enzymes which consist of subunits since such interactions are, as we have seen, weaker as the temperature is lowered, leading to the transition $N_n \rightarrow nR$ in aqueous solutions.

DISSOCIATION OF β-LACTOGLOBULIN IN COOLED MIXED SOLVENTS

It is known that β-lactoglobulin can be dissociated into subunits α and β between pH 6.0 and 9.0 in aqueous solutions at room temperature. Dimeric form (α, β) is stable at acidic pH (3.0–6.0). Above pH 9.0 the dissociation into α and β subunits is irreversible due to a transconformation termed as R → S, leading to aggregation (nS → Sn). These processes can be schematized as follows.

pH	3–6	6–9	9–12	
	N_2	$N_2 \rightleftharpoons 2R$	$2R \longrightarrow$	$2S \cdots \rightarrow Sn$
	Dimer	Reversible dissociation	Transconformation	Aggregates

It has been shown previously that a decrease in temperature from 20 to 0°C induces in aqueous solutions the transition $N_2 \rightarrow 2R$, with a transition pK of 7.7, the dissociation being fully reversible by warming up.

On the other hand, in the presence of 50% by volume of ethylene glycol, there was no modification of the above picture and that the transition pK was of 7.7 (66). However, a decrease in temperature from 20° to −40°C determines a progressive shift of this pK value as shown in Table 8. The

TABLE 8

Temp (°C)	20	10	0	−10	−20	−30	−40
pK of dissociation	7.7	8.1	8.4	8.7	9.0	9.3	9.53

pK change between 20 and −40°C is of 1.85 unit. Thus cooling protects the dimer in the pH range where it is dissociated at room temperature. Between pH 6.0 and 9.0, cooling determines the reassociation, the equilibrium between N_2 and 2R being shifted in the direction $2R \rightarrow N_2$. It may be assumed that some hydrogen bonds linking the subunits, which dissociate above pH 6.0 at room temperature, are protected against dissociation as the temperature decreases, due to a change in the pK of dissociation. Such a process is schematized in Fig. 22.

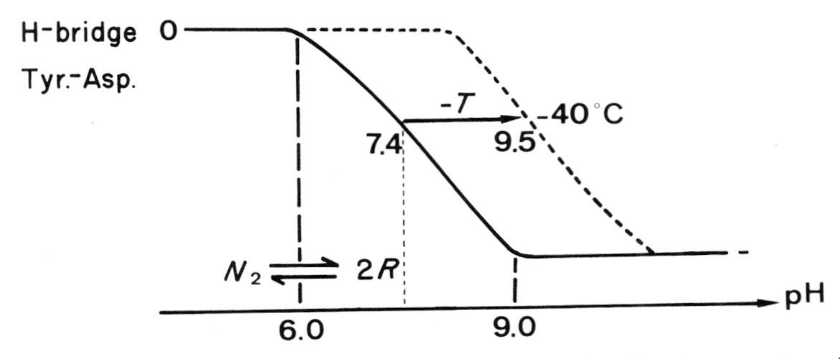

FIG. 22. Scheme depicting the evolution of the hydrogen bonding between tyrosin and aspartate residues upon cooling.

Thus it can be seen that if β-lactoglobulin had a catalytic activity, which is not apparently the case, cooling of its mixed solution would bring about a protection against its inactivation between pH 6.0 and 9.0 and even its "reactivation" by shifting the equilibrium $N_2 \rightleftharpoons 2R$ in the direction $2R \rightarrow N_2$.

A number of oligomeric enzyme systems could be investigated in these conditions of medium and temperature, at specific pa_H values, to discover conditions of protection against dissociation into subunits. The shift in the equilibrium $N_n \rightleftharpoons nR$ as a function of temperature at constant pH values would support the Kavanau hypothesis according to which the non-linearity in the Arrhenius plots would be due to the progressive and reversible formation of catalytically inactive species.

Enzymes which consist of subunits bound by "polar" contacts such as salt bridges between anionic (A^-) and cationic (B^+) residues would also be sensitive to temperature variations, according to the values of the enthalpy of ionization of these groups and to the fact that, according to the pH, they obey the following equilibria: $A^- + H^+ \rightleftharpoons AH$ (neutral acid); $BH^+ \rightleftharpoons B + H^+$. In Fig. 23 which represents these groups involved in a salt bridge for a given pH interval, we can see that any temperature variation will influence the dissociation constant of each group and therefore could determine the breaking of the bond.

It can be assumed that the equilibrium state of cationic and neutral acid groups will be affected by temperature variations according to the value of their enthalpy of ionization. These values are: 1–2 kcal/mol for aspartate and glutamate, 6 kcal/mol for tyrosine, 7 kcal/mol for histidine, 8 kcal/mol

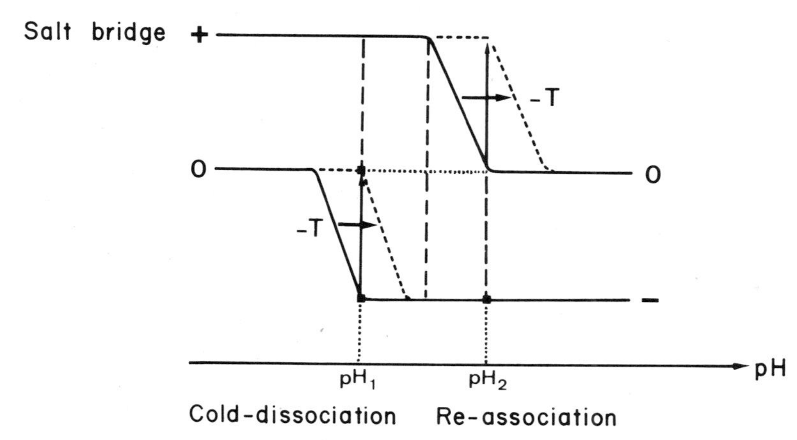

FIG. 23. Scheme depicting the influence of temperature variations on the dissociation of a salt bridge between two subunits. This explains the processes of cold-dissociation as well as of cold-association observed in a number of enzyme systems.

for cysteine and 10 kcal/mol for guanidinium. Cold-dissociation and re-association processes can be expected according to the groups involved in salt bridges and to the value of pH. Further investigations are needed to improve our knowledge about such aspects of the low temperature procedure, but we can see that some fall-offs in the Arrhenius plots could result from such processes.

Conclusion

Each enzyme system to be investigated at subzero temperatures must be tried in different mixtures and at selected temperatures to test its solubility and its catalytic activity, under suitable conditions of ionic strength, pH*, and dielectric constant.

Trials consist of recording reaction rates and pH*-activity profiles in order to find a new pH* value corresponding to the optimal activity in these conditions, to compare it with that observed in aqueous solution and then to check whether the cosolvent effect on such a rate is totally reversible by infinite dilution or dialysis.

Once these requirements are fulfilled, reaction rates can be recorded as a function of temperature over a selected and as broad as possible range, to determine whether the Arrhenius expression is followed, since a fall-off might be due to a modification in the reaction mechanism once some "trivial" causes considered above are eliminated. It is essential to determine any such modifications in these abnormal conditions of medium and temperature and, as we will see in chapter 5, any information obtained about intermediates, their formation and conversion will be used to answer this question.

As we will see, cooling $\rightleftharpoons$ warming cycles can be used to obtain the "temporal resolution" of a number of multistep enzyme-catalyzed reactions, to "quench" them at a given stage and to stabilize corresponding intermediates, and finally to analyze them and to record kinetics of elementary steps. Comparison with data obtained by rapid kinetic techniques and rapid recording in usual conditions of medium and temperature will help to establish whether reaction mechanisms are similar in both types of conditions.

In spite of all these efforts and data, some additional effects on reaction rates and mechanisms could occur in a number of systems as yet not investigated, depending on the change in the solvation of a protein, in the electronic structure and reactivity of coenzymes, prosthetic groups and substrates according to the nature and composition of mixed solvents and temperature. This is why each new system submitted to the present procedure must be tested carefully, with all prerequisite physico-chemical conditions fulfilled. In this respect, we have seen that the recording of pH*-activity profiles of enzyme systems quite sensitive to pH can be characterized both in water and in mixed solvents by curves of identical shape, and that it is possible to gather evidence suggesting that the reaction mechanism of an enzyme-catalyzed reaction is unaffected by the presence of a cosolvent.

Enzyme systems tested in this laboratory appeared to enjoy considerably enhanced stability in such conditions, demonstrating that unexpected changes in bulk dielectric constant and pH* can destabilize the native form of proteins, which then become quite sensitive to a cosolvent which might otherwise have little effect on its conformation. As we have seen, once the destabilizing factors are neutralized, the cosolvent effect on enzyme structure and activity becomes reversible, and weak as compared to the temperature effect in terms of reaction rates, and might be used as we will see in chapter 6, as a tool in molecular biochemistry.

References

1 E. S. Amis. *J. Chem. Educ.* **30**, 351 (1953).
2 E. S. Amis and G. Jaffé. *J. Chem. Phys.* **10**, 646 (1942).
3 E. S. Amis. *Solvent Effects on Reaction Rates and Mechanisms*. Academic Press, London and New York (1966).
4 E. S. Amis and J. F. Hinron. *Solvent Effects on Chemical Phenomena*. Academic Press, London and New York (1973).
5 E. A. Moelwyn-Hugues. *Physical Chemistry*, Chapter XXIV. Pergamon Press, New York (1957).
6 B. Jones. *J. Chem. Soc.* 1006 (1928); 3073 (1928).
7 S. W. Benson. *The Foundation of Chemical Kinetics*, Chapter XV. McGraw-Hill, New York (1960).
8 S. Ono, K. Hiromi and Y. Sano. *Bull. Chem. Soc. Japan* **36**, 431 (1963).
9 K. Hiromi. *Bull. Chem. Soc. Japan* **33**, 1215 (1960); **33**, 1264 (1960).
10 F. Travers. Personal communication.
11 M. L. Barnard and K. J. Laidler. *J. Am. Chem. Soc.* **74**, 6099 (1952).
12 C. E. Clement and M. L. Bender. *Biochemistry* **2**, 836 (1963).
13 M. Mares Guia and A. F. S. Figueiredo. *Biochemistry* **11**, 2091 (1972).
14 P. Strittmatter. *J. Biol. Chem.* **242**, 4633 (1969).
15 D. Findlay, A. P. Mathias and B. R. Rabin. *Biochem. J.* **85**, 139 (1962).
16 M. L. Bender and E. J. Kaiser. *J. Am. Chem. Soc.* **84**, 2556 (1962).
17 M. L. Bender and F. J. Kezdy. *Ann. Rev. Biochem.* **34**, 49 (1965).
18 H. Lineweaver and D. Burk. *J. Am. Chem. Soc.* **56**, 658 (1934).
19 M. Dixon and E. C. Webb. *Enzymes*, 2nd ed., p. 116. Longmans, Green and Co. Ltd., London (1964).
20 H. Gutfreund. *Trans. Faraday Soc.* **51**, 441 (1955).
21 R. G. Bates. In *Solute-Solvent Interactions*, J. F. Coetzee and C. D. Ritchie, Eds., p. 71. Marcel Dekker, New York and London (1969).
22 M. Born. *Z. Physik* **1**, 45 (1920).
23 J. W. Larson and L. G. Hepler. In *Solute-Solvent Interactions*, J. F. Coetzee and C. D. Ritchie, Eds., p. 36. Marcel Dekker, New York and London (1969).

24 R. A. Robinson and R. H. Stokes. In *Electrolyte Solutions*, 2nd ed., p. 538. Butterworths, London (1959).
25 E. Shaw, M. Mares-Guia and W. Cohen. *Biochemistry* **4**, 2219 (1965).
26 P. H. Petra, W. Cohen and E. Shaw. *Biochem. Biophys. Res. Comm.* **21**, 612 (1965).
27 T. Inagami and J. M. Sturtevant. *Biochim. Biophys. Acta* **38**, 64 (1960).
28 M. Castaneda-Agullo and L. M. Del Castillo. *J. Gen. Physiol.* **42**, 617 (1959).
29 P. Maurel, G. Hui Bon Hoa and P. Douzou. *J. Biol. Chem.* **250**, 1370 (1975).
30 P. Maurel and P. Douzou. *J. Biol Chem.* **250**, 2678 (1975).
31 R. Lumry and S. Ragender. *Biopolymers* **9**, 1125 (1970).
32 D. J. Ives and P. D. Marsden. *J. Chem. Soc.* 649 (1965).
33 L. Juliano and A. M. C. Paiava. *Biochemistry* **13**, 2445 (1974).
34 G. C. Pimental and A. L. McClellan. *The Hydrogen Bond.* Freeman, San Francisco (1960).
35 D. E. Koshland and G. Neet. *Ann. Rev. Biochem.* **37**, 404 (1968).
36 M. L. Ludwig, J. A. Harstuck, J. A. Seitz, T. A. Muirhead, J. C. Coppola, G. N. Reeke and W. N. Lipscomb. *Proc. Nat. Acad. Sci. U.S.* **57**, 511 (1967).
37 N. S. Bayliss and E. G. McRay. *J. Phys. Chem.* **58**, 1002 (1954).
38 E. G. McRae. *J. Phys. Chem.* **61**, 562 (1957).
39 T. T. Herskovitz and M. Laskowski, Jr. *Feder. Proc.* PC 56–57 (1960).
40 T. T. Herskovitz and M. Laskowski, Jr. *J. Biol. Chem.* **237**, 2481 (1962).
41 J. W. Donovan. In *Methods in Enzymology*, Vol. XXVII, p. 497. Academic Press, New York (1973).
42 J. Bello. *Biochemistry* **9**, 3563 (1970).
43 H. J. Sage and S. J. Singer. *Biochemistry* **2**, 305 (1962).
44 M. Dixon and E. C. Webb. In *Enzymes*, 2nd ed., p. 116. Longmans, Green and Co., London (1964).
45 P. Maurel and F. Travers. *C. R. Acad. Sc. Paris* **276D**, 3057, 3383 (1973).
46 D. C. Phillips. *Scient. Amer.* **215**, 78 (1966).
47 S. K. Banerjee, I. Kregar, V. Turk, J. A. Rupley. *J. Biol. Chem.* **248**, 4786 (1973).
48 W. Kauzmann. *Adv. Protein Chem.* **14**, 1 (1959).
49 C. Tanford. *J. Am. Chem. Soc.* **84**, 4240 (1962).
50 D. Elbaum and T. T. Herskovitz. *Biochemistry* **13**, 1 (1974).
51 W. J. Crozier. *J. Gen. Physiol.* **7**, 189 (1924).
52 V. P. Maïer and A. L. Tappel. *Analyt. Chem.* **26**, 564 (1954).
53 V. P. Maïer, A. L. Tappel and D. H. Volman. *J. Am. Chem. Soc.* **77**, 1278 (1955).
54 E. W. Westhead and B. G. Malmström. *J. Biol. Chem.* **228**, 655 (1957).
55 I. W. Sizer. *Adv. in Enzymol.* **3**, 35 (1943).
56 G. B. Kistiakowski and R. L. Lumry. *J. Am. Chem. Soc.* **71**, 2006 (1949).
57 J. L. Kavanau. *J. Gen. Physiol.* **33**, 193 (1950).
58 B. F. J. Hofstee. *J. Gen. Physiol.* **32**, 339 (1949).
59 J. Jarabak, A. E. Seeds Jr. and P. Talalay. *Biochemistry* **5**, 1269 (1966).
60 H. N. Kirkman and E. M. Hendrickson. *J. Biol. Chem.* **237**, 2371 (1962).
61 O. P. Chilson, L. A. Costello and N. O. Kaplan. *Fed. Proc. Supp.* **15**, S-55 (1965).
62 K. Shikama. *Sci. Rep. Tohuka Univ.* **29**, 91 (1963).
63 H. T. Meryman. *Cryobiology*. Academic Press, London and New York (1966).
64 W. F. Claussen. *J. Chem. Phys.* **19**, 1425 (1951).
65 M. V. Stackelberg and H. R. Muller. *Z. Elektrochem.* **58**, 25 (1954).
66 G. Hui Bon Hoa, S. Guinand, C. Pantaloni and P. Douzou. *Biochimie* **55**, 269 (1973).

4

Methodology for Experiments at Subzero Temperatures

Preparation of cooled solutions of biopolymers

In most of the early work, cooling was not carried out until the organic solvent had already been added to the aqueous solution of the enzyme (1,2). However, subsequently Bielski and Freed (2) sprayed droplets of the aqueous solutions of the enzyme into the cold aqueous–organic mixture. Under such conditions, enzymes were assumed to dissolve in media not so far from the "critical" value of the dielectric constant (*D* 80). Under this condition rapid denaturation was presumed to be avoided. Unfortunately, the droplets of enzyme solution form small "icebergs" which melted rather slowly when the mixture was stirred (3).

In this laboratory, we synchronized the progressive addition of organic solvent with the cooling of the enzyme solution (4). As was shown in chapter 2 this was intended to prevent any change in the dielectric constant and take advantage of the fact that the decrease of dielectric constant would be instantly corrected by the decrease in temperature. Thus, the dangers of denaturation and effects related to a change in the dielectric constant, as well as the effects of the ionic environment on enzyme activity, are prevented. Under these conditions, the results can be readily evaluated with respect to the eventual effects of both organic solvent and temperature upon enzyme activity.

The apparatus consists of two parts, see Fig. 1.

(1) *Temperature control.* The first part is a thermostatically-controlled refrigerated bath (Air Liquide BRT 90 associated with a thermostat C 1319). A cam-operated programmer permits the cooling of the bath at an average rate of 2°/min in a temperature range of +20° to −100°C. The second part of the apparatus is a programmer for introducing the organic solvent into the aqueous solution.

(2) *Flow programmer.* The general function of this programmer is the following: a series of holes, whose positioning corresponds to the temperature of injection of the solution, are drilled in the temperature cam. These

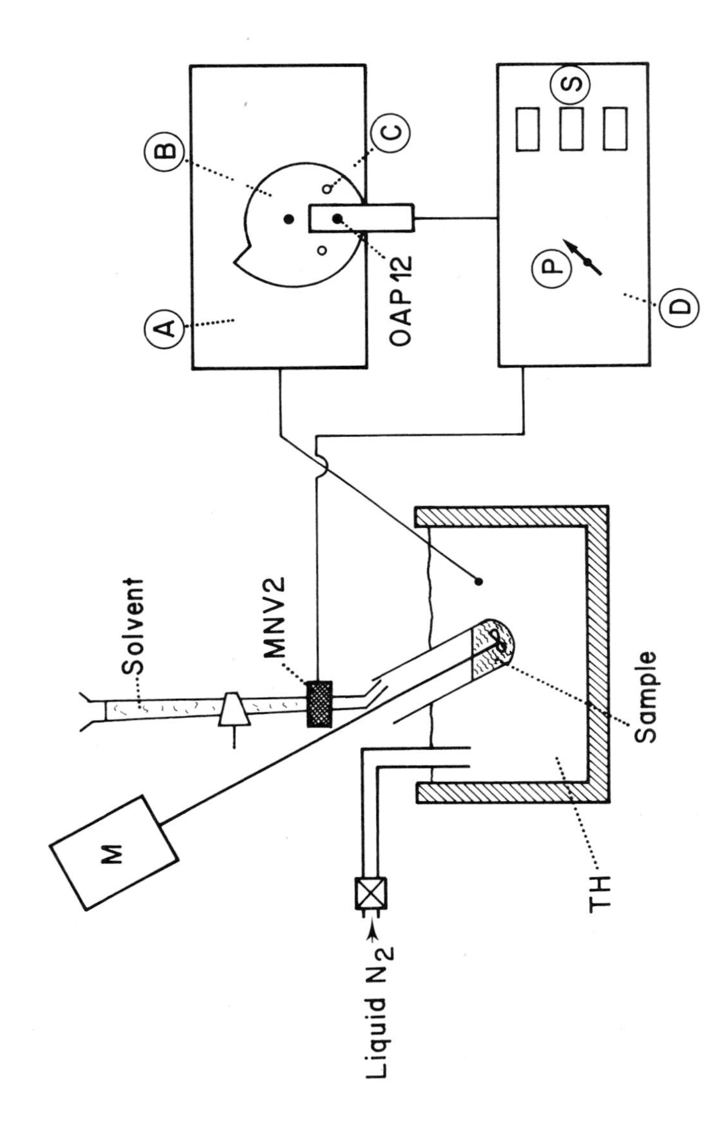

FIG. 1. Device for preparations of cooled aqueous–organic solutions of enzymes. A, regulation of temperature; B, programming came of temperature; C, holes of the injection program; D, stream programming; P, potentiometer; S, selection of capacities; M, motor; TH, thermostat.

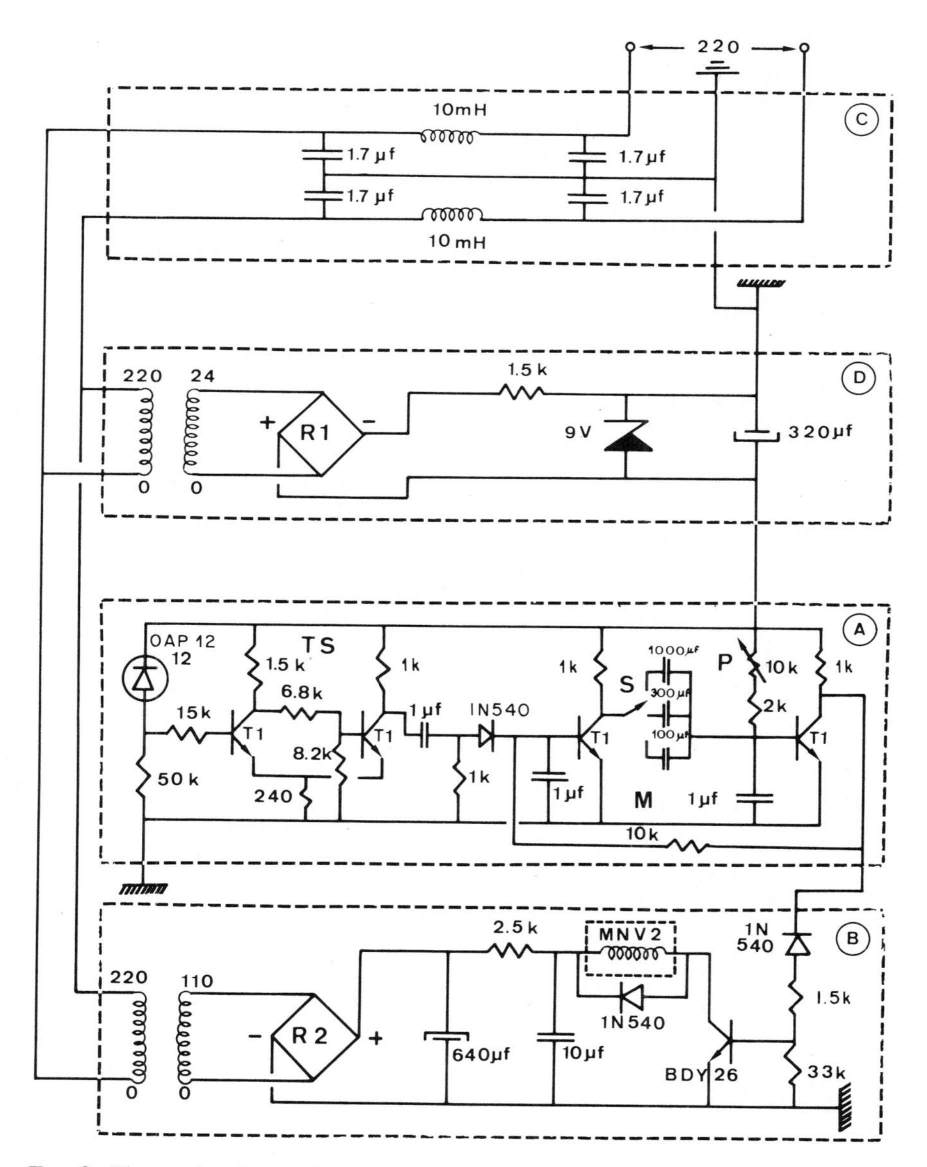

FIG. 2. Electronic scheme: A, monostable M and Schmitt trigger TS; T_1, transistor 2N 1711; P, potentiometer; S, selector of capacities; B, order of the electrovalve; R_2, bridge; C, filter 50 Hz; D, power supply of 9 V; R_1, bridge.

holes are scanned by a photodiode which, on illumination, disengages an electronic device connected to an electrovalve. At each illumination of the photodiode, the electrovalve opens for a set time, allowing the flow of a fixed quantity of organic liquid. The solvent is homogenized by a rotating agitator in a vessel placed in the thermostatic bath.

The electronic device shown in Fig. 2 is composed of three main parts: (*a*) a monostable multivibrator to control the duration; (*b*) an electronic switch for the electrovalve; (*c*) a 50 Hz filter.

(*a*) The monostable is electrically fed (9 V) by a power supply stabilized with a Zener diode and a polarized condenser (D). When the OAP 12 photodiode is illuminated by its passage in front of one of the openings, its resistance diminishes giving a 9 V potential to the base of a 2N 1711 transistor which attacks the Schmitt trigger (TS). The trigger is necessary for the conversion of the relatively slow electric signal delivered by the photodiode to an impulse. The 1 μF condensor generates the signal which attacks the monostable M. This monostable is characterized by a double time preselection: a series of 3 capacitances can be selected within a range of three time scales (S), and a potentiometer (P) permitting continuous adjustment within each of the ranges. The monostable then delivers a square signal (with a time variable from 1 to 20 s) which controls the release of the electrovalve.

(*b*) Electrovalve drive. The electrovalve current is supplied via a rectifier stabilized by a condenser providing a positive potential of 200 V. This tension is recovered by the transitor collector BDY 26 which acts as a commutator.

The apparatus therefore functions as follows: when the monostable is stopped, the transistor BDY 26 is blocked (200 V between the collector and the emittor); that is, no current passes through to the electrovalve. When the monostable is released, the transistor is saturated and a current of 40 mA passes through the electrovalve and disengages it (60 V at its terminals).

(*c*) 50 Hz Filter. Since this system is very sensitive to external random parasitic electrical impulses, a 50 Hz filter, as well as a number of protective capacitors, have been added and are illustrated in Fig. 2.

Cooling ⇀↽ heating and temperature control devices

Depending upon circumstances, low temperatures are obtained either by bubbling gaseous nitrogen through liquid nitrogen, or by cooling methanol or methanol–water mixtures of low freezing point.

The first procedure has been used to perform measurements of physico–chemical parameters and equip optical spectrometers. Cooled nitrogen goes through the cryostat via a low thermal loss guide; the temperature is then controlled by a heating resistance in the path of the cold gas at the entrance of the cryostat. The most efficient regulation is obtained by intermittent and rapid heating, monitored by a pulse generator. Regulation is achieved in most cases with a precision of $\pm 0.1°C$, between $+60°$ and $-100°C$.

The second procedure is applied to fast kinetic apparatus (slow temperature jump, stopped flow); the liquid bath can be alternately cooled and heated, and is fed to the cryostats surrounding the measurement cells. Such a procedure must be used when the observation system to be cryostated presents a high thermal inertia; it has some limitations due to the thermal inertia of the liquid itself and increasing viscosity, which does not permit it to reach temperatures as low as those obtained with gaseous nitrogen (the limit is about $-60°C$).

GASEOUS NITROGEN AS COOLENT

Since many enzyme–substrate intermediates can be detected by their characteristic absorption spectra, we first employed different commercial spectrophotometers (Beckman Acta III, Aminco–Chance DW2, Cary 15) (5,6).

The equipment used consists of three main parts: (*a*) a cryogenic temperature production unit, (*b*) a temperature-regulating device and (*c*) an adapted cell holder and sample compartment. The cell (glass or quartz, 1 cm path length) containing the sample is placed in a metallic cell holder, thermostated by the circulation of gaseous nitrogen, whose temperature is controlled. In order to avoid condensation and ice formation, which appears on the cell walls at low temperatures, the cell holder is isolated in a sample compartment, overpressurized with dry nitrogen.

Cryogenic temperature production unit and regulation device

A detailed description is given in Fig. 3. Dry gaseous nitrogen circulation (from a gas container G), whose delivery can be varied between 0 and 50 l/min. by means of a manometer (M_1), is bubbled into liquid nitrogen, contained in a Dewar flask (D). It is then transferred under vacuum (by vacuum-tube transfer VTT) to the cell-holders (SCH, RCH), and heated as required by an electrical resistance (R), variable between 0 and 200 W, connected via an alternostat (AT) to a temperature regulator (RG). The temperature regulator (Barber Colmann) is designed to accept the output of a standard Chromel–Alumel thermocouple (TC), which measures the temperature of the circulating nitrogen after passing through the cell holder. One then needs only to select a desired temperature suitable nitrogen circulation delivery and heating resistance. The temperatures of the sample

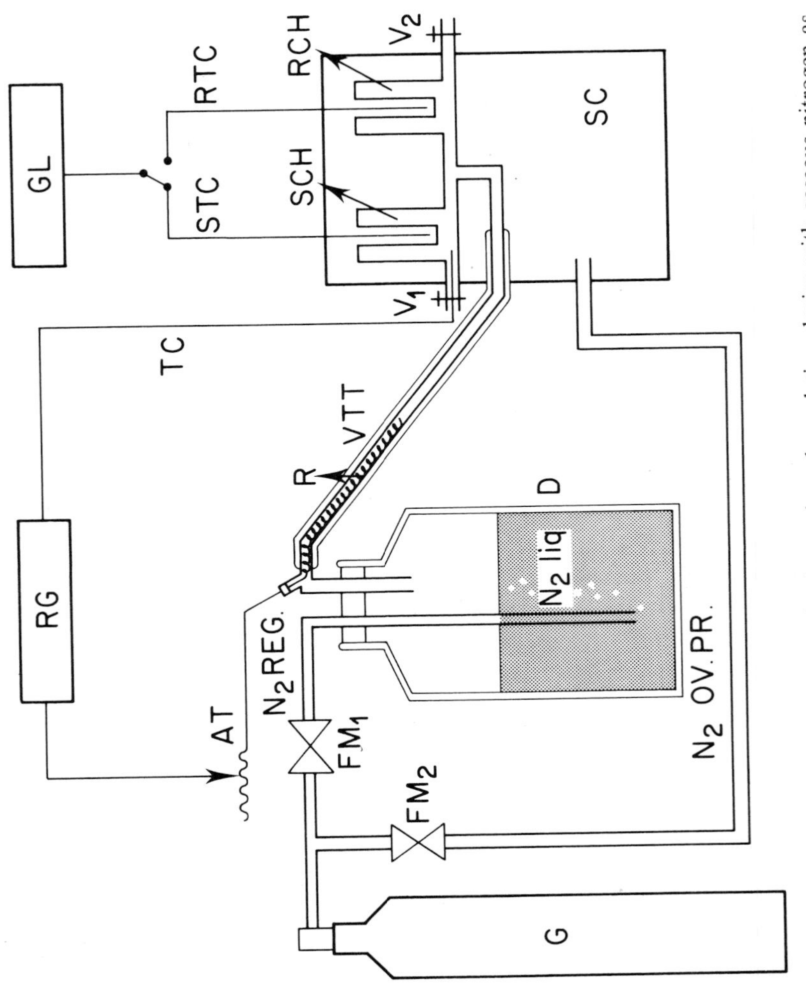

FIG. 3. Cryogenic temperature production unit and regulation device with gaseous nitrogen as coolant.

and reference are directly measured in the cells with Chromel–Alumel thermocouples (STC, RTC) connected to a galvanometer (GL).

Cell holders and sample compartment

The structures of the various commercial spectrophotometers being very different, it is necessary to adapt the cell holder for each particular type of apparatus. Some of these cell holders have been described and published elsewhere (5,6).

Performances

The range of temperature that this apparatus allows extends from −120° to 100°C. Calibration of the thermocouple-galvanometer assembly leads to an absolute error of ±0.1°C in the measured temperature. This precision is sufficient for most experiments. Moreover, we have verified the stability and homogeneity of temperature in both cells by replacing the galvanometer with a Beckman recorder. When temperature regulation is obtained, no temperature variation can be detected, either with time or within the total volume of the sample. Precision is then better than ±0.1°C.

LIQUID NITROGEN AND METHANOL–WATER MIXTURES AS COOLANTS

Liquid nitrogen is transferred under pressure from a container to a heat-exchange coil through electrovalves (e.v.). The coil is immersed in a mixture

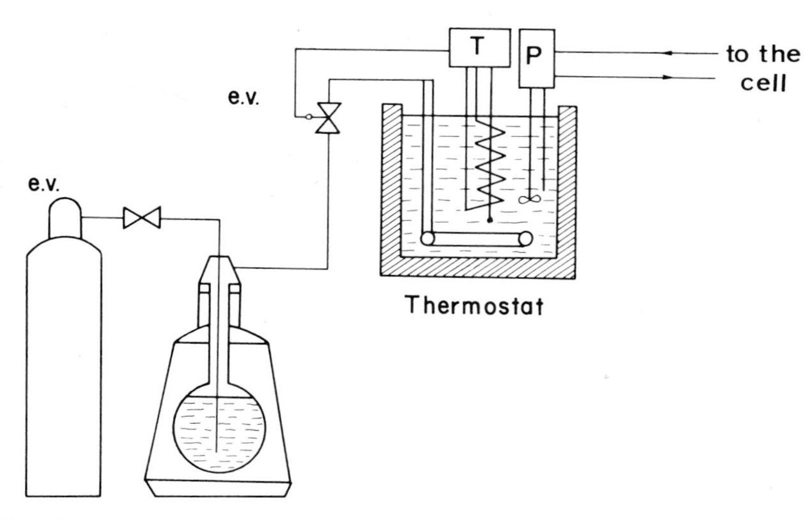

FIG. 4. Cryogenic temperature production unit and regulation device with liquid nitrogen and methanol–water mixtures as coolants.

of methanol–water (70 : 30, v/v) freezing at −115°C. The mixture will be used as coolant. An electronic regulation system automatically controls both the electrovalves and a heating resistance, with a precision of temperature regulation of ±0.01°C between +60° and −60°C. The device is described in Fig. 4.

Such a device is used to thermostat the cells in the fast kinetic apparatus described below.

Fast kinetic techniques at subzero temperatures

FLASH PHOTOLYSIS

Both the flash photolysis and the optical detection systems were designed and built in this laboratory, particularly with a view to working on liquid supercooled solutions. We shall give a brief description of this apparatus and its performance, Fig. 5.

Flash tubes and high voltage generator

Four U-shaped Xenon flash tubes made of Pyrex (height 10 cm, internal diameter 1 cm, distance between electrodes 25 cm, pressure of gas 1 atm) are placed in series around the measuring cell. A cylindrical aluminium reflector (CR) projects the maximum amount of flash light onto the sample. The high voltage generator charges 8 condensors (10 μF each) and flash tubes are fired by a pulse of 12 kV applied to external electrodes around the tubes. This control is synchronized with the detection system. The duration of the flash at half light is 50 μs for a maximum energy of 1 kJ.

Detection

Monochromatic light (from 250 to 700 nm) is provided by a Xenon source (Osram 75 W) and a u.v.-visible monochromator (Jobin and Yvon M 25). The parallel monochromatic light from the exit slit is deflected through 90° by a plane mirror (M) before it enters the measuring cell which has a vertical optical axis. A second monochromator (Baush and Lomb 33-86-25) has the role of optical protection for the photomultiplier (EMI 9558 BQ).

During the flash, the relatively high intensity of illumination saturates the photomultiplier, even with the optical protection of the second monochromator. To eliminate this, we have developed an electronic device which switches off the high voltage applied to the photomultiplier (−1300 V) during the flash (see Fig. 5). The main problems are both the high negative potential required to switch off and the fast dead time during the commutation (less than 25 μs). The control of the duration of the cut off is monitored by a monostable M, adjustable from 0 to 5 ms according to the experimental

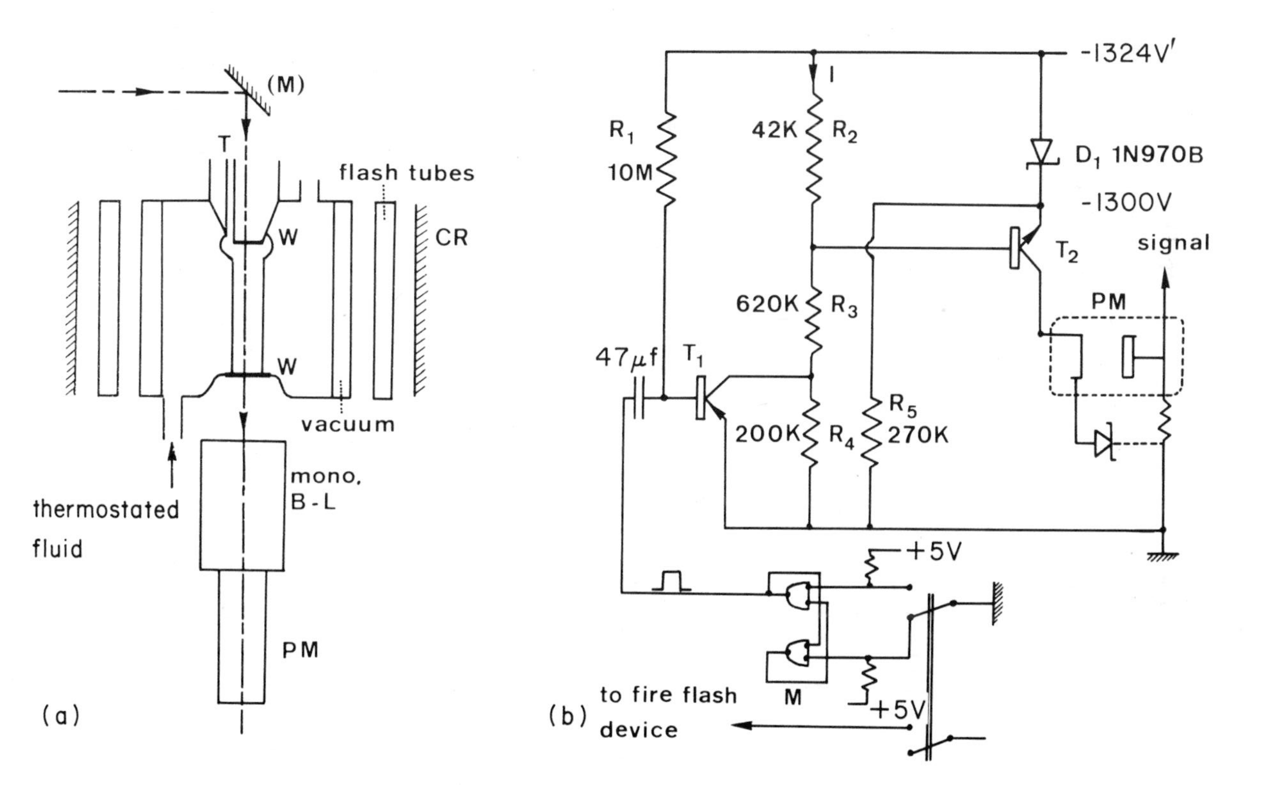

FIG. 5. Schematic diagram of flash photolysis apparatus. (a) measuring cell, (b) electronic device for photomultiplier voltage switch off. (M), mirror; CR, cylindrical reflector; PM, photomultiplier; W, quartz windows.

conditions. The monostable is triggered by a positive front and the duration of the impulse can be varied by a potentiometer. During normal operation, when the high voltage is applied to the PM, the transistor T_2 (BU 105 BE 9 : NPN) is satured and polarized by the current I through the chain of resistance R_2, R_3, R_4 ($I = 2$ mA), the base of T_2 being positive with respect to its emittor. The transistor T_1 (MPSA 92 : PNP) is saturated and thus shunts R_4. T_1 is polarized by R_1 ($Vb_1 = -0.6$ V).

When the positive front of the impulse is applied, through the condensor, to the base of T_1, it renders it inoperative. The current I decreases to 1.6 mA because R_4 is now in the circuit. The base of the second transistor T_2 is then negative with respect to its emittor (-1300 V) and T_2 does not conduct. The zener diode D_1 (24 V) gives two negative reference potentials and is polarized by R_5 (id = 5 mA). The photocurrent of the photomultiplier is amplified with a logarithmic amplifier and read, as optical density, by a storage oscilloscope (Tektronix 5103N).

Measuring cell and temperature control

This cell, made of Pyrex, is composed of an observation chamber of internal diameter 5 mm and optical length limited to 70 mm with two quartz windows (W). The conic shape of the mobile inlet allows accurate positioning of the upper window. Using the tube (T), it is possible to perform experiments in controlled atmospheres. Around this cell is a chamber controlling the temperature of the sample. This chamber is isolated from the surrounding atmosphere by a vacuum jacket. Ice formation on the windows is avoided by a slow flow of dry nitrogen from two inlets. The observation chamber can be thermostated from $+50°$ to $-65°$C with the liquid thermostat described above. Temperature measurements in the sample with a Chromel–Alumel thermocouple (ϕ0.3 mm) indicated a temperature gradient of less than 0.05°C.

"SLOW" TEMPERATURE JUMP

A slow temperature-jump apparatus similar in principle to that designed by Pohl (7,8), but operating between $+60°$ and $-50°$C, was built in this laboratory to overcome the difficulties mentioned above, and also then to record the temperature-dependent kinetics in cases where time constants reaching minutes can be expected at low temperatures (9). Temperature jumps (up and down) are obtained in a microcell by alternating fluids coming from two different thermostats. The microcell is thermostated, and the temperature change is achieved by switching the liquids of the two thermostats to get the required "fast" temperature change of the solution. The thermal capacity of the microcell is as small, and the heat transfer through the walls as large as possible.

Construction details

Thermostat systems. Such a system consists of a stainless steel calorimeter vessel of 10-liter capacity, containing either water or methanol depending on the temperature range employed. Heating is obtained by an electrical resistance, variable between 0 and 500 W, and cooling by liquid nitrogen transferred under pressure from a container to a heat-exchange coil through electrovalves (Asco type 8263 A8P). An electronic regulation system (Unitherm Haake) automatically controls both the electrovalves and the heating resistance. The precision of temperature regulation achieved with such a device is of the order of $\pm 0.01\,^{\circ}\mathrm{C}$ between $+60^{\circ}$ and $-60^{\circ}\mathrm{C}$.

Cell and thermostat chamber. The cell consists of a stainless steel tube, 0.2 mm thick, whose inner diameter is 1.6 mm. Its overall length is 12 mm. Its ends are closed by two quartz windows, made liquid tight by the use of Teflon gaskets. The cell is placed in a thermostat chamber in which circulation of the thermal regulating fluid occurs.

The construction details of this assembly are given in Fig. 6. The thermostat fluid is conducted from baths (T_1 and T_2) to the cell by means of two low-temperature circulating pumps (Lauda type EKS) whose delivery is adjustable. Both pumps are connected to the thermostat chamber by heat-insulated flexible tubes, as short as possible to reduce thermal exchange. Circulation of the thermostat fluid through the cell chamber is controlled by means of four electrovalves (e.v. Asco type 8263 A8L) working in pairs, which allow the cell to be thermostated by either bath T_1 or T_2 as shown in Fig. 7. The electrovalves are directly fixed on the thermostat chamber. This assembly is sufficiently reduced in size so that it can be placed in the sample compartment of any spectrophotometer (the Cary 15, in particular).

Performance

Temperature jump. Two methods can be used to follow the temperature variation inside the cell. (1) The first is based on direct temperature measurement by means of Chromel–Alumel thermocouple (Philips Industrie S.A., type 2ABI 025) placed in the cell; its diameter is 0.25 mm and its response time of the order of 6×10^{-3} s. Recordings of temperature variation with time are made with an X–Y Hewlett Packard recorder, type 9030 AM. While, for a temperature jump from $+20^{\circ}$ to $-40^{\circ}\mathrm{C}$, the time constant of the apparatus is 0.65 s, the overall thermal equilibrium time is 7 s. The same values are obtained when the temperature jump is reversed, from -40° to $+20^{\circ}\mathrm{C}$. (2) The second method is based on the variation with temperature of the pK^* of a given indicator. For this purpose, nitro-3-anilinium is used in a solution made by mixing an aqueous hydrochloric acid solution and methanol (50 : 50, v/v). The pa_{H} of such a solution is known to be independent

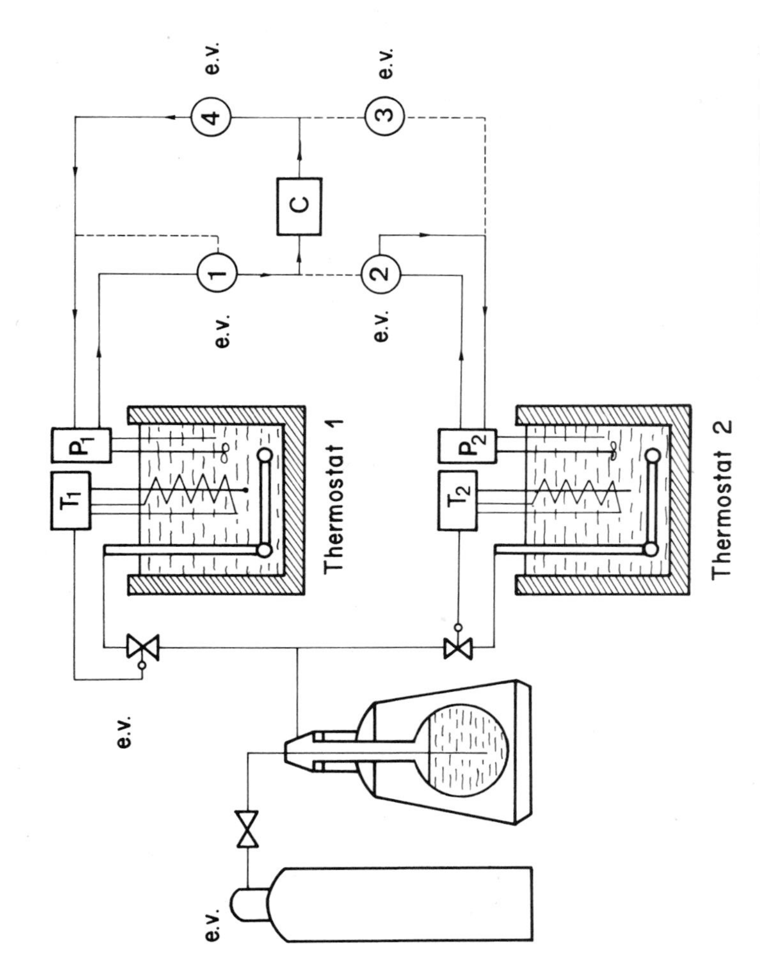

FIG. 6. "Slow" temperature jump device. e.v., electrovalves; T_1, T_2, thermostats; P_1, P_2, pumps functioning between +100° and − 100°C; C, measuring cell.

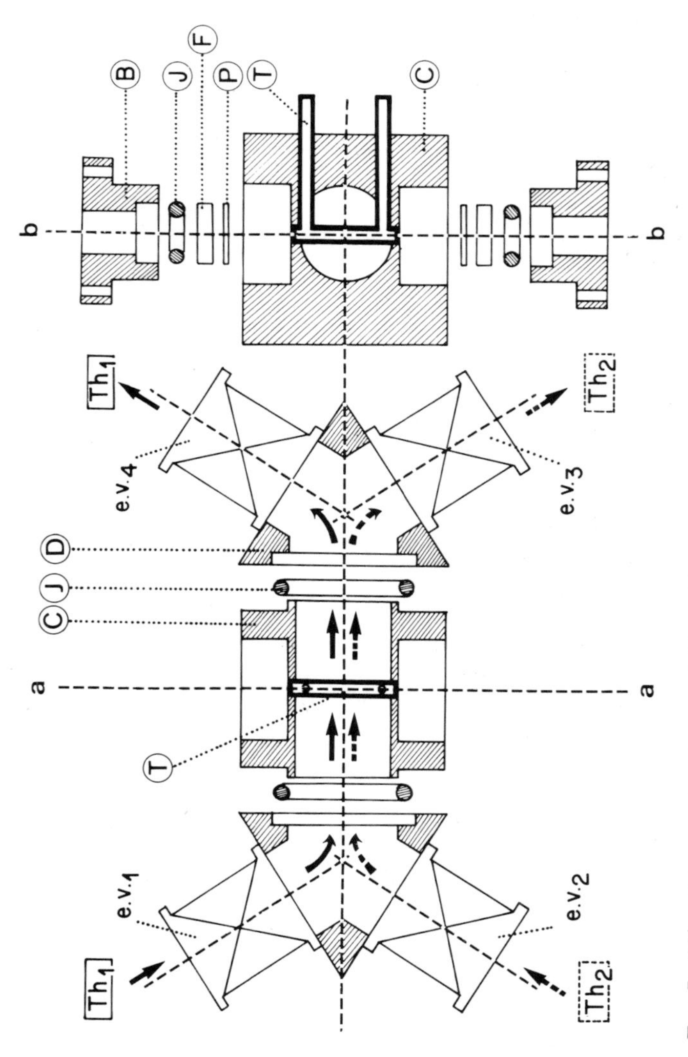

FIG. 7. "Slow" temperature jump cell. T, stainless tube (diam. 0.2 mm); C, insulating block; F, quartz window; Th_1, Th_2, thermostats; e.v.$_1$, e.v.$_2$, e.v.$_3$, e.v.$_4$, electrovalves.

of temperature; in our particular case it was 2.5. The variation of pK^* with temperature for this indicator has been determined under the same conditions (10): it is linear from 1.76 at +20°C to 3.6 at −50°C. Thus measurements of pK variations allow us to determine temperature variations. The pK^* values are given by the equation (see chapter 2)

$$pK^* \simeq pa_H + \log\left(\frac{\alpha}{1-\alpha}\right) = pa_H + \log r$$

where the asterisk means that the values are considered within a hydro-organic mixture.

The concentration ratio (r) of the two forms of indicator is measured at any wavelength by the equation

$$r = \frac{A_{In} - A}{A - A_{InH}}$$

where A is the absorbence measured at the given pH, and A_{In} and A_{InH} are the absorbence for neutral and ionized forms, respectively. Since pa_H is supposed constant, ΔpK is given by:

$$\Delta pK^* = \Delta \log r$$

which can then give ΔT.

The results are given in Table 1, where the variations in absorbance are shown during the temperature jump. Sharp variations occurring at the beginning of the jump are observed and are due to sudden variations in refractive index, induced by the establishment of a temperature gradient and contraction (or expansion) of the solvent. The importance of this phenomenon is proportional to the amplitude of the temperature jump.

TABLE 1. Determination of temperature with indicators: A is the absorbancy at 363 nm at the given pH and temperature; A_{In} and A_{InH} are the absorbences for neutral and ionized forms, respectively, under the same conditions

A	A_{In}	A_{InH}	$\frac{\alpha}{1-\alpha}$	pK^*_{mes}	T (°C)
1.86	2.18	0.22	0.20	1.76	+20
1.79	2.22	0.23	0.27	1.9	+10.5
1.72	2.24	0.24	0.35	2.02	+3
1.52	2.28	0.26	0.60	2.25	−11.5
1.30	2.31	0.27	0.98	2.46	−23

Application: pa_H *jump*

The pa_H reversible jump is a direct application of temperature jump. For a given buffer, a temperature variation induces a pa_H variation. The higher the ionization enthalpy (ΔH°), the larger the pa_H change. Thus, to any temperature change we can relate a corresponding pa_H change for a particular buffer.

Three buffers have been chosen, which are in a hydro–organic solvent, water–methanol (50 : 50, v/v).

$$\begin{array}{lll} \text{Phosphate:} & \Delta H^\circ = 0.94 \text{ kcal/mol} \\ \text{Tris:} & \Delta H^\circ = 10.3 \\ \text{Borate:} & \Delta H^\circ = 3.36 \end{array}$$

As described previously for pK^* change, pa_H change here is followed by the indicator nitro-4-phenol. For this indicator pK^* variation with temperature is known (11). Measurements are made between $+20^\circ$ and -35°C. Variations of pa_H and temperature with time have the same kinetic parameters, i.e., the time constant is 0.65 s, and the total equilibrium time of 7 s. Results are given in Table 2.

TABLE 2. pa_H changes induced by temperature jump. S_1, S_2, S_3 are 10^{-2}M phosphate, Tris and borate buffers, respectively

Temp. before and after jump (°C)	Proton activity	Solutions (methanol-buffer 1 : 1, v/v)		
		S_1	S_2	S_3
$T_1 = +20$	pa_{H_1}	8.38	7.78	8.63
$T_2 = -35$	pa_{H_2}	8.52	9.65	8.05
$\Delta T = -55$	Δpa_H	+0.14	+1.87	−0.58

RAPID MIXING OF REACTANTS: FLOW-STOPPED FLOW

The essential feature of any procedure for rate measurements is that one must be able to obtain observations of the extent of a reaction as a function of time, that is, at different stages of the reaction mixture. This can be achieved in flow systems, and there are an increasing number of designs of equipment in operation, as well as a variety of commercial instruments, for experimentation in the normal range of temperatures. To our knowledge, however, just one apparatus is available for use at cryogenic temperatures. This apparatus was developed by Allen *et al.* (12) and consists of a stainless steel block (8 × 10 × 11 cm) immersed in a thermostat bath. Its volume per run is 3 ml, and its optical path of 2.5 mm is too small for the kinetic studies of many biological reactions. Its thermal inertia is, of course, very important, and the immersed device lacks "flexibility" in changing temperatures.

We built a special piece of apparatus (Fig. 8) to carry out stopped-flow experiments between $+40°$ and $-50°C$ (13). Basically, this apparatus is similar in principle to one designed by Gibson (14). The principle of operation may be followed by reference to Fig. 8. The reagents are placed in 20 ml syringes (RS in Fig. 8), from which they are transferred by correct setting of 2 ml driving syringes (DS). To make a determination, the driving syringes are operated together by the syringe-pushing block (P), using a pneumatic unit (PDP). The reactants go through the mixer and then pass the observation window and flow into the 2 ml stop syringe (SS), driving its plunger upward until the handle strikes the stop (S). Flow is then stopped suddenly. The progress of the reaction is followed by the change in absorbence recorded by a photomultiplier. To repeat the observation, the port valve is opened using a key, allowing the spent reaction mixture in the stopping syringe to drain away through the drain port into a beaker. The port valve is then closed, and the cycle can be repeated by operating the pneumatic unit.

Construction details

To drive the syringes, a pneumatically activated driving block (Hamilton, type PB 600) is used to push the pistons of the parallel driving syringes (Pyrex, medical grade, 2 ml). Syringes and pistons are made to be free-moving and leakproof down to $-45°C$. The air pressure which activates the pneumatic piston is about 40 to 70 psi.

Each syringe is surrounded by a mantle (Hamilton, type 87.204), connected to the circulating system described below. The contents of the two standard syringe reservoirs can be transferred directly to the driving syringe through three-way valves (Hamilton IX P_3). When loaded, the driving syringes are in contact with the plunger of the pneumatic piston, which on activation drives the reactants at high speed through the mixing chamber into the flow cell and out through an exhaust port into the stopping syringe.

Mixing chamber. The stainless steel chamber ($14 \times 16 \times 36$ mm) has a canalization system to allow effective cooling.

The reactants enter the mixing chamber, through two reservoir tubes (r_1 and r_2, length 32 mm, diameter 2 mm), and are then driven into a central tube (length 4 mm) through two jets 0.7 mm in diameter. The jets open tangentially into a central hole of 1.5 mm diameter in the tube, and are so placed as to produce maximum turbulence and hence maximum mixing efficiency.

Observation chamber. The Gilford (model 203-B) stainless steel observation chamber has an optical path of 10 mm and a volume of 0.1 ml. The chamber is perpendicular to the entrance and exit tubes, and consists of a horizontal tube 6 mm^2 in cross-section, closed by two quartz Suprasil windows with

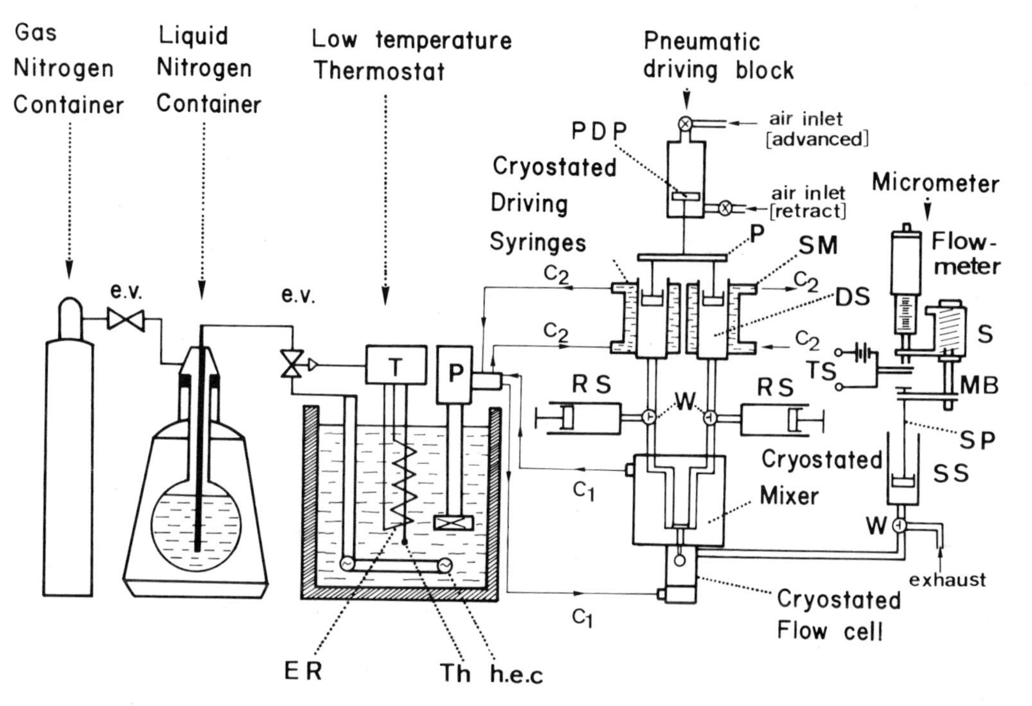

FIG. 8. Schematic diagram of the low-temperature stopped-flow apparatus, showing temperature regulation and circulation of thermostated fluid (C_2, C_1, respectively) in the driving syringes and the mixer and flow cell: e.v., electrovalves; ER, electrical resistance; Th, thermometer; hec, heat-exchange coil; T, electronic temperature regulation block; P, low-temperature circulating pump; PDP, pneumatic driving piston; p, plunger; DS, driving syringes; SM, syringe mantles; RS, reservoir syringes; W, three-way valves; SS, stopped syringe; SP, stopped piston; MB, magnetic bar; S, solenoid; TS, trigger switch.

Teflon gaskets and pressure plates. The observation tube is surrounded by holes to allow accurate cooling. Driving syringes, mixing chamber, and observation chamber are cooled with the same fluid to ensure the uniform temperature.

Temperature control; heating ⇌ cooling system. The liquid thermostat shown in Fig. 8, the temperature of which is controlled by a thermometer and relay, can be alternately heated and cooled.

The liquid bath (methanol–water mixture or methanol) is contained in a stainless steel calorimeter vessel of 10 liter capacity. Heating is obtained by an electrical resistance of 500 W, and cooling by liquid nitrogen, transferred from a container to a heat-exchange coil by pressure through the electro-valves (Asco, type 8263 A 8P). An electronic regulation system (Unitherm, Haake) automatically controls the electrovalves and the heating resistance. The precision of regulation obtained is of the order of $\pm 0.01°C$ between $+60°$ and $-60°C$. Two low temperature circulating pumps (Lauda, type EKS) are used and the output circulating fluids go, in turn, through the syringe mantles, the mixer, and the flow cell before being returned to the main bath.

Control. Mixing and observation chambers can thus be cooled to any temperature down to $-45°C$. They are kept watertight by use of Teflon gaskets and mechanical pressure. The maximum admissible hydrostatic pressure is about 70 psi.

Rigorous temperature control and the maintenance of homogeneity in both chambers are necessary to avoid optical artifacts due to thermal effects in the observation tube. For instance, a difference of 0.5°C between these chambers produces changes in transmittance. These changes, which occur immediately after the flow is stopped, may be positive or negative, depending on whether the temperature of the solution is greater or smaller than that of the observation chamber. Such optical artifacts are due to a gradient of refractive index, caused by a temperature gradient in the stopped solution. By means of differential temperature regulation (using a Philips Chromel–Alumel thermocouple and a Bauer Capa C regulator), it is possible to obtain regulation better than 0.1°C. Finally, water condensation at low temperature can be avoided by careful purging of the chambers with gaseous nitrogen.

Stopping syringe. The vertical stopping syringe (Hamilton, model 1002) is 2.5 ml in volume. It is connected to the exhaust tube of the observation chamber by a high-pressure pipe via a three-way valve (Hamilton I X P_3). Pushed by the flow, the syringe piston moves upward until its motion is stopped by an adjustable micrometer. Before the stopping piston hits the "stop", a microswitch is activated, triggering an oscilloscope. All fluid flow in the system rapidly ceases, and the observation starts. A micrometer displacement of 2 mm corresponds to a volume displacement of 0.1 ml. The displacement of the stopping piston is relayed to a magnetic bar, which moves inside a solenoid. The electromotive inductive force is proportional to the linear rate of displacement of the piston, constituting an electromagnetic flowmeter, the amplitude of its trace on the oscilloscope giving the flow velocity. When the flow stops, the amplitude drops sharply and the time interval between the start and the stopping of mixing gives the transport time of the device.

Optical and electronic systems of detection; light source and monochromator. The source is a high-pressure xenon arc (OSRAM XBO 75 W/2) fed by a commercial, stabilized current supply. The limiting factor in recording a relatively low absorbance change appears to be the occurrence of slow oscillations of intensity. The intrinsic brightness of the xenon lamp is very advantageous. The monochromator (u.v. visible) is a Jobin–Yvon (M. 25); focal length 250 mm and numerical aperture f/3. The resolution is about 0.7 Å, with a slit width of 0.01 mm. In practice, the limit is about 5 Å.

The parallel monochromatic light enters the flow cell and is transmitted to a photomultiplier tube (EMI 6256 B), end-window type. A well-regulated photomultiplier power supply is essential, since the output of the photomultiplier is sensitive to small voltage changes. The photocurrent produced is amplified by a dual-trace differential amplifier (3 A3) of an autotrace storage oscilloscope (Tektronix 564) with a 2867 time-base unit.

The horizontal time-base sweep of the oscilloscope is triggered by the solenoid signal of the flow meter, so that the interval from start through acceleration (5 ms), continuous flow (10 ms), and deceleration to stop (1 ms) is displayed.

Overall performance

For a pressure of 60 psi on the pneumatically driven piston and a volume displacement of 0.2 ml, the "transport time" (time interval of the flow between starting and stopping) is 0.02 s, which corresponds to a volume delivery of 10 ml/s.

The linear flow speed in the observation chamber is then 167 cm/s. The dead volume of the tube (2×4 mm long, d = 1.5 mm) connecting the mixer and the observation tube is about 14 μl compared to the 100 μl volume of the observation chamber. The transit time from the mixer to the middle of the observation window is about 2.8 ms while the transit time through the cuvette is 10 ms. The dead time can be referred to as the time required for a molecule to pass from the mixer to the midway point in the observation cell; it is estimated to be about 8 ms for a first-order reaction.

Direct determination of the dead time can be obtained by measuring the extents of a reaction observed under sufficiently different conditions of rate. Such a reaction can be the reduction of 2-6-dichlorophenolindophenol (DCPIP) by ascorbate, which is a pseudo-first-order reaction whose half-time can be varied by altering the concentration of ascorbate (10^{-3}; 10^{-1} M) while keeping the concentration of DCPIP constant (10^{-5} M). Then a semilogarithmic plot of absorbance vs time for these reactions gives a family of straight lines that intercept at a point whose abscissa is the zero time of the reaction. The difference between this zero time and the time of earliest observation is the dead time of the apparatus, which is about 10 ms.

The mixing efficiency has been evaluated by mixing a solution of HCl (5×10^{-2} M) with NaOH solution (4.8×10^{-2} M) containing indicator (bromothymol blue, 10^{-4} M) and measuring the time of neutralization. Mixing is complete when flow stops at 0.02 s: no optical inhomogeneity indicative of incomplete mixing is observable under these conditions.

Another test of mixing utilizes the reaction of iodine (I_2) with thiosulfate:

$$2S_2O_3^{2-} + I_2 \longrightarrow S_4O_6^{2-} + 2I^-$$

The solution of I^- is colorless, and the solution of iodine absorbs at 430 nm. Any incomplete mixing is easily observed in a way which can be magnified for recording by increasing the concentrations of both reagents, which are prepared in a mixture cacodylate buffer (pH 7.5) and organic solvent. When solutions of iodine and thiosulfate in aqueous methanol (50 : 50, v/v) are mixed at +20°C, no optical inhomogeneity is observable up to a concentration of 10^{-3} M iodine as flow stops (horizontal trace in the scope).

The viscosities of the solvent under consideration are maximum (16 cP) for the aqueous–methanol mixtures at the same temperature. At low temperature (−20°C), where the viscosity is 26 cP, no appreciable deviation of the scope trace is observable under the same conditions. With a more viscous mixture such as water–ethylene glycol (50% v/v), we have succeeded in mixing a solution of 5×10^{-4} M iodine with thiosulfate at +20° and −20°C. Moreover, before each kinetic experiment at any temperature, we mix the colored reactant with the buffer solution alone to control the experimental run of the apparatus and to determine the value of the initial absorbance of the reactant.

The stopped flow has been tested by the reduction of DCPIP by ascorbate at room temperature. This reaction can be made to follow first order kinetics by keeping the initial concentration of DCPIP constant (10^{-5} M) and much smaller than ascorbate (10^{-1} M). Under these conditions the rate depends on the concentration of the reducing agent. In phosphate buffer, pH 7.5 (10^{-1} M), the time constant is about 54×10^{-3} s corresponding to a rate constant of 185 M^{-1} s^{-1} for 10^{-1} M ascorbate.

Reaction kinetics of ascorbate and 2-6-dichlorophenol-indophenol at subzero temperatures. The reaction was carried out in the mixture methanol–water (50 : 50, v/v), the freezing point of which is $-50°C$. Phosphate buffer (10^{-1} M) was replaced by 10^{-1} M cacodylate to avoid any precipitation at subzero temperatures. Under these conditions, the "protonic activity" corresponds to pH 7.5 at $+20°C$ and does not vary widely at subzero temperatures (0.2 units between $+20°$ and $-50°C$).

Reaction kinetics were recorded between $+20°$ and $-35°C$ with three different concentrations of ascorbate (2.10^{-3} M, 5.10^{-2} M, $1.6.10^{-1}$ M) (see Fig. 9). Below $-35°C$ there was a precipitation of ascorbate. As expected, the reactions are slowed in proportion to the drop in temperature, and it can be seen from Table 3 that, for instance, the reduction of the DCPIP in the presence of ascorbate (5.10^{-2} M) has a time constant of 0.18 s $+20°C$ and of 6 s at $-35°C$. Incidentally, let us mention that the performance of the device is practically the same at any temperature.

Problems and difficulties

This apparatus presents the difficulties inherent in any stopped-flow device (air bubbles, leakage, syringe breakage, valve-tip breakage, etc.) and also some additional problems due to the fact that it operates at low temperatures and uses solutions of high viscosity.

TABLE 3. Values of the rate constants k for reduction of 2-6-dichlorophenol-indophenol (10^{-5}M) by ascorbate (2.10^{-3}, 5.10^{-2}, $1.6.10^{-1}$M) obtained at different temperatures with the stopped-flow apparatus

		(ascorb) = 2.10^{-3}M		(ascorb) = 5.10^{-2}M		(ascorb) = $1.6.10^{-1}$M	
T(°C)	($10^3/T$) $°K^{-1}$	τ (s)	k (s^{-1} M^{-1})	τ (s)	k (s^{-1} M^{-1})	τ (s)	k (s^{-1} M^{-1})
20	3.41	4	125	0.18	110		
17	3.44			0.18	110		
16	3.46	5.5	90.5				
11	3.52			0.3	66		
5.5	3.58	6.6	75.5				
0	3.66	11	49.5				
−8	3.77	21.5	23.25				
−16	3.89	35.5	14.1				
−22	3.98			3.8	5.26		
−25	4.04			3.8	5.26		
−28	4.08			5.4	3.68		
−29.5	4.10			5	4		
−33	4.16			6	3.32		
−34	4.17					2	3.12

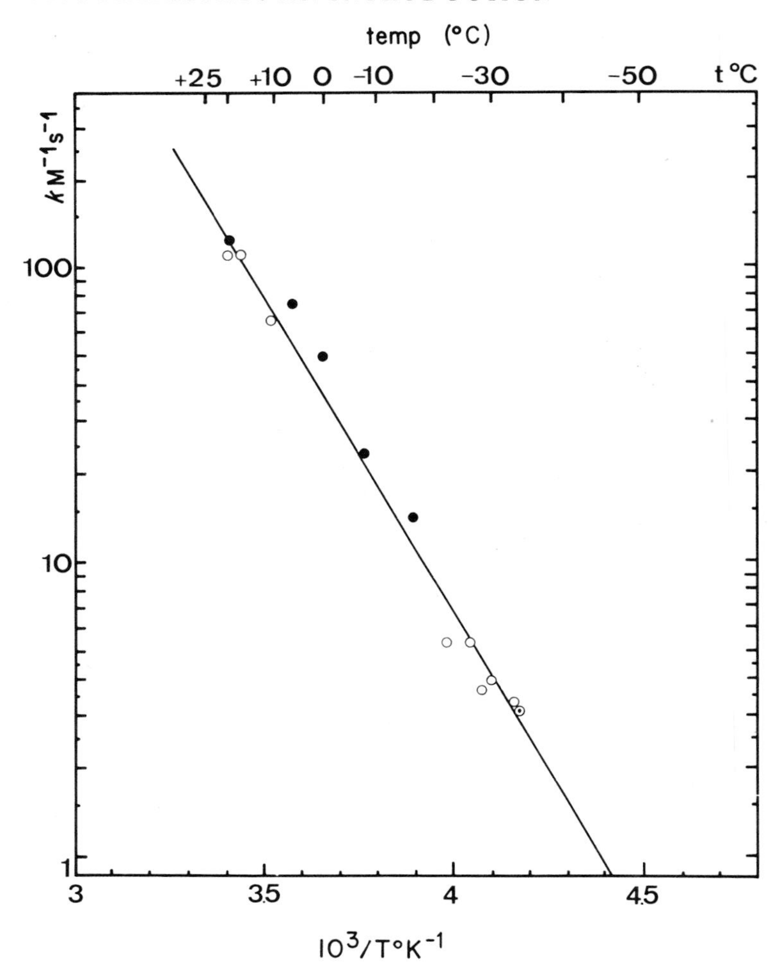

FIG. 9. Arrhenius plot of the rate constants for the reduction of 2-6-dichlorophenolindophenol (DCPIP) by ascorbate at different concentrations. ●, 2×10^{-3} M; ○, 5×10^{-2} M; ⊙, 1.6×10^{-1} M; the concentration of DCPIP is 10^{-5} M; methanol–water solvent is used (50 : 50, v/v); 10^{-1} M cacodylate buffer, pH 7, in water and in methanol–water (50 : 50, v/v); the pa_H at $+20$°C is 7.5 and at -30°C $= 7.6$. The slope of log k against $1/T$ is equal to $E/4.576$, where E is the activation energy (kcal/mol); $E = 9.7$ kcal/mol for the reduction of DCPIP by ascorbate.

Leakage and breakage. As mentioned previously, syringes and pistons are free moving and leakproof down to -45°C, and quite satisfactory results are obtained with the apparatus described above. The pressures used are such that the glass syringe used for stopping the flow can take the stress nearly to its limit. The Teflon tips are somewhat fragile, however, and the valves can fracture at the level of the retaining tip. Repeated changes in temperature (between +40° and −45°C) can lead to "fatigue" of the Teflon tips, which must be replaced from time to time.

Temperature artifacts. When the stainless steel block of the observation chamber is warmer or cooler than the injected solution (by about ±0.5°C), a temperature gradient is quickly established across the liquid, and simultaneously a gradient in refractive index occurs, which is equivalent to a positive or negative lens. This gives rise to a change in optical density for a few seconds; the temperature equilibrium is restored when the chamber and the stainless steel block have the same temperature.

Viscous solutions. Some mixed solutions are highly viscous (more than 50 cP) at low temperatures, and we have seen that it is then advantageous to use a ternary mixture; it is then essential to mix solvents of identical viscosity (i.e. of the same composition and temperature) to avoid inhomogeneities.

Although these different effects have rarely occurred, their possibility must always be borne in mind when anomalous results are recorded.

References

1 G. K. Strother and E. A. Ackerman. *Biochem. Biophys. Acta* **47**, 317 (1961).
2 B. Bielski and S. Freed. *Biochim. Biophys. Acta* **89**, 314 (1964).
3 S. Freed. *Science* **150**, 567 (1965).
4 C. Balny and L. Becker. *J. Chim. Phys.* **68**, 1008 (1971).
5 C. Balny, G. Hui Bon Hoa and F. Travers. *J. Chim. Phys.* **68**, 366 (1971).
6 P. Maurel, F. Travers and P. Douzou. *Anal. Biochem.* **57**, 555 (1974).
7 F. M. Pohl. *Eur. J. Biochem.* **4**, 373 (1968).
8 F. M. Pohl. *Eur. J. Biochem.* **7**, 146 (1968).
9 G. Hui Bon Hoa and F. Travers. *J. Chim. Phys.* **69**, 637 (1972).
10 G. Hui Bon Hoa, R. Gaboriaud and P. Douzou. *C. R. Acad. Sc. Paris* **270C**, 373, 1970.
11 G. Hui Bon Hoa and P. Douzou. *J. Biol. Chem.* **248**, 4649 (1973).
12 C. R. Allen, A. J. Brook and B. F. Caldin. *Trans. Faraday Soc.* **56**, 788 (1960).
13 G. Hui Bon Hoa and P. Douzou. *Anal. Biochem.* **51**, 127 (1973).
14 Q. M. Gibson. *Faraday Soc., Discussions* **17**, 127 (1954).

5

Applications of the Low Temperature Procedure

Study of enzyme-substrate intermediates

It now appears likely that most enzyme catalyzed reactions involve two or three enzyme-substrate complexes acting in the following sequence:

$$E + S \rightleftharpoons E-S \rightleftharpoons E-Z \rightleftharpoons E-P \longrightarrow E + P$$

where $E - Z$ is the "true" transition-state complex, and $E - P$ an enzyme-product complex. An energy diagram for such an enzyme catalyzed reaction is shown in Fig. 1. Furthermore, it should be pointed out that in most enzymic reactions there is more than one substrate molecule and there may be two or more products. It is known that in a reaction with two substrates, S_1 and S_2, there may be three enzyme-substrate intermediates, namely, $E-S_1$, $E-S_2$, and $E-S_1-S_2$. If the reaction has two products, P_1 and P_2, there may be at least three additional complexes, $E-P_1$, $E-P_2$, and $E-P_1-P_2$. Many intermediate steps occur in such reactions, each having its own rate constant. As a result, kinetic analysis of enzymic reactions involving two or more reactants can sometimes be exceedingly complex. The presence of some enzyme-substrate complexes can sometimes be detected by means of sensitive spectroscopic methods, but such direct proof is difficult to obtain because of the instability of the $E-S$ intermediates, which must necessarily decompose rapidly.

One approach to understanding the mechanism of enzyme action might be to characterize these intermediates and to identify them with the different forms involved in the sequential feature of the reaction. Classical kinetic techniques usually do not give any valuable information about intermediates because the enzyme concentrations must be kept very low in order to get the reaction rate within an accessible time scale and consequently the number and nature of the intermediates cannot be determined. Their life-times being very short, a rapid kinetic method is naturally required: but, as mentioned previously, many of the present fast techniques give nothing more than a characteristic relaxation time with no structural information

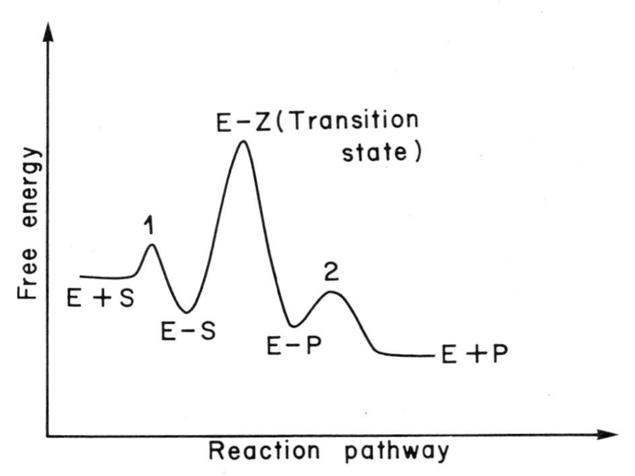

FIG. 1. Diagram for enzyme catalyzed reaction.

about what that time refers to. Very few enzyme mechanisms have been studied sufficiently well to allow any discussion in terms of elementary steps.

The problem is to resolve overall reactions into intermediate stages; progress can be made when a reaction is resolved according to the differences in energy of its stages. Since the magnitude of the activation energies differ from one intermediate step to another, resolution among them would become possible at low temperature according to the Arrhenius relationship.

Reactions too rapid to be resolved at room temperature might fall within the range suitable for measurements at lower temperatures. Freed (1) reported how the relative concentrations of reactants and products vary with temperature and showed to what extent the concentrations of enzyme-substrate complexes increase relative to the concentrations of the separate reactants at low temperature. He also listed the effects of temperature and of activation energy (ΔE) on the ratio of reaction rate constants relative to a reaction rate constant of 1 at 300°K (+27°C): if the activation energy per mole of reactants were 4.6 kcal, the rate would be 46 times slower at 200°K (−73°C) than at room temperature and 2000 times slower at 150°K (−123°C). For an activation energy of 9.2 kcal mol^{-1}, the rate would be 2000 times slower at −73°C, and for an activation energy of 18.4 kcal mol^{-1} it would be 4 million times slower. We can expect such temperature and heat activation effects to be sufficient for intermediate studies based on differences in the energies of reaction stages.

An examination of available data indicates that enzymic reactions may be broken down into elementary steps, but that full explanation of their mechanism at the molecular level requires improved spectroscopic techniques including the most sophisticated ones.

In theory, spectroscopic techniques such as spectrophotometry can characterize the intermediates as well as give their concentrations, but at the present time the sensitivity and speed of response of these methods are, in most cases, inadequate, as the existence of many of the intermediates is so fleeting that their concentration are too low to be detected.

We will see through several striking examples how the low temperature procedure can be manipulated to stabilize some intermediates, and how some of them are or could be isolated and then used as "pure" reactants to study some elementary and essential reactions which cannot normally be carried out by chemical experiments.

Often the characterization of enzyme-substrate intermediates and the recording of their sequential pattern, as opposed to kinetic data obtained under normal conditions by fast techniques, will permit one to correlate reaction mechanisms in both cases, and the superiority of low temperature procedure is its ability to give spectroscopic and therefore structural information about these intermediates.

ENZYME-SUBSTRATE INTERMEDIATES IN PEROXIDATIC REACTIONS

Spectroscopic recording

Peroxidases catalyze the oxidation of a great number of substrates by hydrogen peroxide. The study of reactions involving horseradish peroxidase (HRP) to detect intermediates was pioneered long ago by Chance (2). Using an "accelerated" stopped flow device and rapid spectroscopic recording, Chance found evidence of two consecutive intermediates, complexes I and II, of reasonable stability. The following reaction scheme was then proposed:

$$\text{HRP} + H_2O_2 \rightleftharpoons \text{complex I}$$
$$\text{complex I} + \text{AH} \rightleftharpoons \text{complex II} + \text{A}$$
$$\text{complex II} + \text{AH} \rightleftharpoons \text{HRP} + \text{A} + 2H_2O$$

Absorption spectra of complexes I and II were recorded, as well as rate constant values of these different steps but, as we will see, such observations did not solve the problems raised by peroxidation reactions, particularly in the presence of oxygen which can interfere in a number of derived reactions. However, absorption spectra could provide basic comparison spectra which could be obtained at subzero temperatures when, by quenching the reaction, the intermediates are stabilized and accumulated to a recordable level.

Reactions were carried out in various mixtures: ethylene glycol–water (50 : 50, v/v, freezing point at $-45°C$), methanol–water (50 : 50, v/v, freezing point $-55°C$), dimethylformamide–water (70 : 30, v/v, freezing point $-68°C$), with and without hydrogen donors such as ascorbic acid, in the absence of oxygen.

Reactions carried out in a mixture ethylene glycol–water gave the following resolutions:

$$\text{t } -40^{\circ}\text{C}, \quad \text{HRP}(\text{Fe}^{3+}\text{—H}_2\text{O}) + \text{H}_2\text{O}_2 \rightleftharpoons \text{Fe}^{3+}\text{—H}_2\text{O (complex I)}.$$

This complex was obtained practically "pure" (98 %) and was stable for several hours. With an increase in temperature to about -20°C in the presence of a hydrogen donor (AH) there was a progressive formation of complex II which was stabilized by cooling at -40°C:

$$\text{complex I} + \text{AH} \xrightarrow{-20^{\circ}\text{C}} \text{complex II} + \text{A}$$

By heating to 0°C, complex II decomposed and transformed into HRP:

$$\text{complex II} + \text{AH} \xrightarrow{0^{\circ}\text{C}} \text{HRP} + \text{A} + 2\text{H}_2\text{O}$$

Thus suitably induced cooling ⇄ warming cycles determined the temporal resolution of the whole reaction step by step (3).

The absorption spectra, optical rotatory dispersion and electron spin-resonance spectra of pure complexes I and II were recorded and absorption spectra were found similar in the various mixtures as well as similar to those obtained by Chance under more normal conditions. These different spectra of intermediates I and II obtained in the mixture dimethylformamide–water are given in Figs. 2 and 3.

The differences between experiments carried out by rapid techniques in water and by the low temperature procedure lie in the fact that with rapid techniques it is necessary to use a large excess of substrate to make the reaction and intermediates recordable, while with the low temperature procedure, such a recording can be performed with stoichiometric quantities of enzyme and substrate.

Kinetic studies were performed in the present studies on complexes I and II: complex I, stabilized at -65°C, could be transformed into complex II by heating to -50°, -40° and -30°C in less than 1 min. Under these conditions, it was easy to follow the conversion between I and II as a function of time by measuring the increase in absorbance (A) at 428 nm (Fig. 2). The corresponding rate constants of this conversion were: 1.3×10^2 M^{-1} s^{-1} (-50°C), 3.3×10^2 M^{-1} s^{-1} (-40°C), 11.1×10^2 M^{-1} s^{-1} (-30°C), compared with 2.8×10^6 M^{-1} s^{-1} in aqueous solution at room temperature.

Optical rotatory dispersion spectra. It was found that the ORD spectrum of the enzyme in the dimethylformamide–water mixture (70 : 30, v/v) at -60°C was similar to that recorded in aqueous buffered solution at $+10^{\circ}$C.

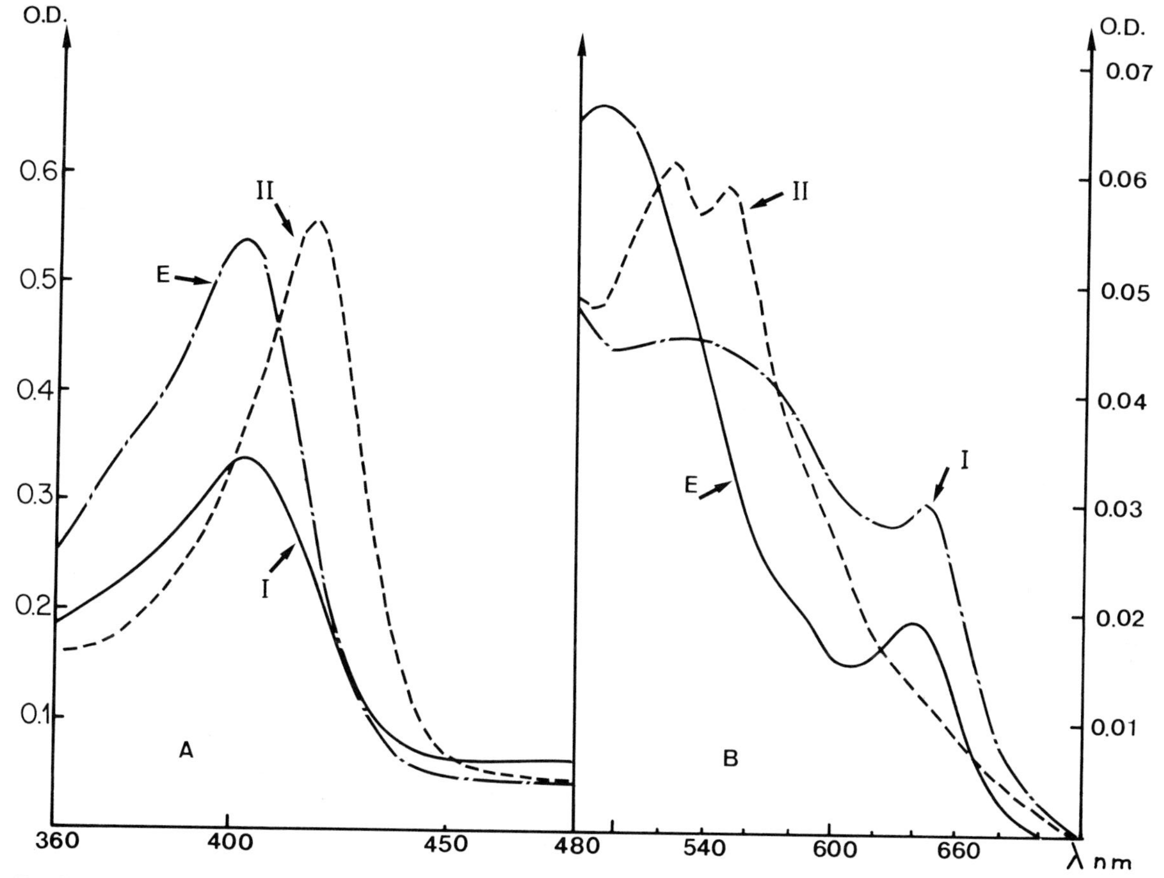

FIG. 2. Absorption spectra of the horseradish peroxidase (HRP) and of complexes I and II in the dimethyl formamide–buffered water mixture (70 : 30, v/v) at −65°C. HRP, H_2O_2 and luminol were 4.45×10^{-6} M. curves A, Soret band; curves B, visible spectra.

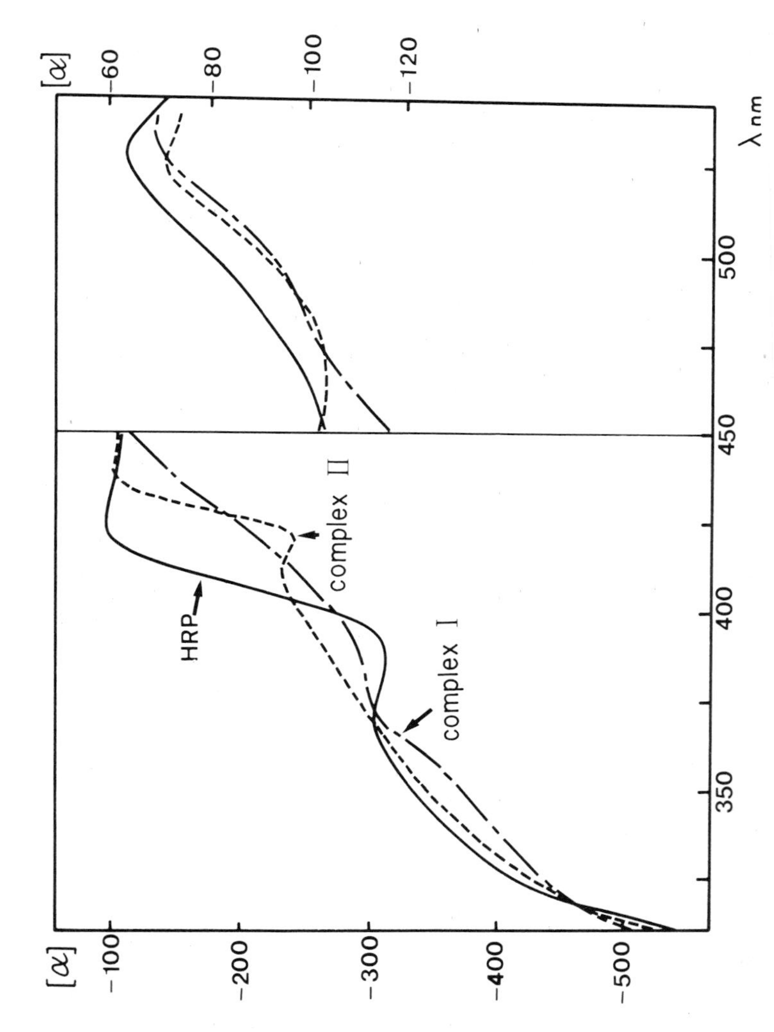

FIG. 3. Optical rotatory dispersion spectra of the enzyme and of the complexes I and II for Soret and visible bands. Temperature, $-60°C$; [α], specific rotation; final concentrations: HRP, 1.17×10^{-5} M; H_2O_2, 1.75×10^{-5} M.

An identical spectrum was obtained in ethylene glycol–water mixture (50 : 50, v/v) at -40°C.

The addition of H_2O_2 to the enzyme solution (dimethylformamide–water at -60°C) resulted in the formation of complex I for which the ORD spectrum was recorded (Fig. 3). Warming up to $+10$°C produced complex II. Upon cooling, complex II was stabilized and its ORD spectrum recorded at -60°C. In both cases, it was possible to check the presence of each complex by monitoring their approximate absorption spectra with the spectropolarimeter.

In Fig. 3, it can be seen that the Cotton effect of the complex I at 403 nm (Soret band) is different from that of the native enzyme and rather similar to that of the apoenzyme. A hypochromic effect is observed in both cases. The Cotton effect appears again with complex II, with a normal wavelength shift and a decrease in magnitude. This result is comparable to data obtained from various ligand forms of peroxidase.

Figure 4a shows the e.s.r. spectra of HRP in a frozen aqueous buffer solution at -196° and in a still fluid mixture of dimethylformamide and buffer (70 : 30, v/v) at -65°C (4). These spectra are very similar. Since the

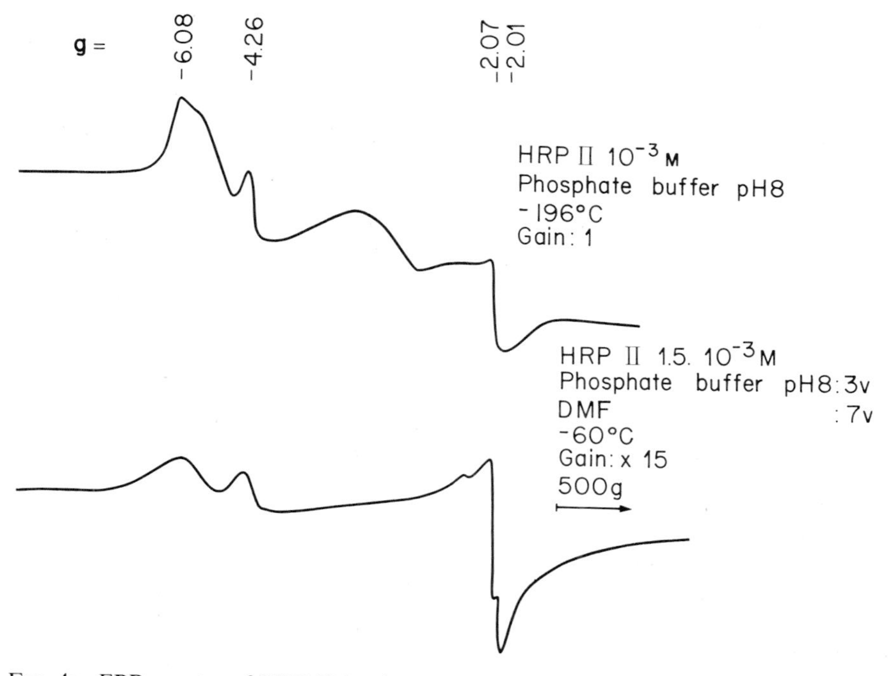

FIG. 4a. EPR spectra of HRP II in phosphate buffer at 77°K, and in the fluid hydro–organic solvent at -60°C.

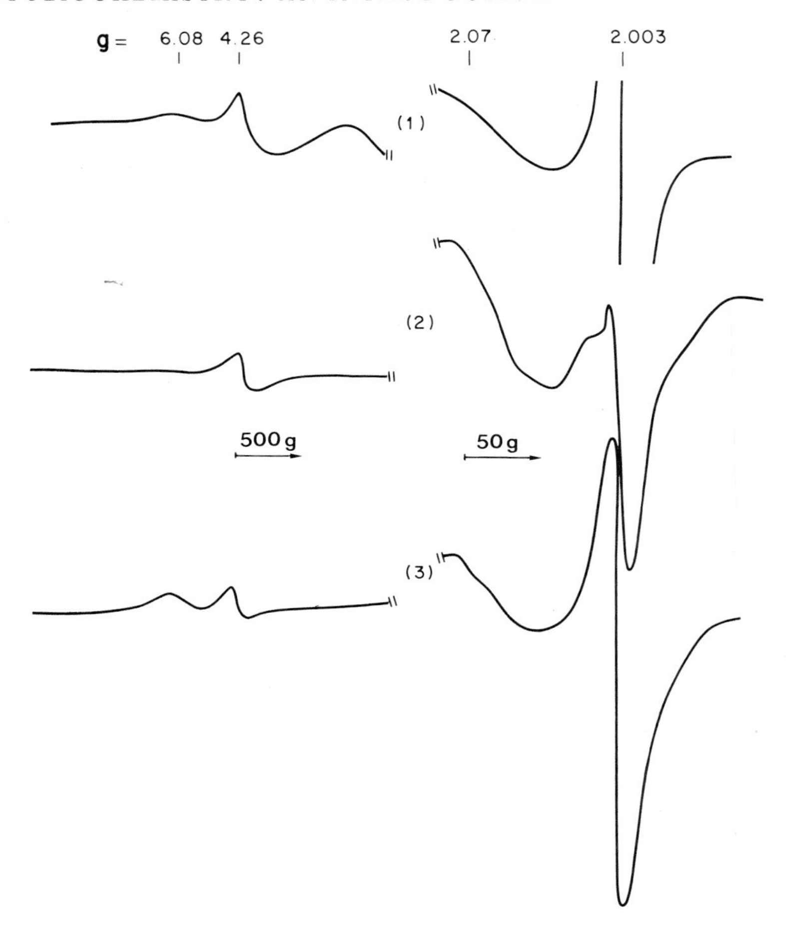

FIG. 4b. (1) HRP II (10^{-3} M) (phosphate buffer pH 8) + H_2O_2 (10^{-3} M): rapid freezing 10 s after mixing; (2) HRP II (10^{-3} M) mixtures DMF–water (70 : 30, v/v) phosphate buffer pH 8) + H_2O_2 (10^{-3} M) and luminol (10^{-3} M) recorded at -60°C; (3) after warming up to -10°C (5 min.) and recooling down to -60°C.

amplitude is temperature-dependent (in fact empirically exponential), the signal from the sample at -65°C is approximately 25 times weaker than that from the frozen solution (where **g** = 6.08).

Figure 4b shows the e.s.r. spectra obtained in the same way during the reaction of HRP with hydrogen peroxide and luminol. Aqueous buffer solutions are frozen 10 s after the mixing of reactants. Hydro–organic mixtures are submitted to cooling ⇌ warming cycles between -65°C and -10°C and the spectra recorded at -65°C.

There are no fundamental differences between the shape of signals recorded in both cases. Nevertheless, due to the temporal resolution obtained in the fluid mixture, it can be seen that the formation of the free radical (**g** = 2.003) appears to occur only during the conversion of complex I to II. The amount of free radical obtained represents in both cases a very low percentage of the original concentration of HRP and also (as seen in the fluid mixture) of the concentration of complex II. Thus, it is confirmed that this long-lived free radical (lasting more than 10 h at $-30°C$ in the absence of free hydrogen peroxide) is a minor component of complex II. Its signal is devoid of any fine structure. At a temperature of $-65°C$ in a fluid solvent, this absence of structure might be due to a localization of the free radical in the protein. Finally, the most useful information gained in fluid solution concerns the evolution of the **g** = 6.08 signal (high spin) during the reaction. From results obtained in the study of the intermediates encountered in catalyses by horseradish peroxidase, we can draw some general conclusions concerning the usefulness of the low temperature procedure applied to enzymic systems.

It should be possible to "resolve" the pathway of a reaction into a number of discrete steps due to the difference in the energy of activation of the various steps as a function of temperature.

The quenching of reactions at a given intermediate stage permits not only the static (non-kinetic) analysis of the individual steps but also a study of the kinetics of two consecutive steps with evaluation of the corresponding activation energy. It is therefore possible to start or stop reactions involving stoichiometric concentrations of enzyme and substrate without having to make the time scale of reactions experimentally accessible by lowering enzyme concentration and increasing the substrate concentration, as in fast techniques. We can thus approach conditions found in nature more closely. The cooling–warming procedure can be used in such a way that a substrate can temporarily behave like an analogue undergoing only part of the catalytic reaction. In fact, the incomplete end product can be ultimately "activated" by a slight increase in temperature to give a completed end product.

Oxidation reactions catalyzed by horseradish peroxidase

If the classical enzyme activity of HRP is the oxidation of donors (AH, AH_2) using H_2O_2 as the acceptor, HRP is also known to catalyze oxidation reactions where O_2 is the acceptor. This remarkable property, first observed in the presence of dihydrofumarate (DHF), and since then with a number of other donors including indoyl acetic acid (IAA), led to the view that peroxidase acts as an aerobic oxidase by means of a ferrous–ferric cycle of valency change as do the cytochromes (5,6,7).

Neither of the three components involved in a peroxidase catalyzed aerobic oxidation, namely O_2, donor and peroxidase, are capable of reacting with each other. It is assumed that the utilization of O_2 as acceptor is due to a reaction with the donor free radical formed in the peroxidation reactions in which

$$\text{complex I} + AH_2 \longrightarrow \text{complex II} + AH^{\bullet}$$
$$\text{complex II} + AH_2 \longrightarrow \text{HRP} + 2H_2O + AH^{\bullet}$$

where AH_2 is the donor and $AH^{\bullet}$ is donor free radical. Thus it is assumed that

$$AH^{\bullet} + O_2 \longrightarrow A + O_2^- + H_2O_2$$

The above reactions, which include, of course, the step

$$\text{HRP} + H_2O_2 \longrightarrow \text{complex I}$$

constitute a chain reaction, chain termination being eventually effected by dismutation of $AH^{\bullet}$

$$2AH^{\bullet} \longrightarrow A + AH_2 .$$

The above reaction scheme composed of chained reactions would require an initial addition of H_2O_2 as free radical initiator.

However, such an addition is not required in the peroxidase catalyzed aerobic oxidation of most donors; the initial reaction rate may be low in the absence of an initiator, but the rate increases with time indicating an autocatalytic reaction. It must therefore be assumed that a reaction step which causes the multiplication of the active intermediates $AH^{\bullet}$ and O_2^- is involved, and could be

$$O_2^- + AH_2 + H^+ \longrightarrow AH^{\bullet} + H_2O_2,$$

O_2^- being generated by the action of some electron donor on oxygen according to an auto-oxidation reaction remaining as yet obscure.

One can consider the above reactions as a branched chain reaction, i.e. as an autocatalytic process presenting a severe obstacle to quantitative kinetic measurements because small unavoidable variations in the initial conditions are amplified so as to cause a very poor reproducibility of the kinetic curves. In a reaction system open to O_2, the peroxidase catalyzed oxidation of DHF or IAA exhibit damped oscillations in the reaction rate as a consequence of the autocatalysis. A synchronous damped oscillation of the concentration of complex III has been found in DHF and also NADH oxidation, but not in indoyl acetic acid (IAA) oxidation.

It is known that peroxidase does not react directly with O_2 to form complex III. The formation of this compound may occur by the following reactions:

$$\text{complex II} + H_2O_2 \text{ (in excess in the medium)} \longrightarrow \text{complex III}$$
$$\text{HRP (ferri. } Fe^{3+}) + O_2^- \longrightarrow \text{complex III}$$
$$\text{HRP (ferro. } Fe^{2+}) + O_2 \longrightarrow \text{complex III}$$

Ferroperoxidase is presumably formed by the reaction

$$\text{HRP (ferri. } Fe^{3+}) + AH^\bullet \longrightarrow \text{ferroperoxidase } (Fe^{2+})$$

Simultaneously, a decomposition of complex III can also take place by the reaction:

$$\text{complex III} + AH_2 \longrightarrow \text{HRP} + H_2O_2 + AH^\bullet \text{ when } AH_2 \text{ is NADH}$$
$$\text{complex II} + AH^\bullet \longrightarrow \text{HRP} + H_2O_2 + A \text{ when } AH^\bullet \text{ is NAD}^\bullet$$

As complex III does not react with DHF or DHF· free radical, there seems to be no reductive way of decomposition in the DHF oxidation but it can be assumed that the rapid decomposition of complex III observed during the depletion of DHF is an oxidative reaction of the type

$$\text{complex III} + H_2O_2 \longrightarrow \text{HRP} + O_2 + H_2O + OH^\bullet$$

or even a reaction between complexes II and III:

$$\text{complex II} + \text{complex III} \longrightarrow 2\text{HRP} + O_2 + H_2O$$

Thus the peroxidase catalyzed aerobic oxidation reaction is mediated by reactive substances such as H_2O_2, O_2^-, and donor free radicals $AH^\bullet$, which are regenerated and multiplied, so that the reaction is autocatalytic, but its intimate mechanism is far from clear.

Support for the above free radical reaction schemes comes from the occurrence of chemiluminescence during the peroxidation of different donors (AH_2) such as NADH, DHF, IAA; chemiluminescence increases strongly in the order NADH > DHF > IAA. It is believed that light is emitted from the intermediates originating from donors, but these intermediates have not been identified. An intense peak of luminescence accompanying the decomposition of complex III during the depletion of DHF reveals an accumulation of excited intermediates during the complex III decomposition.

In spite of a number of observations and suggestions, direct and quantitative data on the reactions involving complexes I, II and II described above are missing because of the difficulty in isolating the reactions of these compounds with various reagents or between themselves.

The low temperature procedure allows one to prepare such compounds as "pure" reactants or, if one prefers, as "precharged" systems which will be triggered by warming up and/or by addition of reagents. As examples of such possibilities, some reactions of pure complex II and III with hydrogen donors, followed by absorption spectroscopy, are reported below:

Complex II. Complex II is prepared from Fe_p^{3+} by action of hydrogen peroxide and IAA in stoichiometric concentrations, in the ethylene glycol-buffer (pH 6.8) mixture (1:1, v/v) at $-10°C$. At the end of several minutes, complex II is free of Fe_p^{3+} and can be considered pure. It is stable at $-30°C$ Under these conditions, the addition of dihydrofumarate (DHF/complex II = 40) results in its decomposition and the formation of complex III (8). The variation of optical density due to the formation of complex III is

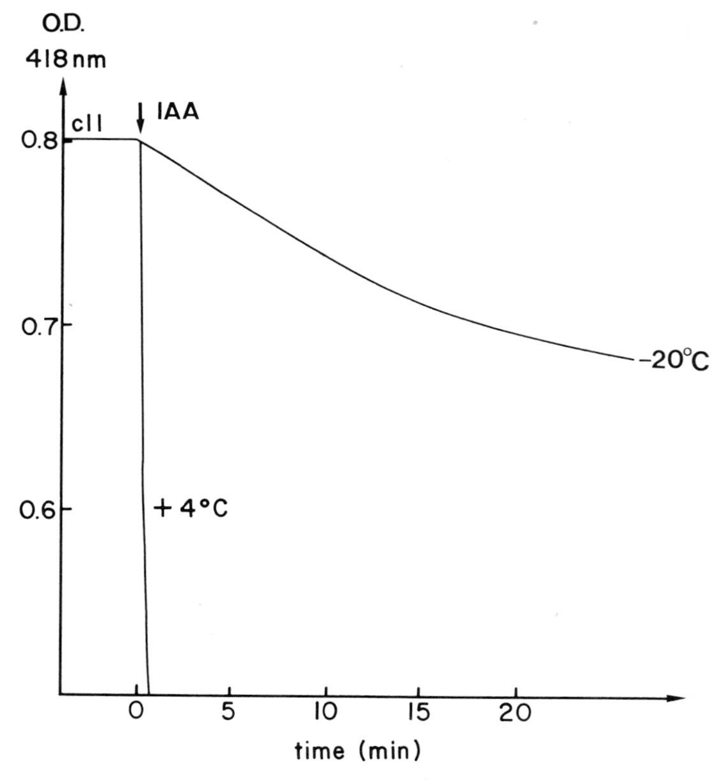

FIG. 5. Kinetics of decomposition of complex II by IAA (ratio IAA/complex II = 100) in ethylene glycol-buffer solution at pH 6.6 (1 : 1, v/v) at $+4°C$ and at $-20°C$ with a constant supply of oxygen.

recorded at 580 nm, and that due to the decomposition of complex II at 526 nm.

Complex II is progressively destroyed upon addition of IAA (ratio IAA/complex II = 50 or 100) at $-20°C$ with a constant supply of oxygen, as shown in Fig. 5. In Fig. 6 the kinetics of decomposition of complex II are shown at different temperatures in the presence of a finite concentration of oxygen in the sample (and in a nitrogen atmosphere in the cell compartment of the spectrophotometer). It can be seen that as the temperature decreases, an increasing amount of complex II is regenerated from the Fe_p^{3+} produced by the initial reaction complex II + IAA + O_2. This process is amplified when an equimolar concentration of Fe_p^{3+} is added to complex II just before the beginning of the reaction (Fig. 7). This is most apparent in the visible spectrum of complex II. The oxygen concentrations

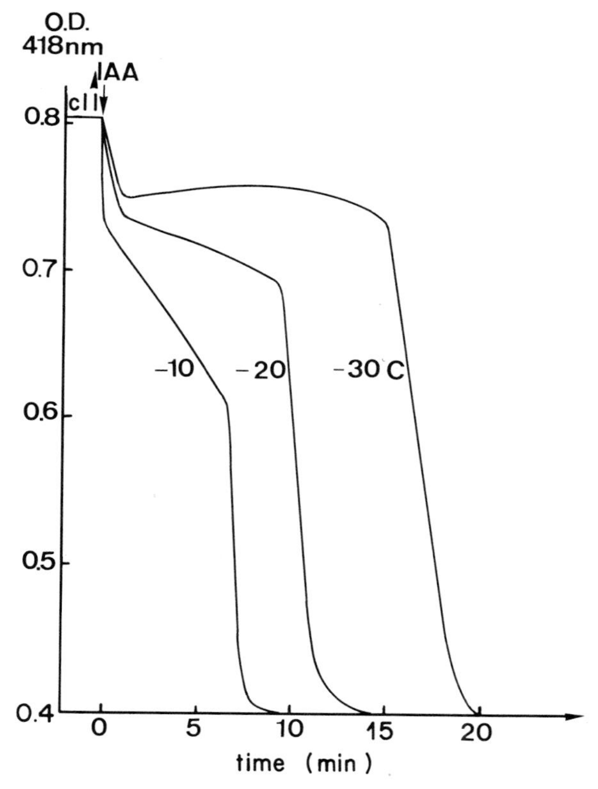

FIG. 6. Decomposition of complex II by IAA (ratio IAA/complex II = 100) in ethylene glycol-buffer solution at pH 6.6 (1 : 1, v/v) at $-10°C$, $-20°C$, and $-30°C$. The solutions initially aerated, are placed in a nitrogen atmosphere.

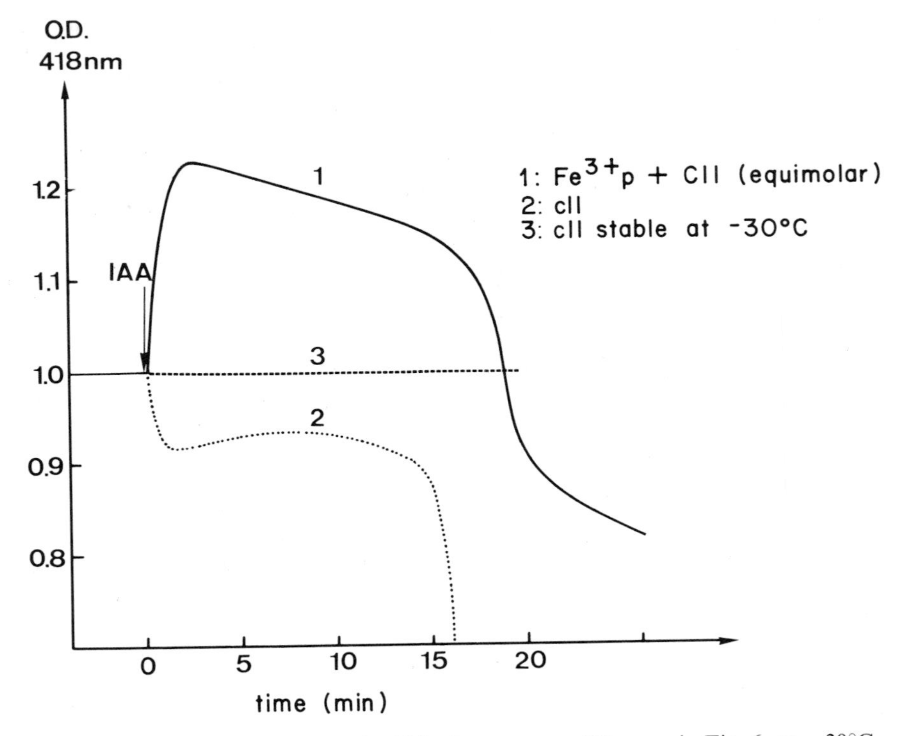

FIG. 7. Decomposition of complex II in the same conditions as in Fig. 6, at −30°C.

were measured at −10°C and −20°C during such reactions in order to confirm that the strange kinetics are mainly due to oxygen consumption. A sudden and brief bubbling of oxygen does, in fact, modify the progress of the reaction, delaying the final and sudden disappearance of complex II.

On the other hand, those kinetics may be useful in luminescence and e.s.r. investigations, for temperatures between −20° and −50°C along the different "segments" of the curves.

Chemiluminescent reactions linked with the decomposition of complexes II and III by hydrogen donors could be studied at selected temperature (−20° → −50°C), for instance, to explore the segments of curves during the reaction

$$\text{complex II} + \text{IAA} \longrightarrow \text{HRP}(Fe^{3+}).$$

Chemiluminescence studies of complex III in presence of DHF and IAA should be continued and could be used as models to explain some bioluminescent reactions which seem to involve a classic peroxidation, or which even require both peroxide and oxygen. For instance the luciferase of a marine hemichordate, *Balanoglossus*, functions as a peroxidase and oxygen is not required; the normal substrate (luciferin) can be replaced by various chemiluminescent components.

Complex III. Complex III of HRP is the most ambiguous with respect to oxidation level and reactivity. This complex reacts both with donors and acceptors, and assuming its structure as (Fe_p^{2+}—O_2), the following mechanisms have been suggested:

$$(1)\quad \text{complex III} \xrightarrow{+el^-} HRP(Fe^{3+}) + H_2O_2 \quad \text{(i.e. complex I)}$$

$$(2)\quad \text{complex III} \xrightarrow{-el^-} HRP(Fe^{3+}) + O_2$$

The best electron donors are ascorbic acid, IAA, phenol, pyrogallol, and the acceptors ferricyanide and *p*-benzoiquinone. DHF is a very poor donor for complex III. It should be noted that one-electron donors such as phenol and ferrocyanide react with complex III as well as two-electron donors. Investigation of equations (1) and (2) should be carried out at subzero temperatures to try to determine some active species such as O_2^- characterized by the reduction of cytochrome c^{3+}, or even luminescence, and presumably involving chain reactions.

Thus, the low temperature procedure applied to complex II provides a method for gathering new data about some reactions which are otherwise difficult to analyze. Complex III can be prepared at −30°C, pH 8.5, in presence of DHF (DHF/HRP = 40) and oxygen.

When complex III interacts with IAA under defined conditions (at −20°C in the ethylene glycol-buffer mixture and in the presence of a finite concentration of oxygen), it is possible to record the rather slow decomposition of complex III which follows a sigmoid-like path and is nearly linear. During the 10 min. of such a reaction, it is easy to take several spectra and to characterize the formation of complex II.

When such experiments are carried out in the presence of a constant stream of oxygen, it is possible to follow the kinetic of decomposition of complex III and the formation of complex II. Complex II can be destroyed by the addition of an excess of IAA (IAA/complex II = 150) and also by progressive warming. At the very end of this reaction a certain proportion of ferriperoxidase (averaging between 10% and 30%) has been destroyed and denatured. In some experiments performed at −30°C, complex II, generated from complex III by the addition of IAA, can be reconverted into complex III by the addition of DHF (DHF/complex III = 50), and so on (9).

INTERMEDIATES OF BACTERIAL LUCIFERASE

Isolation and absorption spectroscopic studies of the oxygenated luciferase-flavin intermediate

An unusual and interesting property of bioluminescent enzyme systems is their ability to create an electronically excited state, and the enzymic intermediates may represent unique states involved in energy storage and conversion, and it is known that in all careful bioluminescent systems the emitting species is an enzyme complex.

Enzymic mechanisms may presumably also be held responsible for the high quantum yields observed, the reaction intermediates being protected in some way against the various types of energy transfer and conversion which would lower quantum yields.

Thus the study of "pure" intermediates would be of interest to explain the mechanisms of the bioluminescence, and Hastings has carried out a number of investigations in this laboratory with a luciferase extracted from luminous marine bacteria.

The reaction *in vitro* of this bacterial system involves the reduced form of the NAD^+, i.e. NADH which reacts with FMN according to the reaction:

$$\mathrm{NADH} + \mathrm{H}^+ + \mathrm{FMN} \xrightarrow{\text{dehydrogenase}} \mathrm{NAD}^+ + \mathrm{FMNH}_2$$

and $FMNH_2$ reacts with O_2 in the following reaction

$$\mathrm{FMNH}_2 + \mathrm{O}_2 \xrightarrow{\text{luciferase}} h\nu \text{ (light)}$$

A long chain aldehyde is a potent stimulant for light emission. The following reaction pathways and intermediates have been proposed for the bacterial reaction; see Fig. 8. Complex II is a long-lived intermediate. In fact, the catalytic cycle, i.e. the time required for a given enzyme molecule to turn over, is unusually long.

When initiated by the rapid mixing of $FMNH_2$ with luciferase in the presence of O_2, the reaction results in an emission which rises to its peak in less than 1 s and then decays exponentially with a half life of 5 to 10 s ($k \simeq 2\ \mathrm{s}^{-1}$ at 25°C). On the other hand the free substrate itself has a lifetime much shorter than the observed emission: free $FMNH_2$ is non-enzymically oxidized within less than 1 s.

Thus the enzyme is exposed to its substrate for a short lifetime compared to the lifetime of the intermediate generated, thus allowing a direct measurement of that lifetime.

The oxidation of $FMNH_2$ by molecular oxygen, catalyzed by bacterial luciferase, involves an enzymic intermediate whose lifetime at 20°C is

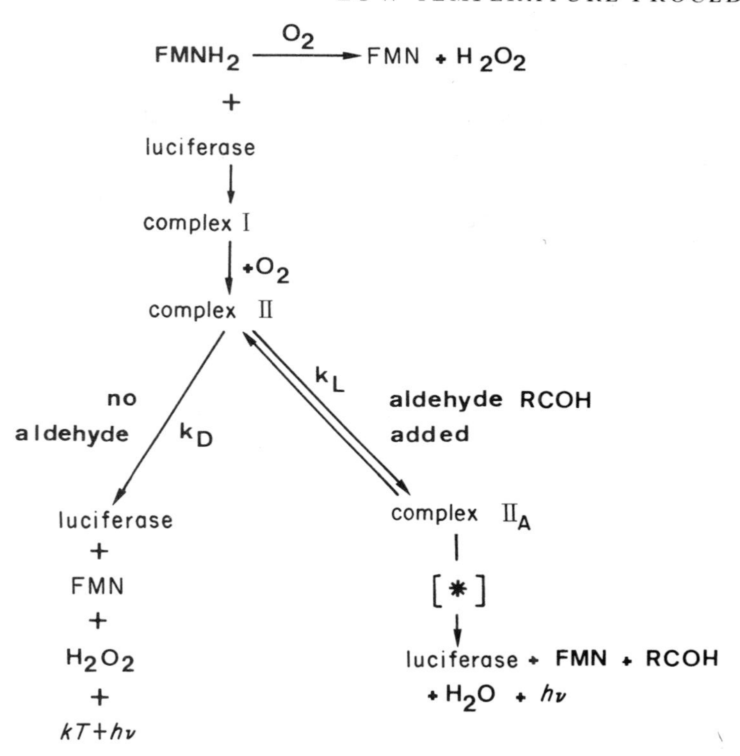

FIG. 8. Hypothetical scheme depicting the pathways and intermediates in the luciferase catalyzed oxidation of $FMNH_2$ by molecular oxygen.

measured in tens of seconds (10). As depicted in Fig. 8, this intermediate (II) reacts further, with concomittent bioluminescence, in the presence of a long chain aldehyde; in its absence, further reaction also occurs, yielding, it is postulated, oxidized FMN and H_2O_2, with little or no light emission. The fact that intermediate II did react with oxygen was deduced from the observation that bioluminescence will occur subsequent to the removal of free molecular oxygen. But in spite of its relatively long lifetime, intermediate II has not been well characterized. Spectral studies are made difficult by the fact that a major fraction of the reduced flavin substrate is oxidized non-enzymatically during the first second, thus remaining in the mixture as background absorption and fluorescence. Even at high enzyme concentrations, for example, where, according to calculation, greater than 90% of the $FMNH_2$ should have been bound, only about 25% was found to be tied up in the complex (11). It was nevertheless concluded that in the intermediate II which is formed, the flavin is not in the fully oxidized state, since concomitant with its emission of light, an increase in both absorption (450 nm) and fluorescence emission (530 nm) occurred.

The most important requirement for a detailed study of the enzyme intermediate is that it should be separated from the free flavin. Since the lifetime of intermediate II at low temperatures ($-20°$ to $-50°C$) is measured in hours or days (12), separation by column chromatography at $-20°C$ was attempted and achieved. Intermediate II was found to have an absorption maximum at 370 nm and a fluorescence emission peaking at 485 nm, consistent with the hypothesis that it involves a flavin-peroxide (13).

The separation of intermediate II from free flavin is shown in Fig. 9. The intermediate was produced by initiating the reaction at $+4°C$, and then trapped by lowering the temperature to $-20°C$. Luciferase (0.8 ml, 10 mg ml^{-1}) without added aldehyde was mixed with 0.2 ml of 2×10^{-3} M catalytically reduced $FMNH_2$ at $+4°C$, both in the 50% ethylene glycol-phosphate buffer. After 10 s, the reaction mixture was cooled to $-20°C$ by transferring it to another tube containing 1 ml of the same buffer at $-35°C$, thereby stopping the reaction and trapping a substantial percentage of the luciferase in the intermediate II stage (12,14). This was promptly applied to the column and eluted with the same buffer at $-20°C$. The isolation and characterization has been repeated many times, with results similar to those described below. The free oxidized flavin (450 nm) is well separated from the protein (280 nm). The latter presumably includes both intermediate II (370 nm) and any unreacted or contaminant protein; the oxidized flavin presumably includes both that which reacted via the non-enzymatic pathway and any which had already been formed as product via enzymatic pathways prior to the time the reaction was stopped.

Aliquots of each tube were assayed for bioluminescence capacity by injecting 0.1 ml into 2 ml phosphate buffer at 22°, pH 7, with 5×10^{-5} M dodecanal. The bioluminescence activity coincides with the protein peak and corresponds kinetically (Fig. 10) to intermediate II, with a decay which is precisely exponential and independent of enzyme concentration (10,15). The absorption spectrum of the peak tube kept at $-24°C$ is shown in Fig. 11. With the exception of the absorption contributed by any protein contaminant or unreacted luciferase, this should represent the absorption of intermediate II itself. According to the postulate of Fig. 8, if the intermediate is then simply allowed to warm to $+20°C$, the reaction goes to completion via the non aldehyde pathway, yielding free oxidized FMN, H_2O_2, and luciferase. The spectrum indicates the production of oxidized FMN, consistent with the postulated scheme.

Subtracting the absorption attributable to the luciferase from these two spectra, one obtains the absorption attributable to the flavin moiety in intermediate II (372 nm peak) and, after warming, the oxidized flavin itself

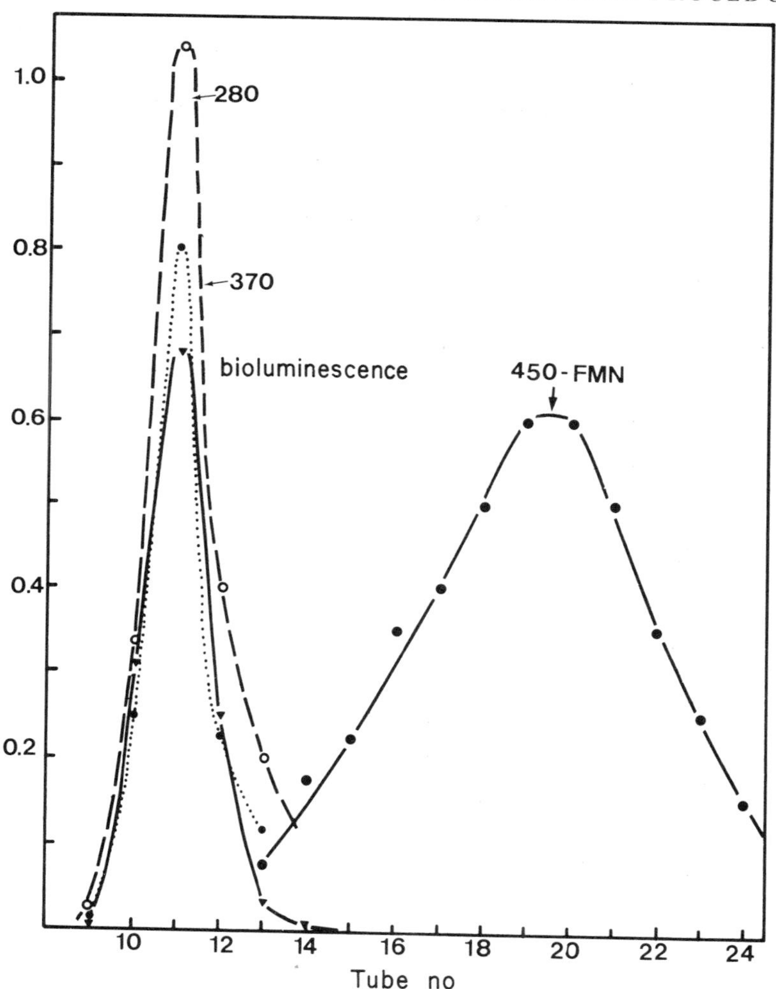

FIG. 9. Chromatography at $-20°C$ of luciferase-flavin intermediate. Ordinate, absorption (280, 370, or 450 nm) and bioluminescence (solid line: multiply by 5×10^{12} to obtain initial intensity in q s^{-1} for a 0.1 ml sample). Abscissa, tube no., 1.9 ml per tube.

(Fig. 12). The spectrum of this flavin is actually nearer that of authentic FMN (16) than was the starting material, which was not a fresh bottle and known to contain impurities, presumably degradation products. Since such impurities presumably do not react with luciferase (17), the isolated product is purified. Based on its absorption at 450 nm (Fig. 12) and assuming a millimolar extinction coefficient of 12.2 (16), the flavin in the peak tube was

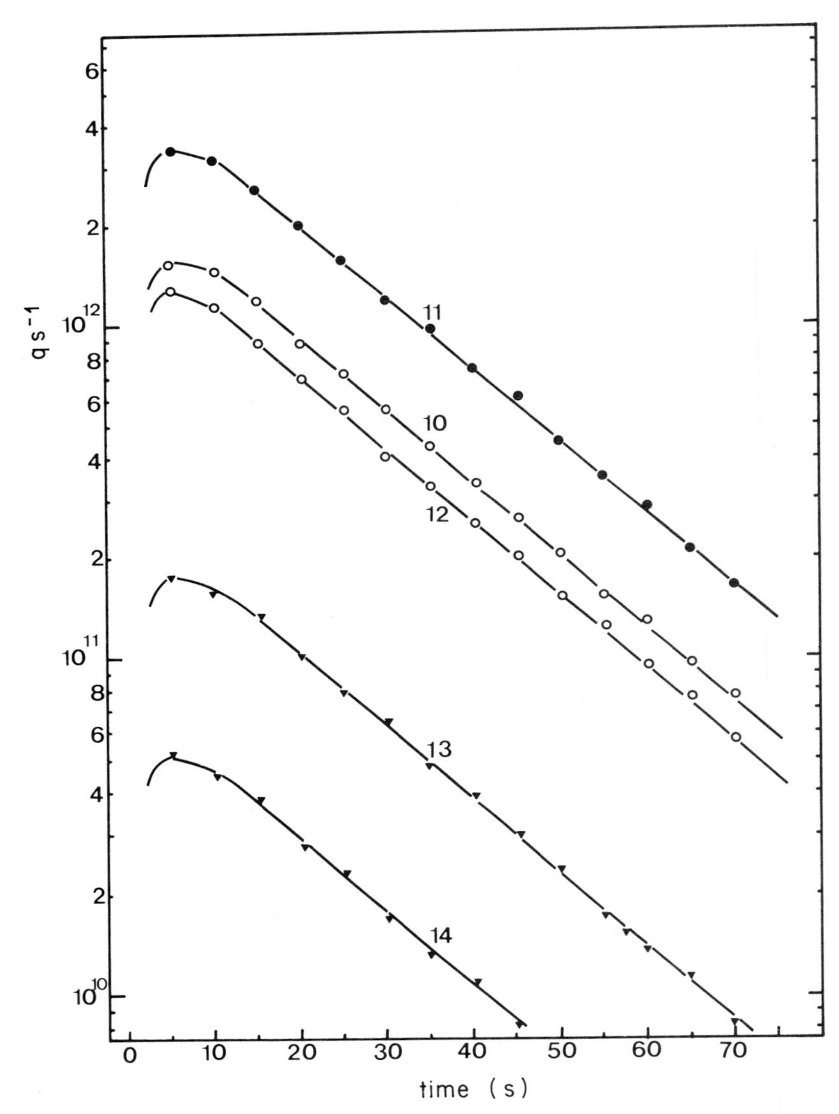

FIG. 10. Time course of the bioluminescence of individual tubes from the column, assayed by injecting into buffer with aldehyde at 22°C.

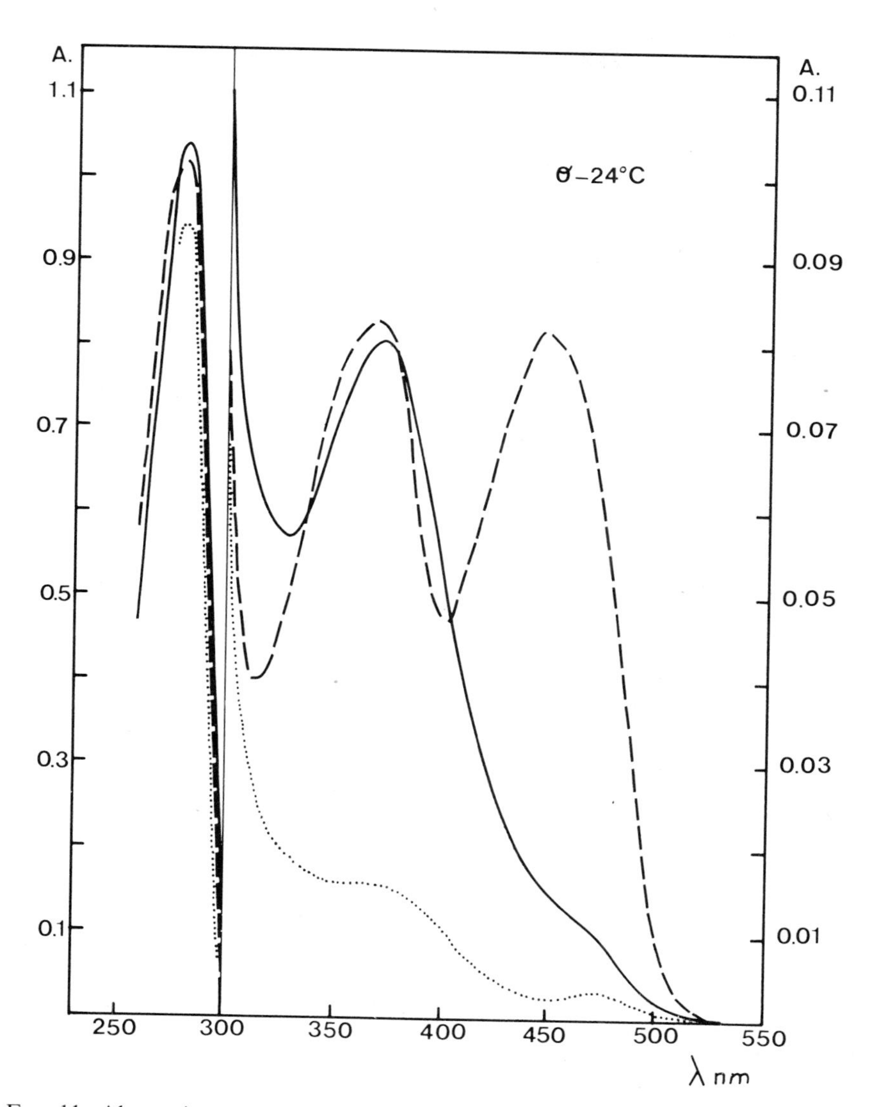

FIG. 11. Absorption spectra, of intermediate II (tube #11) determined at −24°C, both before warming (———) and after warming (– – – –); about 5 min. elapsed between measurements. The absorption of the same amount of luciferase, without additions, at −24°C, is also shown (······).

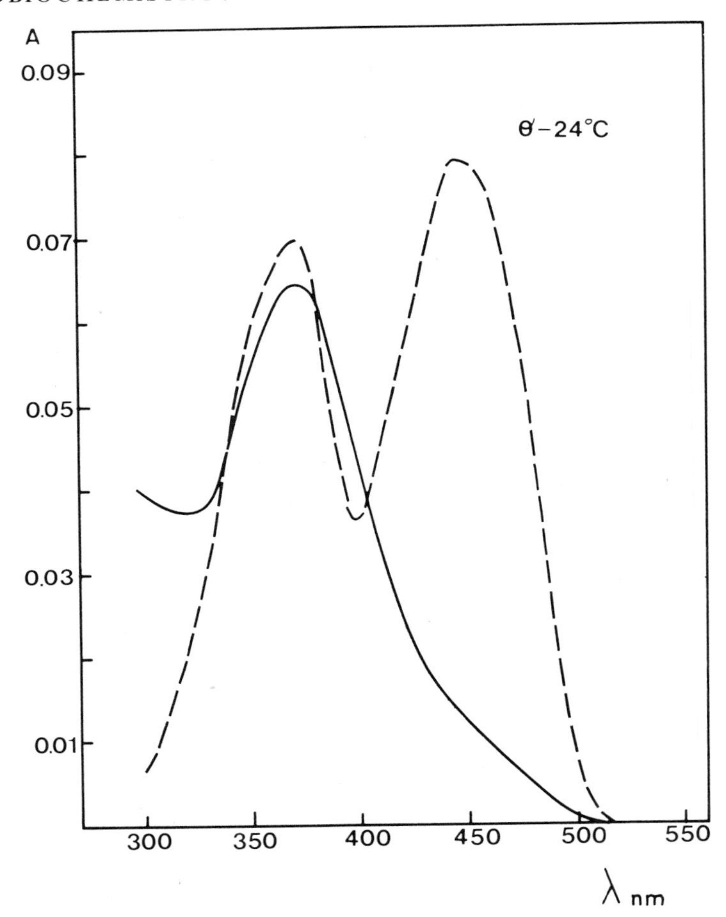

FIG. 12. Difference spectra from Fig. 11 showing the absorption attributable to the flavin moiety of intermediate II (———) peak at 372 nm and the oxidized flavin (− − − −) released after the breakdown of II.

7.5×10^{-6} M. Similar calculations for luciferase, assuming a coefficient of absorption (0.1 %, 1 cm path) of 0.94 (18), and a molecular weight of 79,000 gives 1.2×10^{-6} M luciferase. The excess of luciferase over flavin (1.6 to 1) is reasonable, since unreacted luciferase, or non-luciferase protein impurities in the preparation were not separated from intermediate II in these experiments. Such a separation will be required to determine directly the luciferase–flavin stoichiometry in intermediate II.

The bioluminescence of intermediate II was also slightly better than the starting material, based on a stoichiometry of one flavin per luciferase.

The activity in the peak tube was 3.4×10^{13} q s^{-1} ml^{-1}; based on its FMN concentration (7.5×10^{-6} M) this corresponds to an activity of 5.7×10^{13} q s^{-1} mg^{-1} of luciferase, compared to an activity of 4.5×10^{13} q s^{-1} mg^{-1} for the starting material. Here again one is evidently dealing with a "purified" material, since the calculation is based on only those molecules to which flavin was bound. Adjusting the activity of the starting material for purity, a 20% higher value would give 5.5×10^{13} q s^{-1} mf^{-1}, very close to the measured activity of intermediate II. In a second experiment, the agreement was good but not as close: intermediate II has an activity of 7.8×10^{13} q s^{-1} mg^{-1} luciferase.

If, instead, the activity of intermediate II is calculated on the basis of two flavins per luciferase, which was indicated in the experiments of Lee (19), its specific activity would be 11.4×10^{13} for the first experiment and 15.6×10^{13} q s^{-1} mg^{-1} for the second. A stoichiometry of one flavin per luciferase, as previously reported in other experiments (20,21), is therefore in better agreement with the present work.

The combination of luciferase with reduced flavin was thought to form a complex, intermediate I, which has only a transient existence in the presence of oxygen (10). However, it has recently been suggested by McCapra and Hysert (22) that intermediates I and II are the same, and should both be identified with the luciferase-bound reduced flavin. Therefore, mixtures of reduced flavin and luciferase were prepared, excluding oxygen, and examined spectrally at +24°C and −24°C, to see if the species with absorption at 370 nm, identified above with intermediate II, might in any way be attributable to the luciferase-reduced flavin complex, intermediate I. No indication of this was found. Two ml of a solution containing luciferase (10^{-5} M) and FMN (2×10^{-5} M) with 10^{-3} M EDTA and Dow Corning Antifoam was degassed in an anaerobic cuvette (1 cm × 1 cm) by bubbling with prepurified nitrogen for 40 min, and then subjected to a brief (30 s) irradiation from a 100 W Xenon lamp to bring about the photoreduction of FMN. After absorption measurements at +24°C, the sample was cooled within 3 min to −24°C and similar measurements were recorded. The above procedure was repeated using a lower FMN concentration (4×10^{-6} M) and also by reducing catalytically with hydrogen and platinized asbestos instead of photochemically. In no case was there any indication of a new absorption band in the 300 to 400 nm range. Attempts were also made to determine if there were any changes in absorption when intermediate II is converted to II_A. At −24°C aldehyde was added to a sample from the peak tube (Fig. 9). After 30 min a slight change has occurred, appearing however to be due simply to the production of some oxidized flavin; intermediate II_A appears to react to form product at low temperatures than does intermediate II (12). The sample was then warmed up relatively rapidly, but stepwise, over a period of 15 min. Spectra at several intermediate temperatures ($-5°$, $+8°$, $+15°$C) showed increasing amounts of oxidized flavin. The final spectrum was identical to the one recorded in the absence of aldehyde (Fig. 11). The yield and character of the flavin is thus apparently the same in the two pathways, assuming that most of the intermediate actually reacted via the intermediate II_A pathway. This result also gives no indication of a spectral difference between intermediate II and II_A. But of course, the actual proportion of the intermediate in the II_A form at any given time was not known.

Bacterial luciferase, isolated from *Achromobacter fischeri* strain MAV (15) was purified (23) and stored in the deep-freeze prior to use. Since it was prepared by a shortened procedure, there remained some impurities, especially some with absorbance in the visible range. These were not large enough, however, to interfere with the specific observations. The enzyme's specific activity with $FMNH_2$, determined at the time these experiments were carried out, was 4.6×10^{13} q s^{-1} mg^{-1}, with dodecanal at 22°C. Electrophoresis on acrylamide gels indicated a purity of about 80%. Bioluminescence was measured with a photometer (24) calibrated with the standard of Hastings and Weber (25).

As already described for other enzymes (26), a medium appropriate for studies at low temperatures must be an inert organic solvent mixed with an aqueous system to depress the freezing point, and adjusted to suitable ionic environment and proton activity. For luciferase, a suitable medium in which the enzyme is active and not denatured at low temperatures is 50% ethylene glycol, 0.01 M phosphate buffer, pH 7 (14). Under these conditions it is known that the actual "proton activity" (pH*) of the medium is 7.6 at +20°C and 7.8 at $-20°$C (26) in the range where the quantum yield of the bioluminescence reaction is relatively independent of pH (27).

FMN was obtained from Sigma and was used without further purification. It was reduced catalytically by bubbling hydrogen in the presence of platinized asbestos, in either phosphate buffer or in the ethylene glycol phosphate buffer. Aldehyde (dodecanal) was obtained from Aldrich and dissolved in ethanol to provide a stock solution (0.1% v/v, about 5×10^{-3} M).

Chromatography with Sephadex LH 20 was carried out at $-20°$C in a specially designed column and cold chamber. A general description of the apparatus is given in Fig. 13. The column is provided with a jacket thermostatically controlled by a cooling device. The collector is placed into a cold chamber. Both column and collector can thus be cooled down to $-35°$C. This apparatus is quite different from that designed by Freed and Sack, whose column was permanently surrounded with a bath of diethylketone, maintaining a constant temperature of $-42°$C. The various parts of the apparatus are described in the following sections.

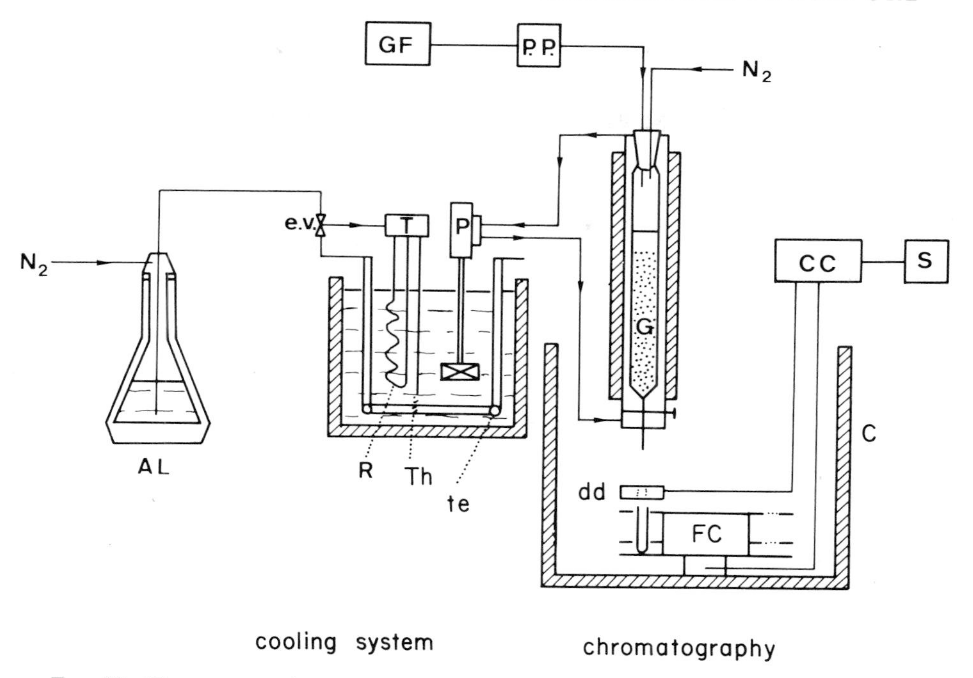

FIG. 13. Chromatography apparatus at low temperatures. AL, liquid nitrogen container; e.v., electrovalves; T, thermostat; R, heating resistance; Th, thermometer; te, thermal exchanger; P, pump; G, gel; GF, gradient flasks; P.P., peristaltic pump; dd, drop detector; FC, fraction collector; CC, collector control; S, security; C, cold chamber.

(*a*) *Column.* The Pyrex column (diameter 2.5 cm, length 35 cm) is provided with a thermostatic jacket within which cooling fluid circulation is allowed. It is surrounded with a glass wool envelope, to minimize thermal exchanges (Fig. 13).

The eluent is a mixture of aqueous buffer and ethylene glycol (volumic ratio 1 : 1). Because of its very high viscosity at −20°C (40 cP), its flow through the column necessitates the use of nitrogen under pressure (1 bar) as well as a peristaltic pump to ensure the required flow rate.

(*b*) *Cooling system.* This device consists of a stainless steel vessel (capacity 10 l) containing the cooling fluid, a methanol–water mixture (volume ratio 7 : 3), whose freezing point is about −100°C. Cooling of the bath is obtained by the transfer under pressure of liquid nitrogen from an appropriate container to a thermal coil exchanger. An electronic regulation system (Unitherm Haake), automatically controls the transfer of liquid nitrogen by means of electrovalves (EV type Asco 8263 A 8P) and a heating resistance. The precision of thermal regulation obtained with such a device is ± 0.1°C inside the column between 0° and -35°C. The temperature within the column is measured by Chromel–Alumel thermocouple and galvanometer.

(*c*) *Fraction collector*. A circular collector, both time and volume controlled (Seive type M 25), is placed in a cold chamber whose temperature is adjusted as required. The temperature regulator is located outside the thermally isolated chamber which contains both column and collector.

The fractions were collected and maintained at -20°C (or lower) during all subsequent manipulations and the transfer to an absorption spectrophotometer (Cary 15) modified as described elsewhere (28), to allow measurements of samples at temperatures as low as -65°C.

The isolation and characterization, using low temperature technology, of the long postulated intermediate II provides support for the proposed reaction scheme and spectral evidence concerning its chemical nature, which is hypothesized to be an oxygenated intermediate in the enzymatic oxidation of reduced flavin. Some years ago, Massey *et al.* (29) suggested that oxidation of reduced flavin might proceed via addition to either its 4a or 1a position. Based on this and other considerations, (13) a luciferase reaction had been postulated with a mechanism starting with oxygen addition to the 4a position to give the peroxide, identified with II of Fig. 8. Evidence concerning the expected spectral properties of such an intermediate (30,31,32) are compatible with the spectra reported here. For example, Jefcoate *et al.* (31) have observed that the spectrum of reduced 1-3-7-8-tetramethylalloxazine is similar to that of reduced FMN; when substituted in position 4a a new band with a peak of 360 nm occurs. This supports (but certainly does not prove) the postulate that the intermediate II with a band at 372 nm is oxygenated in position 4a. It will also be useful to know the fluorescence properties of these model flavin compounds, compared to intermediate II.

Spector and Massey (33) recently reported spectral evidence for an oxygenated intermediate in the reoxidation of the flavin in the reduced enzyme-substrate complex of *p*-hydroxybenzoate hydroxylase. From absorption measurements made during the very early stages in reoxidation (15 ms after mixing), they found a species with an absorption band peaking at about 410 nm, which they believe is the oxygenated intermediate. Since the spectrum is not in agreement with that expected for the 4a position, they suggested that this intermediate could be the oxygen adduct at position 1a. Irrespective of the specific position where oxygen may add, both in this and in the luciferase, if oxygenated species are indeed the intermediate in these two cases, it is quite interesting that the two systems apparently add oxygen in different positions.

With regard to the other postulates concerning the reaction, as set out in Fig. 8, the suggestion that the corresponding acid is produced in the luminescent pathway has received experimental support from several laboratories (22,34,35). The results described here indicate that the product of the reaction is unmodified FMN. This is in agreement with earlier studies showing a yield of more than one photon per FMN in a dynamic system (36). More definitive evidence will need to show that all or the majority of the intermediate reacts via the intermediates II_A pathway, and that the product can be identified unambiguously as FMN.

The results do show unambiguously that intermediate II does not require aldehyde for its formation, and that it can be distinguished from intermediate I. To accommodate the results of McCapra and Hysert (22), one might suggest that in intermediate II $FMNH_2$ and oxygen are bound independently and non-covalently. Although attempts to demonstrate an oxygen binding site on luciferase have not met with success, this cannot be cited as definitive proof, since $FMHN_2$ binding might "induce" oxygen binding. Nevertheless, the spectral band at 370 nm, the fact that intermediate II_A is capable of emission in the absence of free molecular oxygen, and that its combination with aldehyde is reversible, all indicate that oxygen is present in intermediate II in covalent combination, as originally postulated (10), (37).

Use of intermediate II as a "precharged" system to investigate problems of the bioluminescent reaction

Production of O_2^-. It has been found in this laboratory that when an aliquot of a mixed solution of intermediate II stored at $-20°C$ is diluted in an aqueous solution containing oxidized cytochrome c (Fe_p^{3+}), the thermal decomposition of the intermediate yields reduced cytochrome (Fe_p^{2+}). Such a reaction can be explained by the production of superoxide ions, see Fig. 14:

$$\text{intermediate II } (\text{E}\genfrac{}{}{0pt}{}{-\text{FMNH}_2}{\diagdown \text{O}_2}) \longrightarrow \text{E—FMNH}^\bullet + O_2^-$$

and

$$O_2^- + \text{cytochrome } (Fe_p^{3+}) \longrightarrow O_2 + \text{cytochrome } (Fe_p^{2+}),$$

quenching the luminescence. Moreover, E—FMNH$^\bullet$ could react with another molecule of oxygen:

$$\text{E—FMNH}^\bullet + O_2 \longrightarrow \text{E—FMN} + O_2^-$$

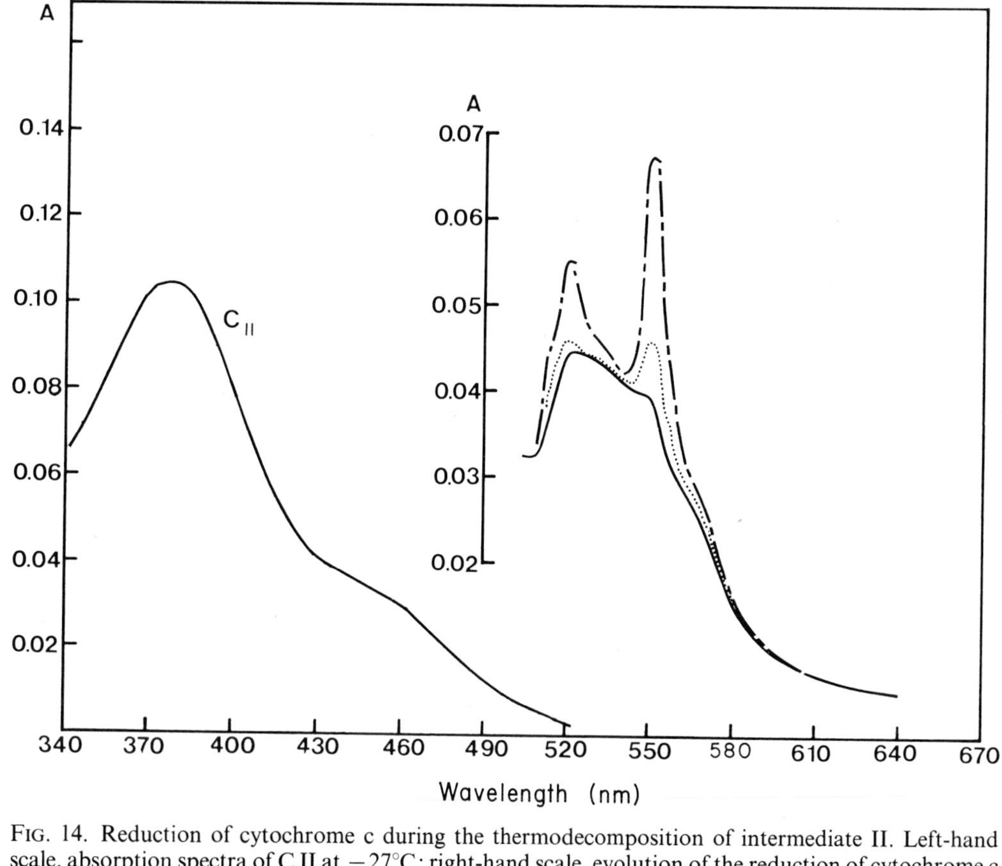

FIG. 14. Reduction of cytochrome c during the thermodecomposition of intermediate II. Left-hand scale, absorption spectra of C II at −27°C; right-hand scale, evolution of the reduction of cytochrome c as a function of time and temperature. ———, after injection of 5 μl cytochrome c 10^{-3} in 1 ml of C_{II}, at −27°C; · · · · · ·, after warming at +22°C; (mixed line), after 15 min. at this temperature.

The production of O_2^- ions could be responsible for the luminescence then observed through the following set of reactions:

(1) dismutation:

$$O_2^- + O_2^- \xrightarrow{2H^+} H_2O_2 + {}^1O_2^*$$

(2) creation of excited molecular pairs of singlet oxygen:

$${}^1O_2^* + {}^1O_2^* \longrightarrow {}^1O_4^*$$

(3) deactivation of the dimer

$${}^1O_4^* \longrightarrow 2\,{}^3O_2 + h\nu$$

Let's mention another possible reaction:

$$E{-}FMNH\cdot + H_2O_2 \longrightarrow E + FMN + H_2O + OH\cdot \ldots \text{chain reactions} \ldots$$

According to the above set of reactions, one can expect a poor quantum yield, even if stoichiometrically

$$E\genfrac{}{}{0pt}{}{-FMNH_2}{\diagdown O_2} + O_2 \longrightarrow E + FMN + 2O_2^- \xrightarrow{2H^+} H_2O_2 + {}^1O_2^*$$

Experiments have clearly shown that there is one, not two, binding sites of $FMNH_2$ per luciferase molecule which consist of two unidentical subunits α and β.

This finding eliminates mechanisms requiring the participation of $2FMNH_2$ molecules for the emission of a single photon. Experiments reported above have shown that intermediate II does not require an additional molecule of oxygen to yield bioluminescence, and it seems therefore that each intermediate II might decompose and give one O_2^- ion and one FMNH· radical. In presence of O_2 and of free $FMNH_2$, these radicals should give autocatalytic reactions similar to those postulated in the case of the decomposition of complex III of HRP:

$$O_2^- + FMNH_2 \longrightarrow H_2O_2 + FMNH\cdot$$

$$FMNH\cdot + O_2 \longrightarrow FMN + O_2^-$$

$$FMNH\cdot + H_2O_2 \longrightarrow FMN + H_2O + OH\cdot$$

net reaction:

$$FMNH\cdot + O_2^- \longrightarrow FMN + H_2O_2$$

These hypotheses could be checked by recording the luminescence yielded by intermediate II decomposed in the absence of free $FMNH_2$ and O_2, in presence of $FMNH_2$ without O_2, and vice versa.

We have seen that long chain aldehyde has a striking stimulatory effect upon light emission: early investigations were convinced that aldehyde is not consumed in the reaction and that its role is catalytic. It was proposed that the effect of aldehyde may be attributed to a conformational change in the long lived intermediate II, thereby favoring the final step reading to the electronically excited state. This proposal was based in part upon experiments carried out at low temperatures.

Since the enzyme intermediate has a relatively long lifetime, it is possible to trap the enzyme intermediate by rapidly cooling the system to the temperature of liquid nitrogen (77°K). The intermediates are stable at this temperature, but upon gradual warming the quanta are released in the form of a glow curve peaking at about $-10°$, similar in pattern to the glow which is obtained in the thermoluminescent emission (38).

If a reaction mixture which has not been exposed to aldehyde is carried through the same freezing and warming procedure, the photon yield is far greater than that obtained when the reaction is carried to completion in the liquid state (12). As this indicated that the physical freezing and warming could in some way spare the aldehyde requirement, the suggestion was made (39) that some analogous physical effect might obtain in the living cell (e.g. luciferase binding to a particular structure), thus meaning that aldehyde is not normally involves *in vivo*.

This interpretation is not easily reconciled with the fact that there are dark mutants which will light up *in vivo* when the colonies are exposed to aldehyde vapor (40). The mutant of Rogers and McElroy was unfortunately lost before a complete analysis could be carried out, but Nealson has recently obtained a number of similar mutants using nitrosoguanidine as a mutagen.

A detailed analysis of the mutants—whether they differ with regard to protein structure or conformation, or aldehyde biosynthesis—should give a valuable new insight into this aspect of the problem.

More recent investigations have shown that aldehyde is converted into the corresponding fatty acid (22). On the other hand, the finding that only a single $FMNH_2$ molecule is involved thereby implicates aldehyde oxidation as an integral part of the bioluminescent reaction, since the oxidation of only one $FMNH_2$ molecule will clearly not provide sufficient energy for photon emission at 490 nm (60 kcal/mol). Consequently a mechanism involving the oxidation of aldehyde and a single $FMNH_2$ molecule would seem likely in the bioluminescent reaction which can be expressed as follows.

$$FMNH_2 + O_2 + RCHO \xrightarrow{\text{luciferase}} FMN + H_2O + RCOOH$$

Thus a model in which aldehyde has only a catalytic role and is not oxidized cannot be reconciled with the energetic requirements in a single $FMNH_2$ mechanism.

However, we must not forget that the photon yield in absence of aldehyde is not negligible, though representing only few per cent, through the freezing and warming procedure.

Further investigation of the intermediate II would be necessary to solve the problem of bacterial luminescence and the experiments suggested above in presence and in absence of free $FMNH_2$ and O_2 might be helpful. Quite recently, it has been found that iso-FMN, FAD and riboflavin, as well as substituted FMN can give intermediate II but with a low photon yield, in the presence of long chain aldehyde. Thus $FMNH_2$ is specifically required to obtain the "real" bioluminescence. These investigations on the long lived intermediate of bacterial luciferase must be considered as only preliminary.

Reaction products. The intermediate II isolated by chromatography at low temperature can be used to investigate the reaction products. As shown in Fig. 8, in the presence of aldehyde the formation of an intermediate designated II_A is hypothesized; in this reaction pathway an excited state (*) is produced, yielding bioluminescence. In the experiments described below, it is possible to measure the amounts of FMN and H_2O_2 produced via the two pathways (dark and light), and then to compare intermediates II and II_A spectrally.

Previous experiments (41) indicated that intermediate II reacts reversibly with aldehyde, the two intermediates (II and II_A) breaking down exponentially at specific rates (k_D or k_L) independent of concentration to yield the different products as indicated. The demonstration that one mole of H_2O_2 is produced per mole of intermediate II upon its breakdown is illustrated in Fig. 15.

H_2O_2 was measured using luminol as an indicator, since under appropriate conditions luminol gives a chemiluminescence proportional to the amount of added H_2O_2 (42). In order to measure H_2O_2 production, aliquots of the individual tubes of intermediates II eluted from low temperature chromatography were rapidly warmed to 20°C, permitting the reaction to go completion without light emission. The concentration of the intermediate which had been present in the peak tube was measured by flavin absorbence at 450 nm after warming (Fig. 15, inset), and found to be 7.5×10^{-6} M, in very close agreement with the amount of H_2O_2 (7.2×10^{-6} M) produced in the absence of aldehyde. In the presence of aldehyde (octanal or decanal), the results show that H_2O_2 is not produced in these conditions.

The flavin products via the two pathways were compared by difference spectrophotometry. Intermediate II at -25°C was placed in each of two matched cuvettes, at this temperature, one in the reference and the other in the sample compartment of an Aminco Differential Recording Spectrophotometer. Difference spectra between the samples after reaction with and without aldehyde gave no evidence of alteration in the flavin, after warming to room temperature, results confirmed by thin layer chromatography on

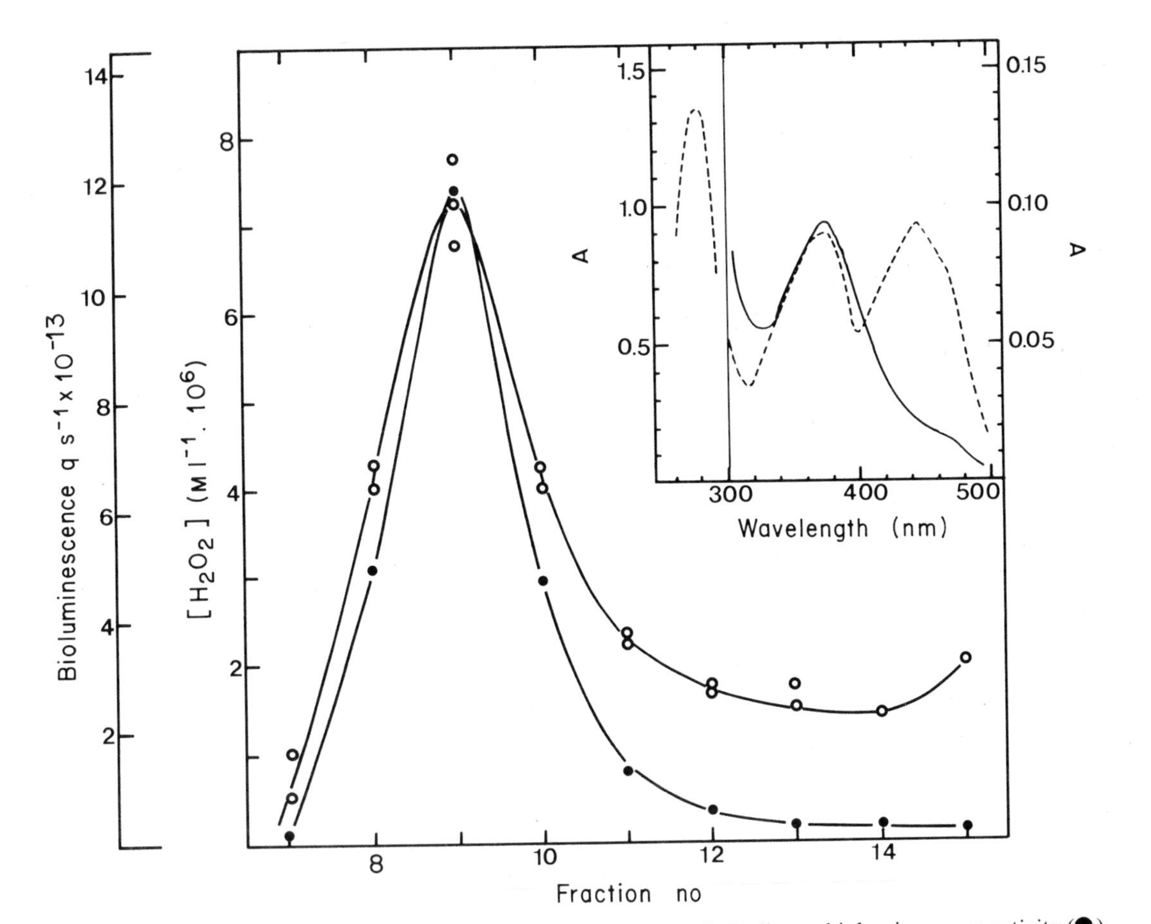

FIG. 15. Production of H_2O_2 after decomposition of intermediate II. Ordinate, bioluminescence activity (●) with dodecanal in units ($\times 10^{-13}$) of q s^{-1} ml^{-1} of sample; abscissa, chromatographic tube number. The spectrum of tube #9 is shown in the inset both before (———) and after (– – –) warming, both measurements at -20°C. The H_2O_2 content of each tube after warming was measured and is plotted (○) in concentration

the flavin produced after reaction. The product of the reaction, both that obtained in the presence and that obtained in the absence of added aldehyde was tested for ability to emit light in the luciferase reaction. The results show that the product is fully capable of turning over a second time.

The demonstration that H_2O_2 is produced by the dark pathway and not by the luminescent one provides direct support for the postulated reaction (Fig. 8). The fact that the luciferase catalyzed oxidation of reduced flavin can occur in either a coupled or an uncoupled mode has an analogy with coupling in electron transport where, as here, the energy in the uncoupled mode is dissipated as heat. The observation that the yield of the product FMN via the two pathways is the same is not surprising, although there is clear evidence that there accumulates *in vivo* an inactive form of luciferase to which some flavin compound is bound (43). The scheme of Fig. 8 postulates the reversible binding of aldehyde to intermediate II, thus the existence of at least two species at this stage, II and II_A. From other studies (44,45,46), it is known that aldehyde and flavin binding have some interdependence. The possibility that aldehyde binding might therefore give rise to spectral alterations in the optical absorbence of II was investigated. At $-30°C$, with intermediate II in each of two matched cuvettes the baseline was adjusted, and ethanolic solutions of octanal or dodecanal were added to the cuvette in the sample compartment of the Aminco Spectrophotometer. The difference spectrum (Fig. 16) of the intermediate in the presence and absence of aldehyde shows that intermediate II_A exhibits a decrease in absorbance which is maximal at 362 nm and an increase which is maximal at 421 nm.

Moreover, it is known that certain other analogs of the fatty aldehydes, including the alcohols, will complete for the binding site on luciferase but are inactive for light production (41). With octanol, no new spectra were observed, so the spectral shift observed with aldehyde must be attributed to interaction between the aldehyde functional group and the protein and/or flavin, and not just to the binding of the fatty residue.

The observations of Nicoli *et al.* (45) and Meighen and McKenzie (46) indicate that the aldehyde and flavin substrates are bound closely together in the active center of the luciferase. The observation of spectral alterations in intermediate II resulting from binding of aldehyde is consistent with this interpretation and affords corroborative evidence for the existence of intermediate II_A species.

On the other hand, the direct isolation of the intermediate with aldehyde (II_A) bound was attempted chromatographically. But, even at $-30°C$, the pure intermediate II_A was not isolated by this procedure (47).

Fluorescence study of the intermediate. The fluorescence emission of intermediate II, measured at $-30°C$, on an Aminco–Bowman Spectrofluorometer

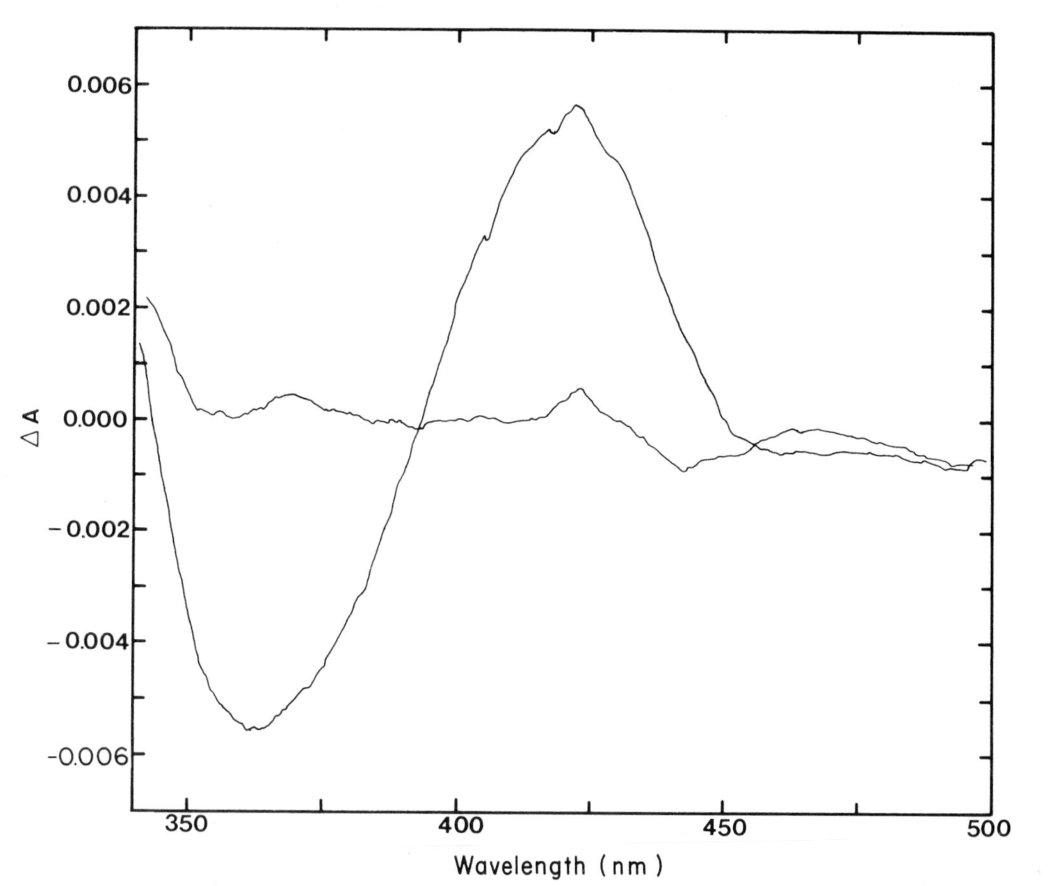

FIG. 16. Difference spectra between intermediates II and II_A. A baseline was established by placing intermediate II in each of two matched cuvettes; octanal final concentration 10^{-4} M was then added to the sample cuvette. After 1 min. the difference in the absorption spectrum of the two was evident.

modified to permit low temperature control of the sample down to as low as $-65°C$ (28), is excited by the single absorption band whose maximum is at about 370 nm. (Fig. 17). The emission spectrum, however, suggests the presence of at least two components, one with a maximum above 500 nm, but not identical with that of oxidized flavin mononucleotide, and the second in the vicinity of 485 nm (48).

With continued irradiation at 370 nm at $-30°C$ the fluorescence emission at 485 nm increases dramatically (five-fold may occur within 10 min.), and the product formed does not decay (rapidly) in the dark. The sample prior to irradiation is referred to as II_x, and afterwards as II_y; are not intended to refer to molecular species, but only to the samples. After irradiation, the emission spectrum shifts to the blue, with a final maximum in the vicinity of 485 nm (Fig. 17). The excitation spectra for II_x and II_y are the same, with a single peak corresponding to the absorption maximum of II. Changes in the absorption spectrum following irradiation at $-30°C$ were thus not expected, and none were observed. The amount of FMN formed was the same whether starting with II_x or II_y upon warming, and the presence of aldehyde does not alter the yield of FMN. The bioluminescence potential of II_x and II_y were also found to be the same, assayed by injection into buffer at 23°C with aldehyde. The addition of aldehyde to II_y at $-30°C$ did not cause a decrease in the fluorescence intensity and did not decrease the bioluminescence potential. The addition of aldehyde at $-30°C$ prior to irradiation to II_x also had no effect upon the bioluminescence potential. Warming the intermediate in the presence of aldehyde, followed by irradiation at 370 nm for 6 min. at $-30°C$, was also directly carried out in the cuvette in the spectrofluorometer. The bioluminescence emission began to occur at about $-20°C$ and rose to a peak at $+4°C$, decaying thereafter. During this period the fluorescence emission at 485 nm declined, while that of the product FMN at 525 nm appeared. The emission spectrum of bioluminescence had been recorded during warming and spectrally it appears to be identical to the fluorescence of II_y (Fig. 18).

The effect of light on II_x is unusual, even spectacular. There are several unresolved questions. For example, II_x is, at the outset, apparently composed of at least two compounds and whether or not one of these is II_y cannot be said. Secondly, it is not known whether II_y occurs uniquely as a result of irradiation, or if it is also a normal intermediate in the pathway to bioluminescence.

The fluorescence of intermediate II is of interest primarily since it is an intermediate in the bioluminescence oxidation of reduced FMN whose identity would be of importance to our understanding of the reaction mechanism. But in addition, its fluorescence emission spectrum after irradiation matches exactly the bioluminescence emission spectrum, so that its

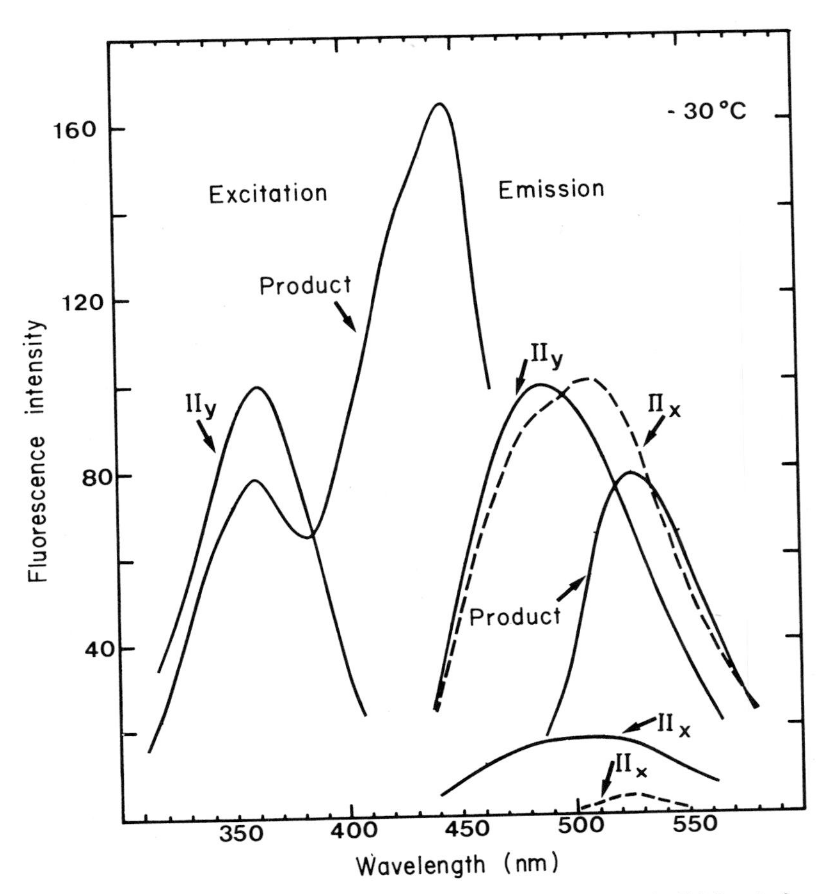

FIG. 17. Excitation and emission spectra for the intermediate II_y (solid lines), formed after irradiation of II_x in the cuvette for 15 min. ($\lambda_{exc.}$ 370 nm $\lambda_{emis.}$ 485 nm). The spectrum of II_x is shown at its correct relative maximum intensity (about 18%, solid line) and also normalized (dotted line) to facilitate comparison with II_y. The fluorescence emission spectrum of product (actually measured by excitation at 450 nm) is plotted normalized to excitation at 375 nm in order to permit a comparison of relative yields of FMN, II_x, and II_y. The product excitation spectrum is also shown. Finally, the amount of 525 nm fluorescence (excited at 450 nm but based on 375 nm) which was present in the initial II_x preparation is shown as a dashed line whose maximum is about 4% of that of the intermediate. This increased to about 6% following irradiation.

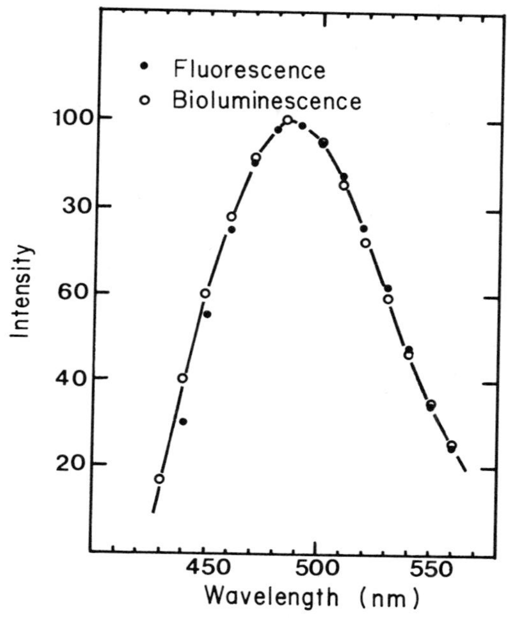

FIG. 18. Emission spectrum of bioluminescence (○), measured directly from the cuvette during warming, plotted together with the fluorescence emission spectrum of II_y (●).

structure is of even greater interest. In spite of the spectral similarity, it seems almost certain that II cannot be identical with the emitting species, for subsequent reaction with aldehydes is required to populate an excited state.

In conclusion one can say that low temperature investigations of bacterial luciferase lead to a number of interesting observations about its function and should be used to investigate other bioluminescent systems.

INTERMEDIATES OF CYTOCHROME P_{450}

Over the past decade, one of the most challenging problems in the field of biological oxidations has been related to the enzymic activation of molecular oxygen. Is O_2 activated by oxidase? What would be the activated form of oxygen and what is the mechanism by which oxygen is activated enzymatically? Or is it that the substrates to be oxygenated are activated by these enzymes and the molecular oxygen merely interacts with the activated substrate?

Spectrophotometric demonstrations that oxygenated forms of enzyme as intermediates in the oxygenase catalyzed reaction exist and might be active forms of oxygen could help to answer the above questions. It can be assumed that in most cases, the enzyme binds with an organic substrate to form an E—S complex and then O_2 binds with the E—S complex to form a ternary complex

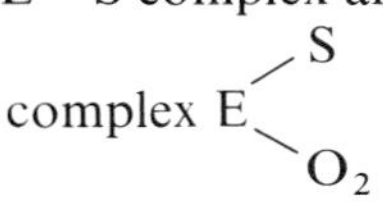

Such a complex must be detected and isolated as a discernible entity to answer the question of whether complex formation and decomposition proceeds in a concerted or a successive manner. The low temperature procedure already successfully applied to peroxidases and bacterial luciferase should be helpful in slowing down the rate of a number of oxidase-catalyzed reactions and to "quench" some of their intermediates. This procedure was applied to the study of microsomal and bacterial cytochrome P_{450}. The reaction of this enzyme presents some striking similarities with the bacterial luminescent reaction and while the detailed mechanism involved is not known, schemes include an "active oxygen" intermediate.

Cytochromes P_{450}, so called because of their strong absorbence near 450 nm when the reduced form of the protein (Fe^{2+}) is exposed to carbon monoxide, bind a substrate, are reduced, and bind oxygen in a cyclic manner and catalyze the specific oxidation of the substrate. Broadly distributed in nature, these b-type cytochromes act as terminal oxidases, and in fact as hydroxylase monooxygenases involved in "detoxification" processes.

The liver cytochrome P_{450}, often simply termed P_{450}, is associated in microsomes with an electron transport chain involving two flavoproteins, the NADPH cytochrome P_{450} reductase and the NADH cytochrome b_5 reductase, a cytochrome b_5 and another possible intermediate termed "X". Whereas the actual pathway of the reactions leading to the hydroxylation of a number of substrates is only postulated, there is no doubt about the formation of transient intermediates at the level of cytochrome P_{450}, including an oxygenated form of the reduced enzyme going to completion specifically through a sequential transfer of two electrons to cytochrome P_{450}.

The bacterial P_{450}, termed $P_{450\,cam}$ for it is specific for the oxidation of camphor, is isolated in high purity from *Pseudomonas putida* with its associated flavoprotein and an iron–sulfur protein. Hydroxylation occurs specifically at carbon-4 of the camphor ketone group to form an exoepimer. Whereas many aspects of the intermediate states of cytochrome P_{450} and putidaredoxin have been described in the literature, the critical questions of oxygen and substrate activation remain largely unclarified. The following reaction pathways and intermediates can be proposed for the bacterial P_{450}, in Fig. 19, where Pd^r and Pd^o are the reduced and oxidized forms of the putidaredoxin, the protein cofactor necessary for the specific oxidation of the substrate camphor.

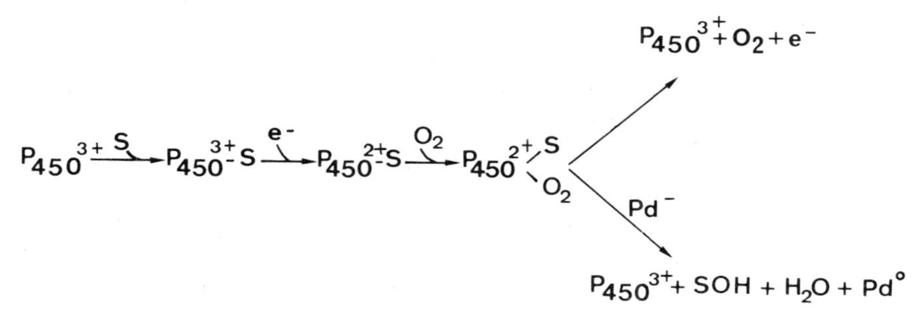

FIG. 19. Reaction cycle of bacterial cytochrome P_{450}.

In spite of its relatively long lifetime, of the order of tens of seconds, intermediate complex II has not been studied as it should be it, that is as a system where decomposition leads to substrate oxidation through short-lived intermediates and/or reactive species. So we decided to stabilize and accumulate intermediate II to use it ultimately as a "precharged" reactant.

Mixtures of ethylene glycol–water and glycerol–water up to 60% (v/v) in cosolvent have been found inert towards cytochrome P_{450} and even protective against the conversion of the microsomal cytochrome P_{450} into cytochrome P_{420}. It has been checked that cooling does not affect the bacterial cytochrome P_{450} and has a protective effect on the microsomal one.

In this laboratory, Debey studied the interaction of carbon monoxide with ferrous microsomal P_{450} (Fe^{2+}) in ethylene glycol–water mixtures at normal and subzero temperatures both by stopped flow (49) and flash photolysis (50), and found that the kinetics below 0°C are in perfect continuity with those observed above 0°C, which are identical in both aqueous solution and the above mentioned mixture, demonstrating the validity of experiments carried out in such unusual conditions.

Intermediates I (Fe^{2+}—S) and II $\left(Fe^{2+}\begin{smallmatrix} \diagup S \\ \diagdown O_2 \end{smallmatrix}\right)$ of bacterial cytochrome P_{450} were then prepared and studied under similar conditions of medium and temperature: kinetic data about the formation of intermediate I were obtained as well as data about the formation, spectra and decay of intermediate II. This intermediate, and the oxygenated form without substrate which is less stable, were "trapped" in cooled solutions. At −40°C, both have lifetimes measured in days, and can be accumulated without any contamination by oxidized P_{450}. Their spectra are reported in Fig. 20. The log of the rate constant of decomposition of the two oxygenated intermediates as a function of $1/T$ is shown in the Fig. 21. It can be seen that activation energies

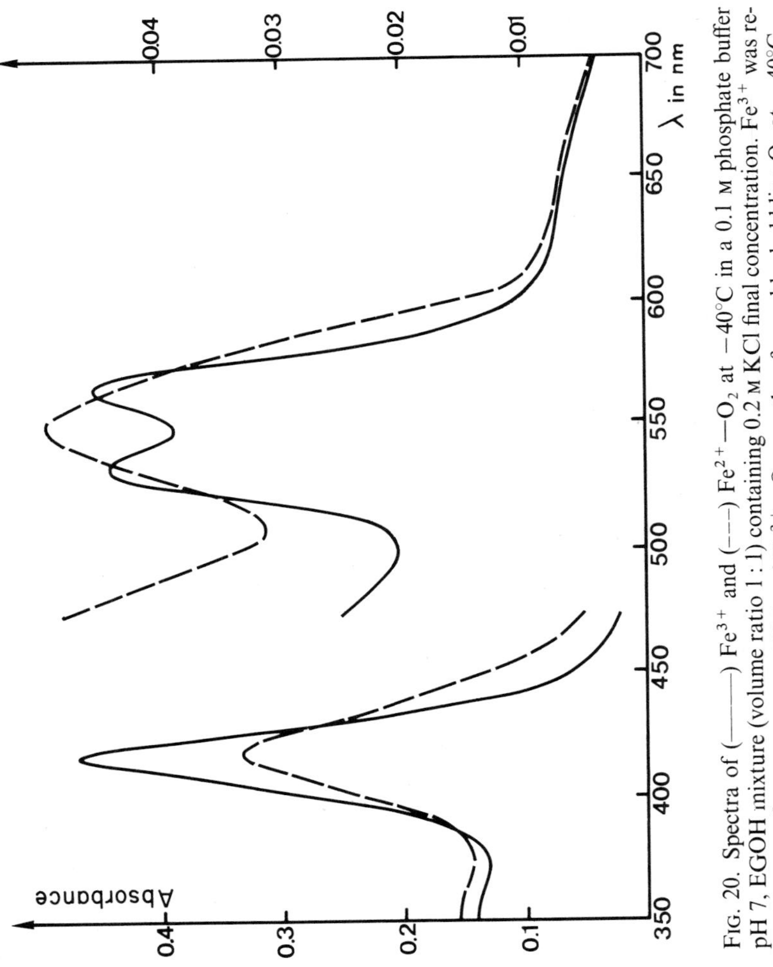

FIG. 20. Spectra of (——) Fe^{3+} and (---) Fe^{2+}—O_2 at −40°C in a 0.1 M phosphate buffer pH 7, EGOH mixture (volume ratio 1 : 1) containing 0.2 M KCl final concentration. Fe^{3+} was reduced by 10^{-5} M $Na_2S_2O_4$ at 10°C and Fe^{2+}—O_2 was thus formed by bubbling O_2 at −40°C.

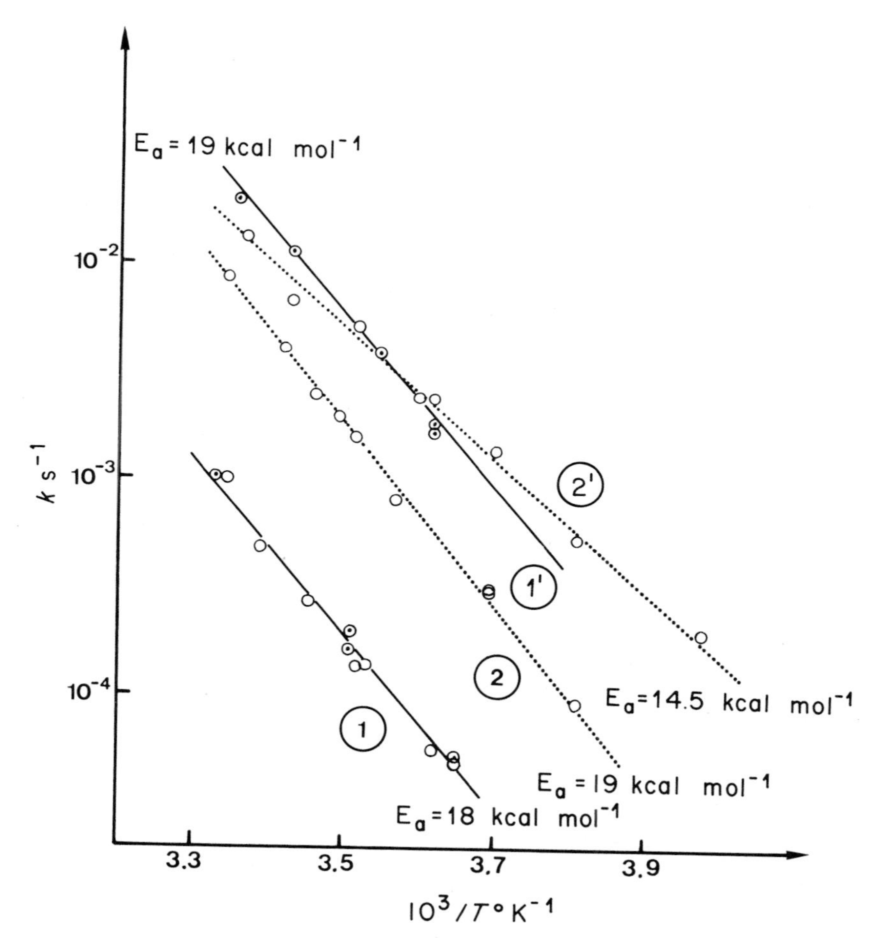

FIG. 21. Arrhenius plots of the autoxidation rate constants k for $Fe^{2+}\langle^{O_2}_{S}$ and $Fe^{2+}—O_2$ in aqueous buffer and EGOH–buffer mixture (volume ratio 1 : 1). (———), (1) $Fe^{2+}\langle^{O_2}_{S}$ in 0.05 M phosphate buffer ○, pH 7.4; ⊙, pH 7 (For clarity, the plot has been arbitrarily translated from one decade lower); (- - - -), (2) Same species in the 0.1 M phosphate buffer pH 7–EGOH mixture (volume ratio 1 : 1, $pH^*_{20} = 7.4$); (———), (1′) $Fe^{2+}—O_2$ in 0.05 M phosphate buffer pH 7.5; (- - - -), (2′) $Fe^{2+}—O_2$ in the mixed solvent described for the curve (2).

of the decomposition of $Fe^{2+}\begin{smallmatrix} S \\ O_2 \end{smallmatrix}$ are practically identical in water and in the mixture (18 kcal mol^{-1}; 20 kcal mol^{-1} respectively) and are only very slightly different from that of $Fe^{2+}—O_2$ (15 kcal mol^{-1}).

The rate of decomposition of $Fe^{2+}\begin{smallmatrix} S \\ O_2 \end{smallmatrix}$ is slightly decreased in the presence of ethylene glycol, an observation which can be compared with that showing that the rate of decomposition of intermediate II of bacterial luciferase $\left(\text{Lase}\begin{smallmatrix} S \\ O_2 \end{smallmatrix}\right)$ is slightly increased in similar conditions and leads to a luminescence in the absence of long chain aldehydes.

In the absence of putidaredoxin, intermediate II $\left(Fe^{2+}\begin{smallmatrix} S \\ O_2 \end{smallmatrix}\right)$ yields superoxide ions O_2^- on decomposing when heated. Let us remember that superoxide ions were found during the thermal decomposition of intermediate II of the bacterial luciferase in absence of long chain aldehydes.

There are a number of problems remaining, such as the eventual formation of a complex between the oxygenated intermediate and putidaredoxin, the kinetics of the reaction yielding oxidized camphor, the formation of oxidative species involved during the evolution of the oxygenated intermediates; they are now under close examination using the low temperature procedure.

CONCLUDING REMARKS

We have seen that a number of enzyme-substrate intermediates can be stabilized in fluid mixed solutions at subzero temperatures and can be analyzed by all available techniques. Other product intermediates can be stabilized in this way and behave like abortive compounds as long as a low temperature is maintained. Let us mention in this context the case of hydrolytic enzymes such as α-chymotrypsin (51,52,53) and lysozyme, the study of which will be described in this chapter.

However, it may happen that the half-time of the reaction is not sufficiently long for physico-chemical studies of the intermediates and transient oxidizing species involved in this last step of the catalytic "cycle" to be feasible. It seems that freeze trapping of such intermediates and transient species will be necessary and a procedure developed recently by chance *et al.* (54), which combines the use of cooled fluid mixed solvents and rapid freezing, might be applied to the trapping of a number of species with short half-times and in too small a concentration during the reaction.

The procedure (low temperature trapping) has been applied to the study of cytochrome oxidase oxygen intermediates and is in fact a "triple trapping" process in which the first temperature drop is to $-30°C$, where the mixed medium is still fluid and in which oxygen added does not exchange with carbon monoxide bound to cytochrome oxidase at a significant rate. Moreover, it has been confirmed that the mitochondrial integrity is adequately preserved in these conditions.

The second stage of the trapping process is to a storage temperature where the sample can be maintained pending continuation of the experiment. Then photolysis of the CO-compound leads to an oxygen reaction which is faster than the subsequent electron transfer reaction below $-90°C$. Rates of electron transfer and oxygen reactions are sufficiently slow between $-50°C$ and $-120°C$ for trapping by rapid cooling to be feasible, leading to trapped states of the system appropriate to studies with "conventional" spectroscopic techniques.

A number of enzyme-catalyzed reactions such as that involving cytochrome P_{450} and some bioluminescent processes could be investigated by a similar procedure, since it is possible to begin from the "last" oxygenated intermediate and then synchronize its mixing with its "effector", the rapid freezing then being able to trap some unstable species which lead to final products.

Kinetic studies of elementary steps

The temporal resolution of a number of enzyme-catalyzed reactions like those mentioned in previous section allows us to record the kinetics of their pre-stationary state and of some of their elementary steps by rapid mixing of reactants and (or) intermediates with a flow-stopped flow device adapted to low temperature measurements and described in chapter 4. Let us give some examples of the kind of information which may be obtained with this procedure.

FORMATION OF COMPOUND I OF THE HORSERADISH PEROXIDASE

So far, very few kinetic studies of enzyme-substrate complexes are available because the half-time of their reactions are too short to be measured by ordinary spectrophotometers even when they show intense and characteristic absorption spectra. Hemoprotein enzymes meet the spectral conditions and some intermediates can be detected by special procedures.

Chance followed the kinetics of the horseradish peroxidase (HRP) reaction by fast techniques and established the rate constant values of its consecutive steps. The rate constant k_2 (complex I $\rightarrow Fe_p^{3+}—H_2O + H_2O_2$) was calculated from k_1 ($Fe_p^{3+}—H_2O + H_2O_2 \rightarrow$ complex I) and from the dissociation constant K of complex I. The determination of the latter lacks sufficient precision except when hydrogen peroxide is replaced by methylperoxide which combines much more slowly with HRP. Moreover, complexes I and II are simultaneously present in the reaction and their respective spectra were more or less contaminated one by the other. At subzero temperatures (for instance between $-30°$ and $-40°C$), it has been possible to isolate complexes I and II.

Using the flow stop–flow device described on p. 175 we analyzed the kinetics of the first step:

$$HRP\ (Fe_p^{3+}—H_2O) + H_2O_2 \underset{k_2}{\overset{k_1}{\rightleftharpoons}} Fe_p^{3+}—H_2O_2\ \ (\text{complex I}) + H_2O$$

As shown in Fig. 22, with stoichiometric concentrations of enzyme and hydrogen peroxide, the reaction can be sufficiently slowed at the appropriate temperature and almost 100% of complex I can be obtained.

The dissociation of complex I is practically imperceptible at low temperatures, as is the formation of complex II from the endogenous donor. Further investigations of the rate constant k_1

$$(E_{HRP} + H_2O_2 \xrightarrow{k_1} \text{complex I})$$

have been carried out for various mixed solvents at different volume ratios and selected temperatures. An example of the results obtained is reported in Fig. 23. It can be seen that k_1 decreases with the concentration of methanol and that this effect is enhanced by lowering the temperature. In the experiments, suitable pH values have been adjusted, it has been checked that there was no influence of the bulk dielectric constant; on the other hand, k_1 variations were not due to viscosity since it is decreasing proportionately with the increasing concentration of methanol.

Activation energy (E) and the enthalpy of activation (ΔH) are shown from the results below.

Methanol	30%	50%	70%	Ethylene glycol 50%
E	4.1	6.4	7.9 kcal mol^{-1}	9.6
ΔH	3.5	5.8	7.3 kcal mol^{-1}	8.9

The second step of the reaction

$$(\text{complex I} + AH \longrightarrow \text{complex II} + Aox)$$

is now being studied as well as the last step

$$(\text{complex II} + AH \longrightarrow Fe_p^{3+}—H_2O + Aox)$$

under the same conditions in order to attain a full description of the reaction pathway with the respective values of heats of formation and activation energies.

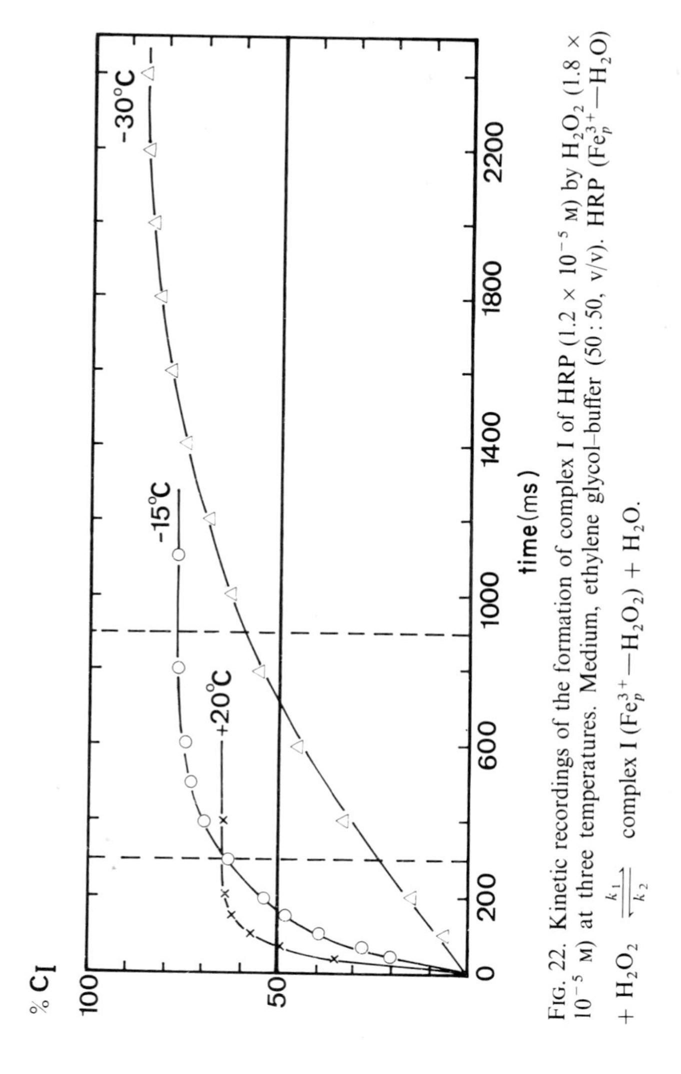

FIG. 22. Kinetic recordings of the formation of complex I of HRP (1.2×10^{-5} M) by H_2O_2 (1.8×10^{-5} M) at three temperatures. Medium, ethylene glycol–buffer (50 : 50, v/v). HRP (Fe_p^{3+}—H_2O) + H_2O_2 $\underset{k_2}{\overset{k_1}{\rightleftharpoons}}$ complex I (Fe_p^{3+}—H_2O_2) + H_2O.

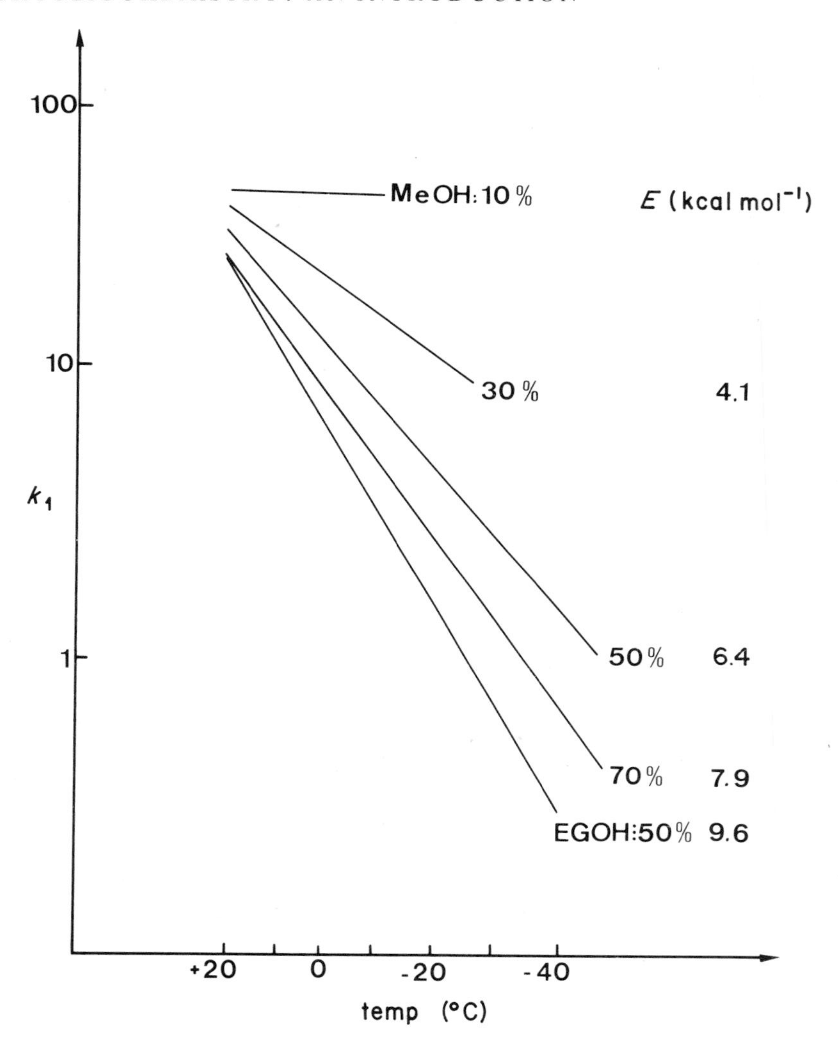

FIG. 23. Evolution of the rate constant k_1 in selected mixtures, as a function of temperature.

SUBSTRATE BINDING BY *Pseudomonas putida* CYTOCHROME P_{450}

As seen in the previous section, the hydroxylation of camphor (S) by activated oxygen occurs in *Pseudomonas putida* via cytochrome P_{450} (Fe^{3+}) which is the terminal oxidase of a three-component electron transport chain. The first step of this hydroxylation mechanism, i.e. the formation of the enzyme-substrate complex (Fe^{3+}—S), was studied by Griffin and Peterson (55), who determined kinetic and equilibrium constants for the reaction

$$Fe^{3+} + S \underset{k_{-1}}{\overset{k_1}{\rightleftharpoons}} Fe^{3+}\text{—}S$$

and found that in aqueous solutions these parameters varied with temperature in an unexpected manner. However, owing to the very rapid binding of camphor, the direct measurement of the association rate constant was possible only at 4°C and its values at higher temperatures were deduced from indirect determinations. Direct measurement of the association and dissociation rate constants of camphor (S) with cytochrome P_{450} (Fe^{3+}) has been carried out in this laboratory at subzero temperatures (56).

The binding of camphor was studied between +30°C and −20°C in the mixed solvent; Fig. 24 shows kinetic traces obtained at 418 nm at 4°C and −18°C, the final concentrations of camphor being respectively 7.13×10^{-5} M 1.25×10^{-4} M, which are large enough, compared to the concentration of cytochrome P_{450} to give apparent first-order kinetics. Furthermore in these conditions of solvent and temperature, the kinetics were slow enough

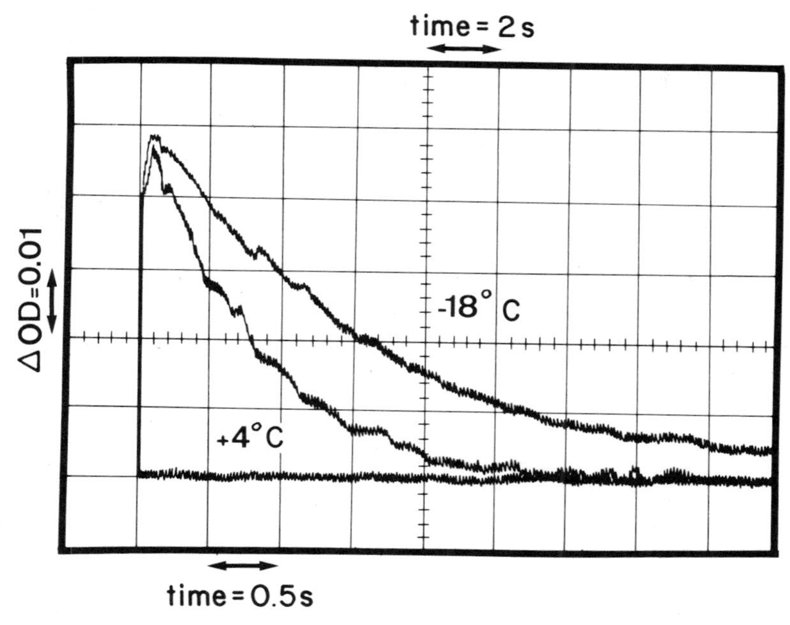

FIG. 24. Kinetic traces of the camphor binding to ferric cytochrome P_{450} in 100 mM phosphate buffer pH 6.5–EGOH mixture containing 100 mM KCl, final concentration. The formation of the P_{450}–camphor complex is measured by the decrease of absorbence at 418 nm. $[P_{450}] = 1.25 \times 10^{-6}$ M, $[S]_{+4°C} = 7.13 \times 10^{-5}$ M, $[S]_{-18°C} = 1.25 \times 10^{-4}$ M.

TABLE 1. Experimental values of association and dissociation rate constants and calculated values of the equilibrium constant for camphor binding to cytochrome P_{450} at different temperatures in 100 mM phosphate buffer pH 6.5–EGOH (volume ratio 1 : 1); [KCl] = 100 mM final concentration.

temp (°C)	+30	+25	+20	+15	+10	+5	0	−5	−10	−15	−18
$k_1 \times 10^{-3}\ \text{M}^{-1}\text{s}^{-1}$	170	120	88	64	45	29	13.2	6	3	1.2	0.86
$k_{-1} \times 10\ \text{s}^{-1}$	26	20	14	9.6	6.6	4.5	2.7	1.7	1.18		
$K_{eq} = k_1/k_{-1} \times 10^{-4}\ \text{M}^{-1}$	6.54	6	6.3	6.67	6.8	6.45	4.88	3.53	2.54		

compared to the dead time of the stopped-flow apparatus (5 to 10 ms), to permit the direct measurement of the binding constants over our entire temperature range. The results are summarized in Table 1; at 4°C the rate constant in 50% EGOH is $1.85 \times 10^4\ \text{M}^{-1}\ \text{s}^{-1}$ which is approximately 200 times lower than the corresponding value in water:

$$(k_{1\ \text{water}} = 4.1 \times 10^6\ \text{M}^{-1}\ \text{s}^{-1}).$$

The dissociation kinetics of camphor from the Fe^{3+}—S complex were obtained by trapping the free ferric cytochrome with metyrapone (MP) which competitively binds to it in a very fast and essentially irreversible reaction.

Then the reaction sequence is:

$$Fe^{3+}\text{—S} \xrightarrow{k_{-1}} Fe^{3+} + \text{S}$$

$$Fe^{3+} + \text{MP} \xrightarrow{k_2} Fe^{3+}\text{—MP}$$

the first reaction being rate limiting.

Figure 25 shows the first order kinetics of the formation of Fe^{3+}—MP (absorbence increase at 422 nm) after mixing the saturated Fe^{3+}—S complex

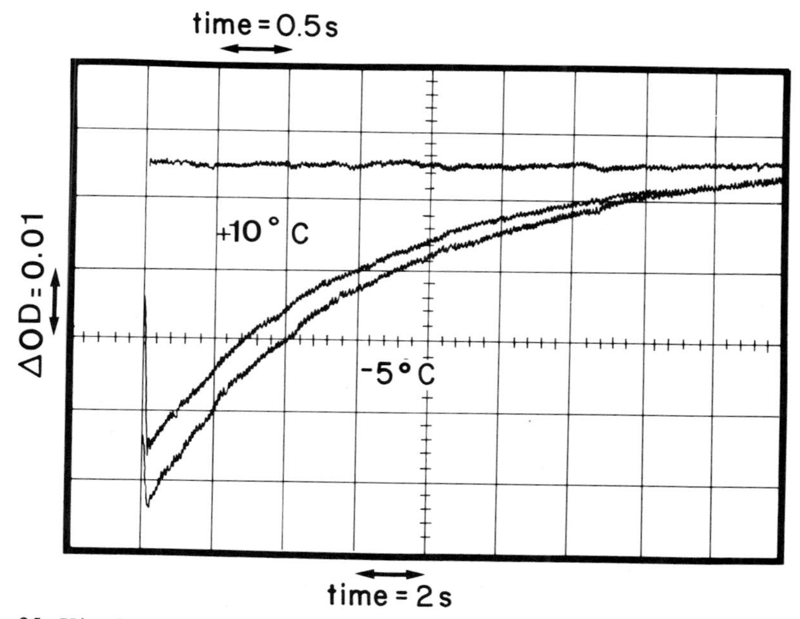

FIG. 25. Kinetic traces of the camphor-dissociation reaction in the same solvent as Fig. 24 in presence of an excess of metyrapone (10^{-1} M). The formation of the Fe^{3+}–MP complex is followed by the increase of absorbence at 422 nm. $[P_{450}] = 10^{-6}$ M $[S] = 10^{-4}$ M at +10°C and −5°C.

with an excess of MP (10^{-2} M to 10^{-1} M) in the mixed solvent. We have verified that the reaction order and rate constant are independent of MP concentration between 10^{-2} and 10^{-1} M under our conditions of medium and temperature.

The Arrhenius plots of the experimental rate constants are shown in Fig. 26. The logarithm of k_{-1} depends linearly on $1/T$ from $+30°C$ to $-20°C$ with $\Delta H^{*}_{-1} = 12.5 \pm 0.5$ kcal mol^{-1}. However the Arrhenius plot for k_1

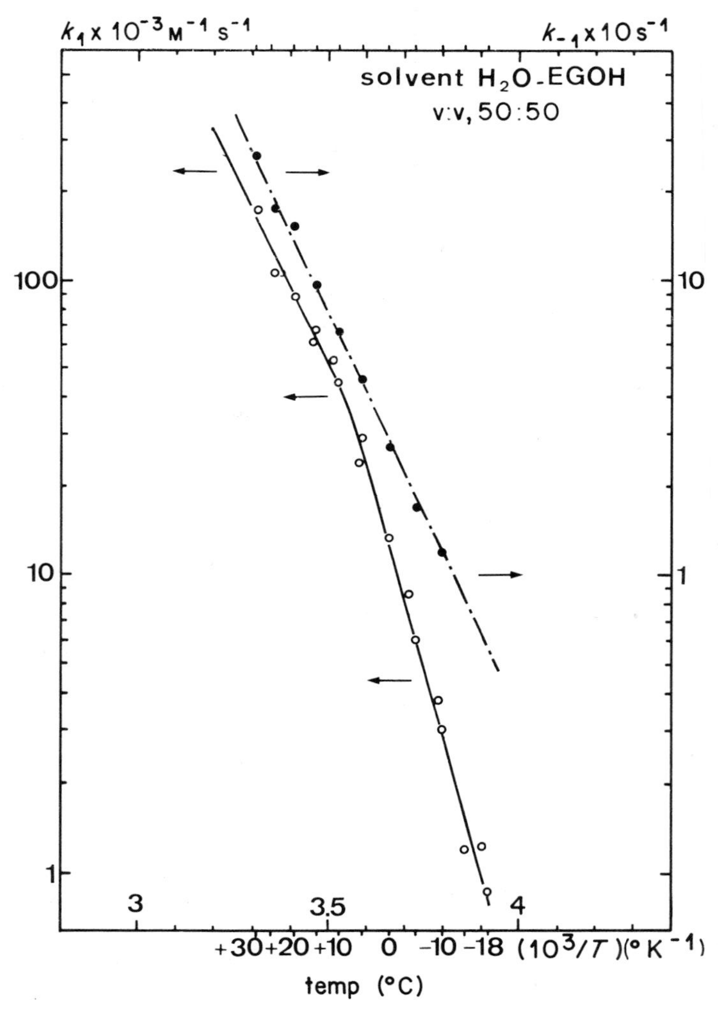

FIG. 26. Arrhenius plots of the experimental rate constants for the association and dissociation of camphor and ferric cytochrome P_{450} in the mixed solvent (see Fig. 24).

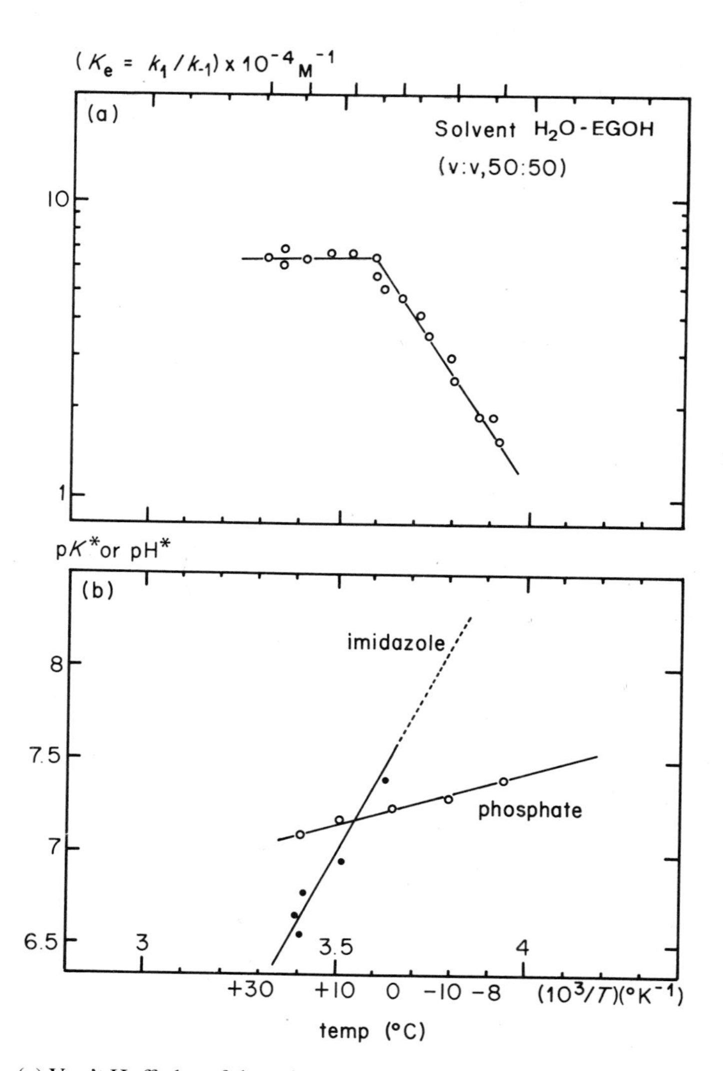

FIG. 27. (*a*) Van't Hoff plot of the calculated equilibrium constant for camphor binding to cytochrome P_{450}. (*b*) Temperature dependence of the pK* of imidazole (p$K^*_{+20°C}$ = 6.65) and the pH* of phosphate buffer (p$H^*_{+20°C}$ = 7.1) in the 1 : 1 volume ratio water–EGOH mixture. The enthalpies of ionization of imidazole and phosphate are respectively 16 kcal mol^{-1} and 2.6 kcal mol^{-1}.

consists of two linear portions of different slopes, with a break near 8°C. The calculated values of ΔH_1^* above and below the break are respectively 11.8 ± 0.5 and 21 ± 0.5 kcal mol^{-1}.

The equilibrium constant $K_e = k_1/k_{-1}$ may be calculated from the experimental values of the rate constants. As shown by Fig. 27, it is exponentially dependent on $1/T$ from -18°C about $+8$°C; the enthalpy of the reaction is 8.3 kcal mol^{-1}. There is a break in the Van't Hoff plot near $+8$°C, above which the equilibrium constant is temperature independent. This unusual temperature dependence of the equilibrium constant deduced by calculation from experimental rate constants was confirmed by direct measurement of the equilibrium constant at different temperatures and in the same conditions of medium. However, the results obtained by these two methods differ in the position of the temperature break (-5°C) and the value of ΔH at low temperatures ($\Delta H = 5.6$ kcal mol^{-1}) (56).

The above observations show that the most immediate effect of EGOH on the interaction between cytochrome P_{450} and camphor is to decrease both the association and, to a lesser extent the dissociation rate constants. If the driving force for camphor binding is the removal of camphor from an aqueous environment to an hydrophobic region of the protein, the higher hydrophobicity of EGOH as compared to water should decrease the free energy of solvation of camphor, and thus decrease the driving force for association. This could explain the observed reduction in the association rate and the affinity for camphor cytochrome P_{450}. On the other hand both association and dissociation rate constants have very similar temperature dependence in the presence and absence of organic solvent, suggesting that the solvent does not affect the conformation of the hydrophobic pocket of the cytochrome. From a practical point of view, the reduction of k_1 by a factor of about 200 at 4°C in the presence of 50% EGOH allows a direct measurement of the rate constants between 30°C and -20°C.

Kinetic investigations of a number of elementary steps are made possible by the temporal resolution of overall reactions at subzero temperatures, and should lead to a full description of the reaction pathway, with the respective values of heat of formation and activation energies. Direct observation of the rate of some reactions at the molecular level is therefore foreseeable with the low temperature procedure.

Protein and enzyme-substrate crystallography at subzero temperatures

D. C. Phillips was the first to realize that a detailed knowledge of the physicochemical properties of mixed solvents would allow conditions to be chosen to maximize the similarity between the microenvironment of protein crystals and that provided by the normal mother liquors; and further that cooled fluid mixtures would allow the diffusion of substrates into the crystals in the cold with subsequent stabilization. Direct observation of productive enzyme-substrate intermediates could then be carried out.

Earlier attempts had been made to set up X-ray crystallography of proteins at subzero temperatures, including some assays by Low and colleagues (57) with orthorhombic insulin crystals carefully cooled to $-13°C$ in 0.7 M citrate and by Haas (58), who added glycerol to orthorhombic hen egg white lysozyme crystals in 2% NaCl at pH 9.5, but since the crystals dissolved when exposed to 10% glycerol, it was necessary to cross-link with glutaraldehyde to render them insoluble. Moreover, the relatively high freezing point of the mixture meant that a glass state was nearly always obtained below $-20°C$ and therefore substrate diffusion was impossible. Sucrose has been selected as a compound which would be likely to prevent ice formation (59) but unfortunately caused some intensity changes, and on freezing, the crystal unit cell dimensions decreased by an amount roughly proportional to the sucrose concentration. For phase determination in protein crystallography by nuclear γ-resonance scattering (60) at 77°K, Mössbauer and his colleagues attempted to overcome the formation of ice I in the crystal lattice by freezing spermwhale myoglobin crystals at a hydrostatic pressure of 2500 atm. They reasoned that the ice III phase, which exists at that pressure, could be formed without damage to the crystal since its volume is less than water. After immersing the crystals in isopentane to remove surface water they succeeded in freezing them at 77°K in a hydraulic press. They were able to collect data at ordinary pressure since the "ice" phase formed was metastable at low temperatures. The crystals diffracted strongly at 2.2 Å with no apparent increase in mosaic spread (61).

Despite the attraction of this purely physical technique, it poses considerable problems as a general method. The equipment required is complex and difficult to handle. A fluid condition does not result so substrate diffusion experiments are impossible, and the process produced some unit cell dimension changes. Furthermore, the frozen crystals were totally destroyed on warming to 120°K at atmospheric pressure, prohibiting cooling–warming cycles without pressure cells. Finally a number of questions remain about this technique, chiefly the exact nature of the mother liquor when frozen. The X-ray photographs taken by these workers do not show the characteristic diffraction lines of high-pressure ice crystals, suggesting that rather than ice III, some amorphous phase is formed. The effect of this phase on protein conformation and enzyme activity is still unknown.

A new approach which allows carrying out protein crystallography at subzero temperatures without major changes of unit cell dimensions and allowing substrate diffusion experiments has been set up by Petsko (62). The basic feature of this technique is the replacement of the normal crystal mother liquor with a cryoprotective salt-free aqueous organic mixture of high cosolvent concentration and low freezing point, using the physicochemical data now available to control any changes in the essential properties of the system under investigation.

PREPARATION OF CRYO-PROTECTED CRYSTALS

Proteins crystallized from alcohol–water mixtures

These crystals are already in a "cryoprotective" mother liquor of low freezing point, but to achieve very low temperatures without freezing, the concentration of alcohol must be high, i.e, >50%. For proteins crystallized from alcohol–water, the alcohol concentration could be raised to any desired level without cracking the crystals by coupling the progressive addition of alcohol to the progressive lowering of the temperature. The relative amounts of each were chosen to keep the dielectric constant of the mother liquor as close as possible to its original value. This experiment could be performed in a cold room, or on the diffractometer by the use of a polystyrene flow cell, the crystal being held in place by Sephadex chunks.

Proteins crystallized from salt solutions

These crystals must be transferred from their normal mother liquor to salt-free aqueous–organic mixtures. Usually an alcohol concentration of 75% or higher is necessary to prevent dissolution of the crystals. The transfer is carried out in the cold, as close to the freezing point of the salt solution as possible, to minimize differences between the dielectric constants of the salt and mixed liquids. The crystal is selected and scooped up with a Pasteur pipette; as much salt solution as possible is removed from the crystal with a filter paper or a micro-pipette, and the crystal is then plunged directly into the alcohol–water mixture. The best alcohol and its concentration must be determined empirically for each protein; studies in solution are particularly valuable as a means of ruling out certain alcohols without the sacrifice of possibly scarce crystals. In the transfer experiments particular attention should be paid to whether the crystal cracks and/or shatters (alcohol concentration probably too low). In practice, a difference in alcohol concentration of 5% by volume is often critical. Precise control of pH is particularly important to prevent crystal destruction on transferring to mixed solvents. Buffers must be selected with regard to their solubility in mixed solvents at very low temperatures; this often necessitates the replacement of phosphate by cacodylate, for example. Even more important is the effect of alcohol and temperature on pH. In practice, buffers were prepared in aqueous solutions at pH values such that, at the temperature of transfer and at the alcohol concentration required, the pH* would equal that of the high salt mother liquor. The tables given in chapter 2 were used. Care was also taken to select those buffers whose pH* would change the least with subsequent lowering of temperature.

Although Petsko has never encountered a crystal that did not fit either of the previous two categories (i.e. which could not be transferred to some aqueous–alcohol mother liquor of high alcohol concentration) he has developed a third procedure which works for all of the crystals we have examined and which may allow subzero temperature crystallography on such difficult cases. The procedure is to replace the precipitating salt in the normal mother liquor by ammonium acetate at equivalent ionic strength. Ammonium acetate is extremely soluble (>20 M) even at very low temperatures (-70°C), and is not precipitated by cryoprotective additives such as glycerol or ethylene glycol. Therefore one has two options: one can do crystallography at subzero temperatures in ammonium acetate mother liquors down to about -25°C without fear of precipitation of the salt, or one can double the concentration of ammonium acetate and then add 50% glycerol or ethylene glycol; the crystal when transferred to this mother liquor can be cooled to below -70°C. The effect of the additive on the pH of the ammonium acetate solution must be taken into account, as before. Flow cells are ideal for this procedure.

X-RAY CRYSTALLOGRAPHY

Principles

The principles governing the investigation of enzyme catalyzed reactions in fluid media have been applied to stabilizing crystalline enzyme at low temperatures. Studies have been extended to mixed solvents where the alcohol is of high molecular weight (MPD^+, propanediol) and therefore better able to keep the protein out of solution. In addition to this necessary property, high molecular weight alcohols are less volatile than, say, methanol or ethanol and are therefore easier to control.

The physico-chemical data reported in chapter 2 made it possible for Petsko to do crystallographic experiments at subzero temperatures, because he could understand and control the changes in the crucial properties of the system.

Proteins crystallized from alcohol–water mixtures. Considering that structures have been determined for several proteins crystallized from alcohol–water mixtures, it is surprising that no detailed low-temperature studies have been done on them. One difficulty is that the percentage of alcohol required to precipitate most proteins is lower than that required to prevent freezing at very low temperatures. Often, increasing the concentration will crack or disorder the crystals. This may be due to the sudden change in dielectric constant and pH that accompanies the addition of alcohol; whatever the cause, such disordering can be avoided by coupling a progressive addition of organic solvent to a progressive lowering of the temperature. The two are adjusted so as to keep the dielectric constant of the mixed solvent equal to its original value in the native mother liquor. At the same time, the pH* of the medium must be rigorously controlled, as the addition of alcohol will cause considerable shifts.

As an example, bovine pancreatic ribonuclease A has been chosen, which is crystallized from 40% ethanol–60% water, pH* 5.5 with 0.02 M acetate. The crystals are grown at room temperature, and in their normal mother liquor cannot be cooled below −50°C without severe disruption. Increasing the alcohol concentration above 60% causes cracking of the crystals, but if the proper sequence of steps is followed they can be transferred to over 90% ethanol without damage. A series of vials is set up with different

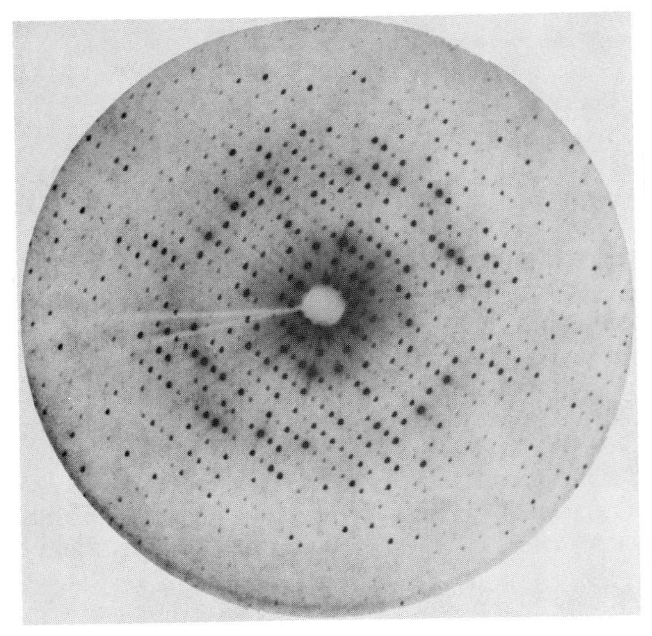

FIG. 28. Precession photograph of the hOl zone of a bovine pancreatic ribonuclease A crystal in 80% ethanol–20% water, after 10 cycles of cooling to −130°C and warming to room temperature, data to 1.9 Å resolution are shown. This demonstrates that crystals may be cooled without damage when they are in aqueous–organic solvents. (Reproduced by courtesy of G. Petsko.)

concentrations of ethanol, from 60–80% in 5% increments. Each is maintained at a different temperature, from 0°C to −40°C in steps of −10°C. The buffers for each are made up to allow for the increase of 0.07 pH* unit for every 5% of added ethanol, and the 0.034 pH* unit increase for every 10°C lowering of temperature.

Figure 28 shows a precession photograph of the hOl zone of a bovine pancreatic ribonuclease A crystal in 80% ethanol–20% water after 10 cycles of cooling to −130°C and warming to room temperature. This demonstrates that crystals may be cooled without damage when they are in mixed solvents. Figure 29 shows two hOl photographs of ribonuclease A at 1.9 Å resolution. Figure 29(*a*) shows the crystal in 40% ethanol–60% water at 20°C, Fig. 29 (*b*) shows the same crystal after transfer to 80% ethanol–20% water and after 15 cycles of cooling to −130°C.

Proteins crystallized from salt solutions. According to Petsko, there are two distinct problems that must be overcome if crystals grown by "salting out" are to be studied at subzero temperatures. The first is the problem of freezing the liquid, although it is difficult to understand why it has been assumed that a normal ice phase is formed within the crystal lattice. The extremely small diameter of the "pores" in a protein crystal, and the presence in them of hydrophilic amino acid side chains, suggests that liquid in the interstices might have physical properties quite different from those in bulk solvent. We believe the second problem, that of salt precipitation as the temperature is lowered, is likely to be more serious. One could conceive of

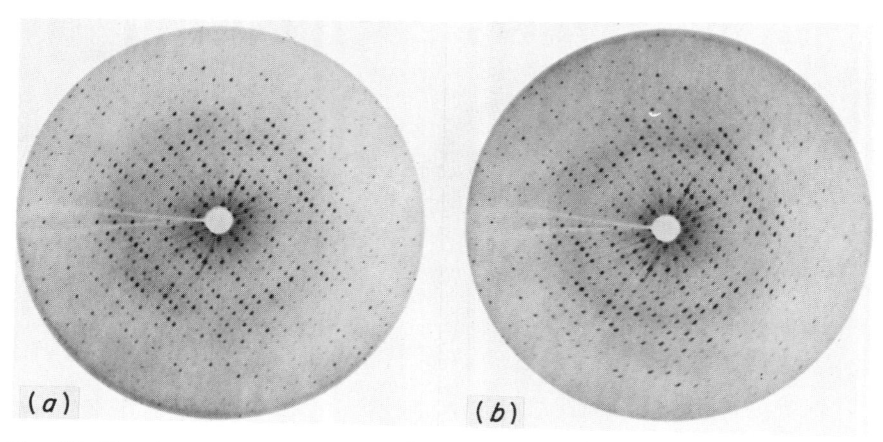

FIG. 29. Two hOl photographs to 1.9 Å resolution of ribonuclease A. (*a*) the crystal in 40% ethanol–60% H_2O at 20°C; (*b*) the same crystal after transfer to 80% ethanol–20% H_2O and after 15 cycles of cooling to −130°C. (Reproduced by courtesy of G. Petsko.)

TABLE 2

Crystalline protein	Normal mother liquor	Cryo-protective mother liquors	Highest resolution observed (Å)	Lowest temp. reached (°C)
Hen egg white lysozyme	5% NaCl, pH 4.7 with acetate	70% MPD–30% H_2O; 1 M (NH_4) Ac; 1 M (NH_4)Ac–50% glycerol; 1 M (NH_4)Ac–50% ethylene glycol; 74% propylene glycol–26% H_2O (all pH* 4.7 with acetate)	2.0	−75
Human leukaemic lysozyme	7 M $(NH_4NO_2$, pH 4.7 with acetate	80% MPD–20% H_2O; 7 M (NH_4)Ac (both pH* 4.7 with acetate)	2.3	−90
Chicken triose phosphate isomerase	3.1 M $(NH_4)_2SO_4$, pH 7.8 with triethanolamine or pH 7.5 with Tris	85% MPD–15% H_2O; 88% propylene glycol–12% H_2O; 72% ethylene glycol–28% H_2O (all pH* 7.5 with phosphate)	2.5	−75
Rabbit triose phosphate isomerase	50% $(NH_4)_2SO_4$, pH 5.6 with acetate	70% MPD–30% H_2O, pH* 5.6 with acetate	2.5	−85
Human deoxyhemoglobin	2.5 M $(NH_4)_2SO_4$, pH 6.5 with $(NH_4)_2HPO_4$	70% MPD–30% H_2O, pH* 6.5 with phosphate	2.8	−75
Bovine pancreatic ribonuclease A	40% ethanol–60% H_2O, pH* 5.5 with acetate	80% ethanol–20% H_2O, pH* 5.5 with cacodylate; 85% isopropanol–15% H_2O pH* 5.8 with cacodylate	0.9	−120

TABLE 2 (*continued*)

Crystalline protein	Normal mother liquor	Cryo-protective mother liquors	Highest resolution observed (Å)	Lowest temp. reached (°C)
Bovine pancreatic ribonuclease S, form W	40% $(NH_4)_2SO_4$, pH 6.6 with acetate	65% MPD–35% H_2O, pH* 6.6 with acetate	2.0	−75
Bovine thrombin	2 M $(NH_4)_2SO_4$, pH 5.8 with acetate	75% MPD–25% H_2O, pH* 5.8 with cacodylate	2.8	−100
Pig heart cytoplasmic malate dehydrogenase	65% $(NH_4)_2SO_4$. pH 5.0 with acetate	75% MPD–25% H_2O, pH* 5.0 with cacodylate	2.5	−75
B. Subtilis subtilopeptidase A (Carlsberg)	11% Na_2SO_4, pH 5.6 with cacodylate	60% ethylene glycol–40% H_2O 75% MPD–25% H_2O 65% isopropanol–35% H_2O 1.4 M $(NH_4)Ac$ + 50% glycerol all pH* 5.6 with cacodylate	1.6	−100
Human serum prealbumin	55% $(NH_4)_2SO_4$, pH 7.1 with phosphate	75% MPD–25% H_2O pH* 7.5 with phosphate	2.5	−69
Porcine stomach pepsin	20% ethanol, pH* 1.8 with HCl	60% ethanol or methanol pH* 1.8 with HCl	2.5	−50

eliminating the freezing problem by doubling the salt concentration and then adding 50% by volume glycerol, ethylene glycol, or some other "antifreeze", but the relative insolubility of most precipitating salts in alcohol–water mixtures or even in water at low temperatures would lead to crystal destruction by salt precipitation long before the freezing point of the mother liquor was reached. (A corollary of this argument is that careful removal of bulk liquid from around the crystal might enable some protein crystals to be "supercooled" in their normal mother liquor, a possibility now under investigation.)

Petsko's solution to these problems is to replace the normal crystal mother liquor with salt free aqueous–organic solvents of low freezing point. The protein crystal must be transferred, in one step, to a mixed solvent of high enough alcohol concentration to prevent dissolution. If this concentration is too low for the temperature required, it may be increased by the procedure described above.

Although it is not certain that "dielectric shock" is responsible for crystal disordering on transfer to mixed solvents, it has been observed that the transfer operation proceeds much more smoothly in the cold for all proteins examined. The present procedure is to cool the crystal in its salt solution to just above its freezing point; the mixed solvent is cooled to the same temperature and the transfer proceeds as outlined on pp. 238–240. In preparing the buffer for the mixed solvent, the changes in pH* on addition of alcohol and lowering of temperature are taken into account. Generally one selects buffers whose pH* values show little change with temperature; thus, Tris and borate are avoided while acetate and phosphate are used frequently. Once the transfer has been performed in the cold, the crystal may be warmed up slowly to room temperature and is then stable. In each case the crystals may be cooled and warmed repeatedly without damage. Twelve different crystalline proteins have been examined at low temperatures (Table 2); in each case neither disorder nor cell dimension changes have been observed. Intensity changes have been minimal and for the most part confined to very low resolution. In many cases, the first mixed solvent system tried was successful; in others (e.g. human lysozyme) the solvent listed was the only useful one found after over fifty experiments. The table is not meant to imply that the solvents listed are the only possible ones for any given crystal, nor that the listed temperatures and resolutions are the limits observable.

Special cases. Although no exhaustive survey has been made, it has been found that many protein crystals can be transferred from their normal salt mother liquor to ammonium acetate solutions of equivalent ionic strength. The extreme solubility of ammonium acetate even at very low temperatures in the presence of high concentrations of alcohols makes it possible to use this salt to keep the crystal out of solution when alcohol alone will not do the job. It has been observed that there is no precipitation problem with ammonium acetate even down to $-100°C$ at 15 M, so the only problem is to prevent freezing. This can be done by making a cryoprotective mother liquor of 50% "antifreeze"–50% water, with the required ionic strength of ammonium acetate. Preliminary studies indicate that glycerol and ethylene glycol are useful, mild "antifreeze" additives for this procedure.

Figure 30 shows the diffraction pattern at 2.8 Å resolution of the hkO zone of hen egg white lysozyme, which is crystallized from 5% NaCl, pH 4.7 with acetate. Figure 30(*b*) shows the same zone of the same crystal after transfer to 50% glycerol–50% water, 1 M ammonium acetate, pH* 4.7 with acetate. The crystal has undergone 10 cycles of cooling to $-69°C$ and warming to 20°C. Note that the change in pH* of acetate on cooling is very small in this solvent system (less than 0.15 units), and the change on addition of ethylene glycol has been accounted for in preparing the buffer. The transfer in the cold ($-13°C$) is done as well but without any particular sensitivity to temperatures.

Advantages of and problems with the technique

As stated by Petsko:

> Considering the deleterious effects of 100% alcohol on many proteins it may seem surprising that this technique has worked so well. Regarding this, it is interesting to examine the molecular composition of mixed solvents. A mixture composed of 50% ethylene glycol–50% water corresponds to a molar ratio of 8 moles of water to 1 mole of alcohol; a 70% methanol–30% water mixture has a 1 : 1 mole ratio. Thus there are always large numbers of water molecules in these solvents, and the firmly bound first layer of water ("solvation shell") about the protein should remain essentially unchanged. Provided the pH* of the solvent is controlled, and (possibly) sudden changes in *D* are minimized, mixed solvents can be very mild for proteins. Indeed, we believe that this technique has several advantages over those others which have been tried.

The author also mentions that we have never observed significant unit cell dimension changes at the temperature of interest, probably because at that temperature, *D* and pH* are as close to their values in the original mother liquor as possible. Occasionally small unit cell changes in mixed solvents are observed at room temperature, but it has been found that cooling always restores them to their proper values. The absence of significant cell dimension changes means the powerful and relatively rapid difference Fourier technique can be used, and difficult *de novo* solutions are not necessary. A second advantage of this technique is that the crystals are not in any way disordered, even after cooling–warming cycles. Since, in general, the mixed solvents used are fluid at low temperatures, diffusion of

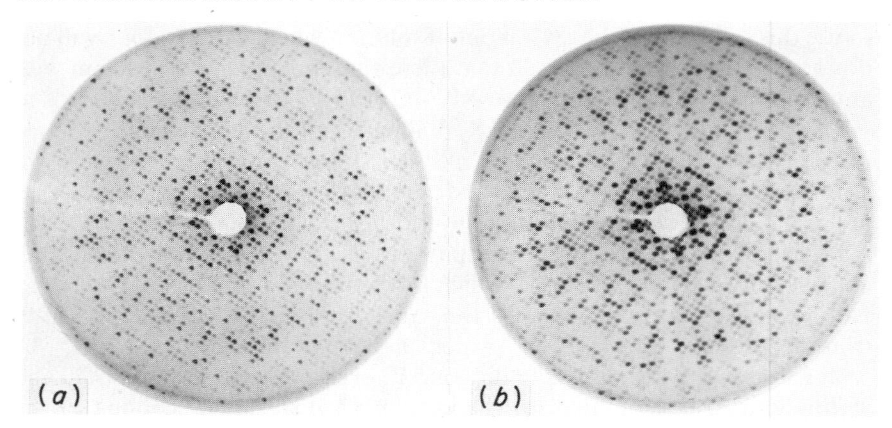

FIG. 30. Two hkO photographs of hen egg white lysozyme at 2.5 Å resolution. (*a*) lysozyme in 1 M NaCl at 20°C; (*b*) after transferring the crystal to 1 M ammonium acetate + + 50% glycerol, after 10 cycles of cooling to −25°C. (Reproduced by courtesy of G. Petsko.)

substrates into the crystal is possible if the viscosity is not too high. No special equipment is needed for this technique, although thermoelectric cooling blocks are convenient for examining different solvents at different temperatures. However, it is believed that the most significant aspect of this technique, one not shared by either the cross-linking or high pressure methods mentioned in chapter 1, is that the mixed solvents have known physico-chemical properties. Their effect on protein conformation and enzymic activity is known for a large number of enzymes and can be studied for any enzyme of interest. One is not working in the dark with these systems. Careful tests in solution can establish which alcohols have the least effect on the structure of the protein, and when one or more of these have been found to stabilize crystals of the protein in question, the activity of the enzyme in the chosen mixed solvent can be studied. This permits precise determination of the temperature at which the reaction is slow enough for X-ray data collection. If the alcohol or the cooling process changes the rate-determining step of the reaction, this can then be detected by such solution studies. Attractive though this technique may be, it is not free from problems which may be serious or even prohibitive for some proteins—although we have not yet found this to be so.

Certainly the biggest drawback of this method is the necessity of finding the exact working conditions for each protein. Although some general recommendations are suggested by our work, it is quite possible that in future work no suitable cryoprotective mother liquor can be found for some proteins and, unfortunately, a failure in an attempted transfer experiment is fatal to the crystal. This may therefore not be the technique of choice when only a small number of crystals are available. Another possible problem is the effect of the mixed solvent on the protein conformation, although in all cases examined so far the diffraction pattern of the protein crystal was essentially unchanged. Another difficulty of the low temperature procedure may be in obtaining correct diffusion of substrates into crystals suspended in highly viscous mixed solvents. To avoid this, Petsko performed pH* "jumps" between room and reduced temperatures. Let us illustrate such an elegant procedure by the following example: a crystalline serine protease in MPD–water, poorly buffered by Tris at pH* 5, is exposed to substrate at 20°C. At this temperature diffusion is possible, so the substrate binds but is not turned over to product because the pH* is too low. The crystal is then mounted and rapidly cooled. As the temperature reaches −50°C the pH* reaches 8, and the enzyme becomes active. But at −50°C the rate of reaction is infinitesimally slow, so data can be collected on this stable, productive enzyme–substrate complex. Such experiments are now under way for the serine proteases, thrombin and subtilopeptidase A (Carlsberg). They should also be possible for other types of enzymes.

Possible generalizations

Our experience with this technique for a wide variety of crystalline proteins enables us to make a few generalizations which, we hope, will make its application by other groups easier.

Careful experiments in solution—e.g. the study of methanol and other solvents on lysozyme, on p. 248, are a necessary prerequisite for all crystallographic studies. They allow the effects of alcohol on the conformation and activity of the protein to be determined; they enable a number of alcohols to be ruled out immediately, and they establish the precise temperature and pH for crystallographic studies. Choice of cosolvent should be dictated by two factors, ease of handling and viscosity. It has been found that MPD is the organic additive most likely to stabilize the crystal and, as it is of high molecular weight, it is easy to handle. However, its extreme viscosity will be a problem if flow cell studies are contemplated, so a less viscous but more volatile alcohol like ethylene glycol or isopropanol may be preferable. In general, experiments to find suitable cryoprotective solvents should start with MPD at 50% and higher, and methanol or isopropanol at 60% and higher. DMSO is to be avoided since the pure organic liquid has a freezing point of +17°C; therefore, mixtures of it are "supercooled" at low temperature and often freeze spontaneously. Ternary mixtures of low viscosity described in chapter 2 are now being investigated.

Unless the protein is more soluble in the cold than at room temperature, all transfer experiments should be done at as low a temperature as possible (usually around $-15°C$). The reason for the better results achieved in the cold has not yet been definitely established. When preparing buffers, the changes of pH on addition of alcohol and change in temperature must be allowed for. Unless pH*-jump experiments are contemplated, buffers with small thermal coefficients should be chosen. These include oxalic and chloracetic acid buffers from pH* 2 to 4.5; acetate from pH* 4.5 to 6; cacodylate or phosphate from pH* 6 to 8.

X-ray diffraction procedure

Diffractometer data were collected either on a Hilger–Watts 4-Circle diffractometer specially modified to measure five reflections quasi-simultaneously (63,64,65) or on a Syntex $P2_1$ diffractometer extensively modified to give very low background counts (66). The former was equipped with a protein crystallography low-temperature device (67) designed to reach temperatures around $-50°C$. For measurements on the Syntex machine the Syntex LT-1 Low Temperature Attachment was used; it permitted temperatures below $-120°C$ to be obtained. Certain modifications were required to this device for efficient protein crystal work (66). Photographs were taken with a Buerger Precession Camera having a specimen-to-film distance of 6 cm. A sealed fine-focus copper X-ray tube run at 50 kV $\times$ 30 mA was used. Photographs were taken at room temperature on Ilford Industrial G X-Ray Film.

Protein crystals were mounted in glass or quartz capillary tubes in the usual way except that columns of mother liquor were not usually placed at the ends of the tubes before sealing. This was because a thermal gradient along the capillary tube often caused distillation of liquid to or from such reservoirs when they were present.

Radiation damage

The effect of cooling on the radiation damage to the crystals has been studied by Petsko and one example is shown in Fig. 31, which indicates that crystals do show less damage under X-ray exposure as the temperature is dropped.

THE COMPLEX BETWEEN LYSOZYME AND A "TRUE" SUBSTRATE

In this laboratory we carried out tests allowing precise determination of the medium conditions protecting the soluble proteins, and of the temperature at which reactions are slow enough for X-ray data collection. We were able

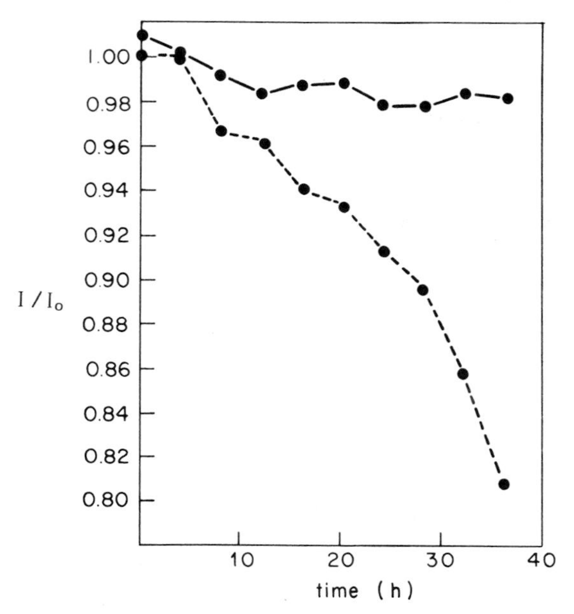

FIG. 31. The effect of cooling on the radiation damage to human hexagonal aldolase crystals, the dashed curve is at 20°C, within 40 h the crystal has lost 20% of its scattering power, the solid curve is at −4°C, the crystal shows less than 2% damage after 40 h. Each curve is the average of 50 reflections at 4 Å resolution. (Reproduced by courtesy of G. Petsko.)

to determine the conditions in which productive complexes of lysozyme and oligosaccharides, of ribonuclease and oligonucleotides, can be obtained and stabilized over hours or days. To ascertain the best conditions for the structure determination of a productive lysozyme-substrate complex, the hydrolysis of bacterial cell walls and oligosaccharides was investigated both in high salt solutions and in mixed solvents. Let us examine the main observations recorded in such cases.

The three-dimensional structure of crystalline hen egg white lysozyme has been determined in atomic detail using the X-ray method by Phillips *et al.* (68) and similar investigations were carried out to find the structure of a reactive complex between lysozyme and its polysaccharide substrate, in the hope that the active groups of atoms in the enzyme could be recognized and their function understood.

Because of the velocity of the catalyzed hydrolysis of polysaccharide molecules and the necessity of a rather long exposure of crystals to X-rays, and because some small amino sugar molecules, freed by the breakage of

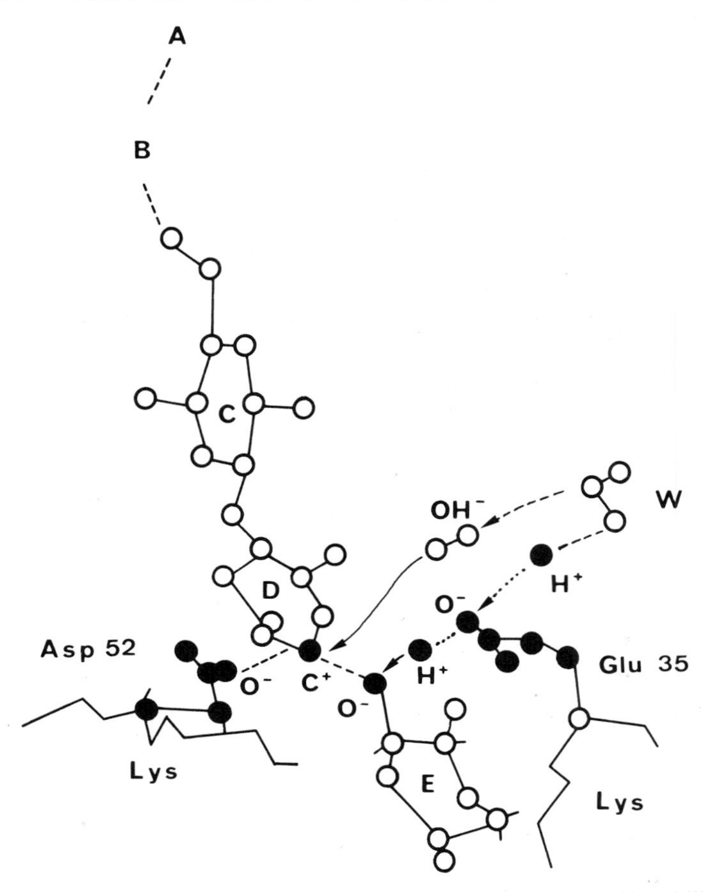

FIG. 32. Mechanism of the splitting of substrate by lysozyme as assumed by Phillips (68).

polysaccharides, act as competitive inhibitors of the enzyme's activity, it was not possible to use an oligosaccharide filling the active site; this contains the six subsites termed A, B, C, D, E, F (see Fig. 32) and undergoes hydrolysis by breakage of the linkage between sugar residues D and E. Simple amino sugar molecules, including the monosaccharide *N*-acetylglucosamine (NAG) and a trisaccharide comprising three NAG units (tri-NAG), which act as competitive inhibitors, were used as substitutes for substrates, since they bind to the enzyme in the way that the true substrate does, stopping the enzyme from working normally by preventing the substrate from binding to the enzyme.

Crystals containing these amino sugar molecules were prepared by adding the sugar to the solution from which the lysozyme crystals had been grown and in which they were kept suspended, the small molecules diffusing into the crystals along the channels filled with water that run through the crystals. Studies at 2 Å resolution have shown that the trisaccharide fills the top half of the cleft (schematized in Fig. 32).

The electron-density map showing the change in electron density when tri-NAG in bound in the protein crystal showed clearly that parts of the enzyme molecule have moved with respect to one another. These changes in conformation are largely restricted to that part of the enzyme structure to the left of the cleft, which appears to tilt more or less as a whole tending to close the cleft slightly. Such an observation supports the "induced fit" theory of the enzyme-substrate interaction discussed by Koshland (69).

Since the complex formed by tri-NAG and the lysozyme is not the enzyme-substrate complex involved in catalysis, because it is stable and a rather complex enzyme-inhibitor, it is difficult to find out how the enzyme works just by analysis of static X-ray images. A model has been built in which another three sugar residues can be added to the tri-NAG bound to the upper half of the cleft so that there are satisfactory interactions between the atoms in the proposed substrate and the enzyme.

Finally these authors delineated the interactions of the hexa-NAG substrate with enzyme and found the amino acid residues involved in catalysis and located between subsites D and E where the carboxyl groups of Asp 52 and Glu 35 were found (see Fig. 32).

Studies of the hydrolysis of various polysaccharides by Rupley (70), and Sharon (71) and their co-workers confirmed the above model. It was deduced by Phillips and co-workers that Asp 52 located in a network of hydrogen bonds, might easily be ionized and would carry a negative charge stabilizing the formation of a carbonium ion involved in the bond cleavage; and that Glu 35, located in a hydrophobic environment, would have the p*K* of its carboxyl group shifted, the group remaining protonated in the optimal pH range for lysozyme.

Thus X-ray crystallographic studies of lysozyme-inhibitor complexes have been rewarding (72,73); the same studies with true substrates (polysaccharides) cannot be carried out since the rate of their catalyzed hydrolysis increases with increasing substrate chain length. On the other hand, we have seen that the formation of non-productive complexes involving three strong-binding subsites with small oligosaccharides and polysaccharides is competitive with substrates undergoing catalytis. One possible method of stabilizing active enzyme-substrate complexes would be to reduce the temperature but such a possibility raises a number of problems which were analyzed in this laboratory. We will give here the main results obtained both in aqueous salt solutions and in mixed solutions of lysozyme.

It must be pointed out that the lysozyme was crystallized from a solution containing sodium chloride and that the environment of each molecule in the crystalline state is not significantly different from its natural environment in the living cell. The crystals include some 35% by weight of mostly watery liquid of crystallization, and the effect of the surrounding water on the lysozyme conformation is thus likely to be much the same in the crystals as it is in solution.

Lytic activity of lysozyme

Use of increasing amounts of cosolvent as "antifreeze" could perturb the conformation and thus the activity of lysozyme. We carried out a number of experiments to try to determine precise conditions for investigating lysozyme reactions and lysozyme-substrate intermediates in cooled mixed solvents, as a preliminary to similar investigation by X-ray diffraction on crystals. Work began with an estimate of the solubility of the enzyme and of several oligosaccharides in various mixtures as a function of temperature, and the assay of lysozyme activity was carried out by recording the cleavage of the polysaccharides of bacterial cell walls of *Micrococcus lysodeikticus*. Let us recall that the important part of these cell walls, as far as lysozyme is concerned, is made up of long polysaccharide chains containing the glucose-like amino sugars, *N*-acetylglucosamine (NAG) and *N*-acetylmuramic acid (NAM), occurring alternately in the chains and connected by bridges that include an oxygen atom (glycosidic linkage) between carbon atoms 1 and 4 of consecutive sugar rings. Lysozyme has been shown to break the linkages in which carbon 1 in NAM is linked to carbon 4 in NAG but not the other linkages. This process brings about the dissolution of bacterial cell walls, which can be measured turbidimetrically.

It has been assumed that the events leading to the rupture of bacterial cell wall probably take the following course. First, a lysozyme molecule attaches itself to the bacterial cell wall by interacting with six exposed amino sugar residues. In the process sugar residue D is somewhat distorted from its usual conformation. Second, glutamate residue 35 transfers its terminal hydrogen atom in the form of an hydrogen ion to the glycosidic oxygen, thus bringing about cleavage of the bond between the oxygen and carbon atom 1 of sugar residue D. This creates a positively charged carbonium ion (C^+) where the oxygen has been severed from carbon atom 1. Third, this carbonium ion is stabilized by its interaction with the associated aspartic acid side chain of residue 52 until it can combine with a hydroxyl ion (OH^-) that happens to diffuse into position from the surrounding water, thereby completing the reaction. The lysozyme molecule then falls away, leaving behind a punctured bacterial cell wall.

Kinetics of lysozyme-catalyzed lysis in mixed solvents at room and subzero temperatures

The solvents used to investigate the above reaction were mixtures of methanol–water, 2-methyl-2-4-pentanediol–water, 7 M ammonium nitrate, and 5 M sodium chloride.

The methanol–water mixture was in the volume ratio 40 : 60; the freezing point is about −35°C. When adding the 40% methanol to a solution of buffer, the protonic activity (pH* or pa_H^*) increases and the difference $\Delta pH^* = pH^* - pH_{H_2O} = 0.5$ and 0.8 respectively in the case of acetate and phosphate buffer. At −50°C the increase of pH* does not exceed 0.22 for acetate and phosphate buffers in 50% methanol; the extrapolated enthalpy of ionization of these buffers in 40% methanol can be estimated then to have the values $\Delta H_i^0 = 0.7$ and 0.9 respectively for acetate and phosphate buffers: so the increase of pH* is very small at low temperatures and can be neglected. The second solvent used is the mixture of 2-methyl-2-4-pentanediol–water with volume ratios of 50 : 50 and 70 : 30. Their freezing points are around −30°C and −90°C. The solution of 2-methyl-2-4-pentanediol (MPD) has an absorption band at 275 nm and is not transparent below 260 nm; near their freezing points thse solvents are very viscous, more than the mixture composed of 50% by volume of ethylene glycol.

The last two solvents used are solutions of high salt concentration: 7 M ammonium nitrate (NH_4NO_3) and 5 M sodium chloride (NaCl). Their freezing points are about −20°C. When studying protein spectrum, 5 M NaCl solution is preferred, because of its transparency in the u.v. region. The ammonium nitrate solution absorbs strongly in the u.v.

Mixed solutions are prepared by progressive addition of the organic solvent to aqueous buffered solutions of lysozyme cooled at 0°C, to prevent any possible denaturation.

All kinetic studies of the lysis of *Micrococcus lysodeikticus* cells by lysozyme are carried out on the Cary 15 Spectrophotometer equipped with a low temperature cryostat and an electronic temperature control device. The temperature regulation is better than 0.05°C from +20° to −100°C. The enzymic activity of lysozyme was measured turbidimetrically by following the decrease in absorption at 711 nm or 500 nm of a mixture of *Micrococcus lysodeikticus* (0.1 mg/ml) and lysozyme (8 μg/ml); at low temperature the induction phase can take a few minutes. A plot of the logarithm of optical density against time gave a straight line indicating that under the experimental conditions the reaction is first order. The main results obtained in these conditions were as follows:

Solubilities. We checked the solubility of lysozyme and oligosaccharides in the different mixtures and we found that they were soluble up to 0.2

mg/ml in 50% by volume of methanol and of methylpentanediol (MPD) down to $-15°C$. In 70% MPD each component is still soluble separately but their concentration must be lowered to 0.01 mg/ml when mixed to be tested down to $-20°C$.

Hydrolysis of bacterial cell walls (Micrococcus lysodeikticus) *at room temperature.* The overall rate constant of hydrolysis in aqueous phosphate buffer (2×10^{-2} M; pH 6.0) is $k_{H_2O} = 17.8 \times 10^{-3}\ s^{-1}$ for $\lambda = 500$ nm at 20°C.

In the methanol–water mixture (40 : 60): When 40% methanol is added to aqueous solutions buffered by phosphate at pH 6.0, the proton activity increases up to pH* 6.8, and the rate constant $k_{methanol\ 40\%} = 7.8 \times 10^{-3}\ s^{-1}$ at $\lambda = 500$ nm and at 20°C. The maximum of the pH* profile (pH^*_{max}) is shifted towards alkaline values (7.5) and the corresponding k_m becomes $11.5 \times 10^{-3}\ s^{-1}$.

In the MPD–water (50 : 50) mixture: In similar pH conditions to those above, $k_{MPD\ 50\%} = 4.6 \times 10^{-3}\ s^{-1}$ at $\lambda = 500$ nm and 20°C. The pH* rate profile indicates a maximum value of the rate constant k_{max} of $7.9 \times 10^{-3}\ s^{-1}$ when the aqueous buffered solution is at pH 7.0.

In the MPD–water (70 : 30) mixture, there is no detectable hydrolysis at 20°C—mainly due to the action of MPD on *Micrococcus.*

Salt solutions: There is a noticeable inhibition of the hydrolysis, insensitive to pH changes.

7 M Ammonium nitrate: The rate constants observed at pH 4.7, 6.0 and 8.0 are respectively $1.13 \times 10^{-4}\ s^{-1}$, $1.9 \times 10^{-4}\ s^{-1}$ and $1.3 \times 10^{-4}\ s^{-1}$ at 20°C.

5 M Sodium chloride: The same results are obtained, and e.g. $k = 4.2 \times 10^{-4}\ s^{-1}$ at pH 4.7. (pH values were estimated using a pH-meter with glass and calomel reference electrodes. Values recorded are indepenent of the salt concentration between 1 M and 7 M but depend on the nature of the buffer.)

Temperature-dependent hydrolysis. Rupley and Gates (70) have determined the temperature-dependent cleavage of penta-NAG by lysozyme in aqueous buffered solutions (pH 5.3) at saturating substrate concentration between 20 and 60°C and have found an activation energy of 17 kcal mol^{-1}.

In the media mentioned above, we have studied the hydrolysis of bacterial cell walls between 20° and $-30°C$. The rate constants lie in a straight negative line in the Arrhenius plot ($\log_{10} k = f(1/T)$). Activation energies are of 16.9 kcal mol^{-1} in 40% methanol, and of 20 kcal mol^{-1} in 50% MPD. The velocity in salty solutions is much more sensitive to salt action (inhibition) than to temperature; in the presence of 7 M NH_4NO_3, the rate factor

k salt/k water $\simeq 9 \times 10^{-3}$, and the calculated activation energy is $\simeq 5.3$ kcal mol^{-1}.

Reversibility of the solvent inhibition. The inhibitory effect of the solvent and of high salt concentration is entirely reversible by infinite dilution; these solutions can be stored for hours at 0°C and for days below 0°C, and then diluted without any loss of activity.

Thermal activation of "quenched" reactions. Lysozyme-catalyzed lysis of bacterial cell walls is practically "quenched" in mixed solvents at about −25°C. By warming up, the rate constant and activation energy values observed during cooling at selected temperatures are restored.

Conclusions. Investigating the kinetics of lysozyme-catalyzed lysis of bacterial cell walls, we have shown that the presence of an organic solvent inhibited the overall reaction rate reversibly. The ratio of the maximum rate constant of lysis in water and in mixed solvent was:

$$\frac{k_{H_2O}}{k_{\text{methanol } 40\%}} = 1.7$$

$$\frac{k_{H_2O}}{k_{\text{MPD } 50\%}} = 2.5$$

It had been previously shown that there was no macroscopic conformation change of the lysozyme in presence of 40% methanol, and we have checked that there is no change of the levorotation at 436 nm and just a slight increase in the extinction coefficient at 292 nm in presence of 50% MPD.

The reversible reduction of reaction rates in mixed solvents is small compared to that obtained in salt solutions where

$$\frac{k_{H_2O}}{k_{NH_4NO_3\ 7\,M}} = 110$$

The temperature effect on the rates of hydrolysis, recorded between 20 and −25°C, shows that the rate constant k is reduced by a factor 200 in presence of 40% methanol in such a range of temperatures, and a factor 510 in the presence of 50% MPD. Such reductions correspond respectively to activation energies of 16.7 kcal mol^{-1} and 20 kcal mol^{-1}.

Let us notice that the first numerical value is similar to that obtained in water for the hydrolysis of the penta-NAG by lysozyme (E = 17 kcal mol^{-1}), and such a similarity would mean that the mechanism of breakdown of the complex E—S to form product is identical in both media.

The reduction of the rate of hydrolysis in the solution containing 7 M NH_4NO_3 is

$$\frac{k_{H_2O}}{k_{NH_4NO_3\ 7M}} = 4.78.$$

with an activation energy of 5 kcal mol^{-1}.

Study of the complexes between lysozyme and oligosaccharides

The above observations carried out on lysozyme in mixed solvents as a function of temperature demonstrate that lysozyme-catalyzed lysis can be performed in such abnormal conditions, and that the reaction can be "quenched" at some subzero temperatures and resumed by heating. The problem is now to check whether one can obtain an enzyme-substrate complex (in this case lysozyme–oligosaccharide) stabilized at low temperature. Such a complex can be detected by differential absorption spectroscopy. Difference spectra in the u.v. region (240–320 nm) were recorded using a Cary 15 and an Aminco-Chance DW2 Spectrophotometer with quartz cells of 1 cm light path. This apparatus is equipped with a low temperature cryostat and control device. Both the reference and sample cells are maintained at the same temperature. The reference cell contains a solution of lysozyme (1.39×10^{-5} M) and the sample cell contains a solution of lysozyme at the same concentration plus substrate, which is here hexa-NAG (1.66×10^{-4} M). The buffer used is acetate buffer pH 5 plus organic solvent and other pH had been used. Also difference spectra have been determined in the presence of a small chain saccharide like di-NAG (4.83×10^{-4} M) and tri-NAG (3.27×10^{-4} M). These substrate and inhibitor concentrations are sufficient to nearly saturate the enzyme.

Lysozyme—hexa-NAG complex at room temperature. An equimolar solution of lysozyme and hexa-NAG, acetate buffer pH 4.7 (corresponding to the maximum of activity) temperature 20°C, is recorded by absorption spectrophotometry against a solution of lysozyme. A specific difference spectrum is obtained (see Fig. 33). The spectrum shows 3 positive peaks (277, 287, 294 nm) and 2 negative peaks (280, 291 nm). The magnitude of the largest peak (at 294 nm, with a shoulder at 298 nm) attains its maximum during mixing of the enzyme and substrate. This confirms previous results obtained by Hagashi *et al.* (74) with neuramidase–glycol–chitin complex. The spectrum is relatively time-independent for almost 30 min.

It can be seen in Fig. 33 that similar spectra are obtained in salt solutions at neutral pH, and that spectra are slightly modified in mixed solvents, due to modifications in pK. Note the transformation of the shoulder at 298 nm into a peak in methanol (40%), MPD (50%), and a much more pronounced perturbation in 70% MPD.

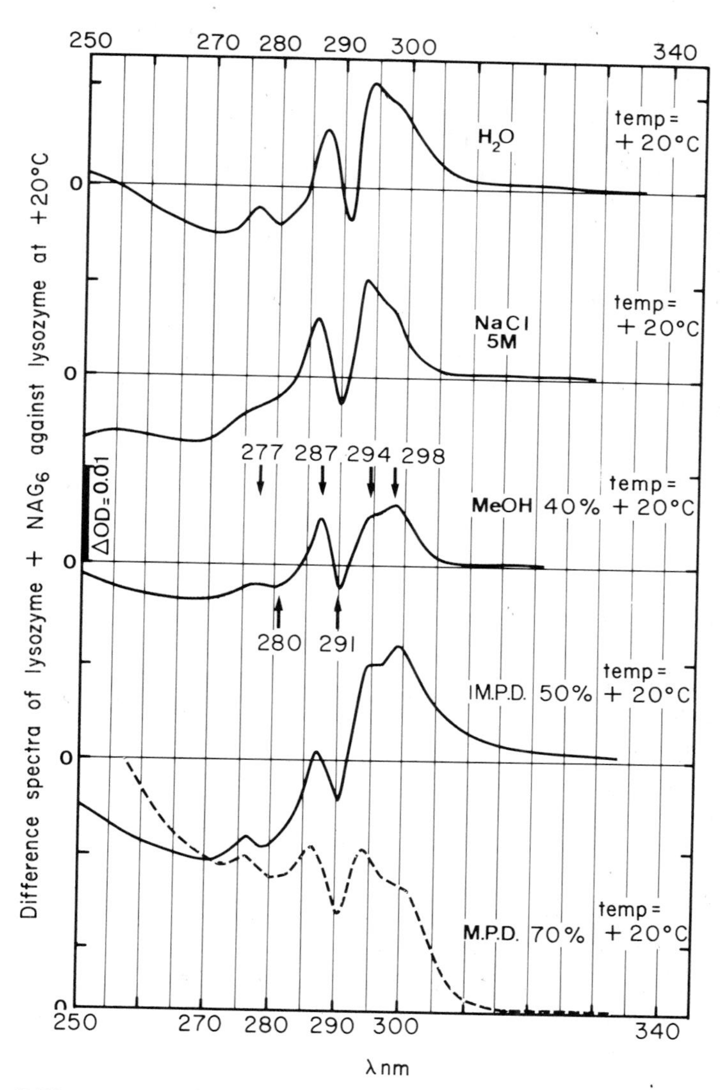

FIG. 33. Difference spectra of lysozyme + $(GL_cNA_c)_6$ against lysozyme in various solvents at 20°C.

Lysozyme—hexa-NAG complex at subzero temperatures. At −20°C in mixed solvents and in the same conditions of concentrations as above, similar difference spectra are obtained (Fig. 34), except that there is a real peak at 298 nm, and that the magnitude ΔOD_{294} is slightly reduced. Obviously, the spectra are more sharply defined than at 20°C. They are quite stable over several hours. Difference spectra have been recorded in 40% methanol between pH* 2.0 and 10.6 at −20°C and are shown in Fig. 35. There is a real peak at 298 nm between pH* 4.7 and 6.1, and this could correspond to the carboxy ionization. The amplitude $\Delta OD_{294} = OD_{294} - OD_{291}$ is pH* dependent (see Fig. 36). In this figure, we can see the profile of substrate binding in 40% methanol at −20°C. This profile shows a maximum of binding at about pH* 4.7, which can be compared to the results obtained in water by Banerjee and Rupley (75).

Lysozyme—di-NAG and—tri-NAG complexes. In water and in mixed solvents, lysozyme (1.38×10^{-5} M) and di-NAG (4.83×10^{-4} M) or tri-NAG (3.27×10^{-4} M) give difference spectra similar to those preceding, except that their magnitude is smaller; ΔOD_{294} increases from di-NAG to tri-NAG and hexa-NAG—lysozyme complexes. Spectra have been recorded between 20 and −20°C. The spectrum of the di-NAG and tri-NAG—lysozyme complex is without structure.

We can now draw the following conclusions: Difference spectra obtained on mixing lysozyme and hexa-NAG in mixed solvents at room as well as at subzero temperatures indicate the formation of an enzyme-substrate complex and it appears that such a complex is indefinitely stabilized at −25°C. Results observed on the temperature-dependent cell wall hydrolysis can be extrapolated to hexa-NAG hydrolysis; note that the respective pH profiles are different.

The shape of the difference spectra attributed to the lysozyme–hexa-NAG complex at low temperature (−20°C) is pH* dependent, and should depend on the ionization of chromophores near the binding site. The maximum amplitude of these spectra corresponds to the pH of the maximum activity of lysozyme in the hydrolysis of hexa-NAG. Decrease in amplitude in the presence of tri-NAG and di-NAG confirms the observations of Blake *et al.* (76) showing that these substrates are bound to the upper part of the cleft (subsites A, B, C).

Conclusions

In conclusion to these preliminary studies we can say that:

1. It is possible to obtain a lysozyme-catalyzed lysis of substrates in mixed solvents. There is a partial and reversible inhibition showing that any organic solvent acts as a modifier of the enzyme specific activity.

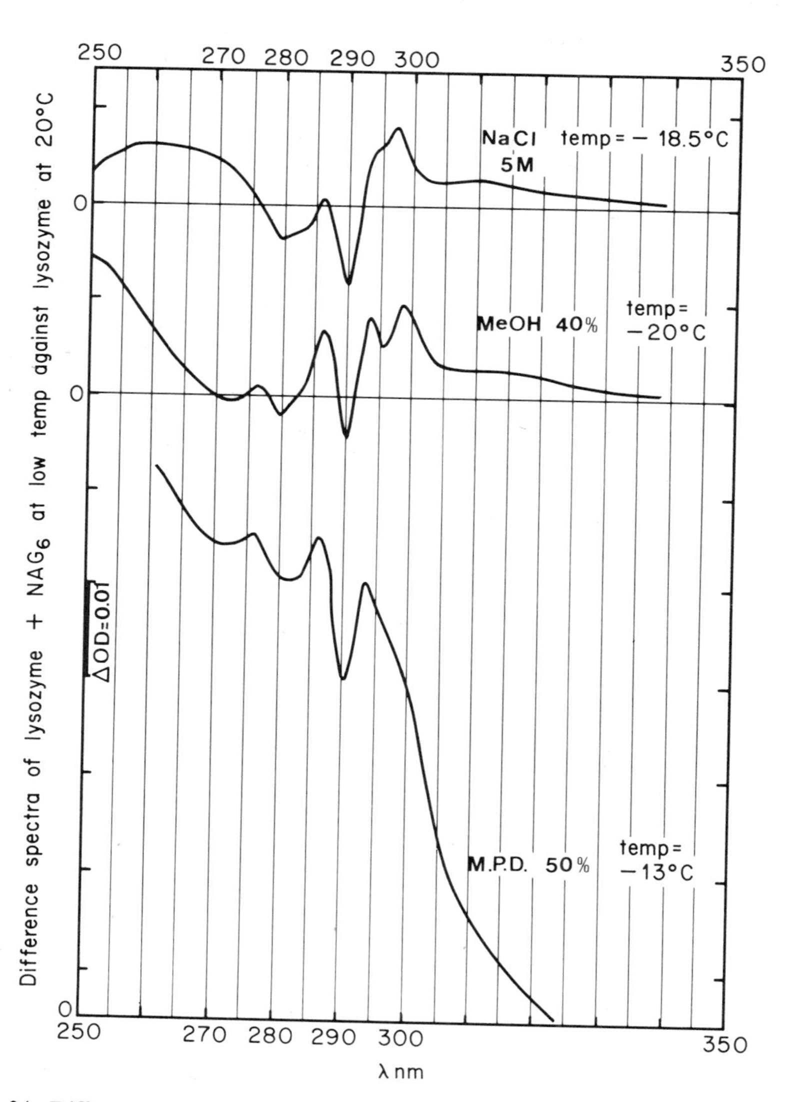

FIG. 34. Difference spectra of lysozyme + $(GL_cNA_c)_6$ at low temperature against lysozyme at 20°C.

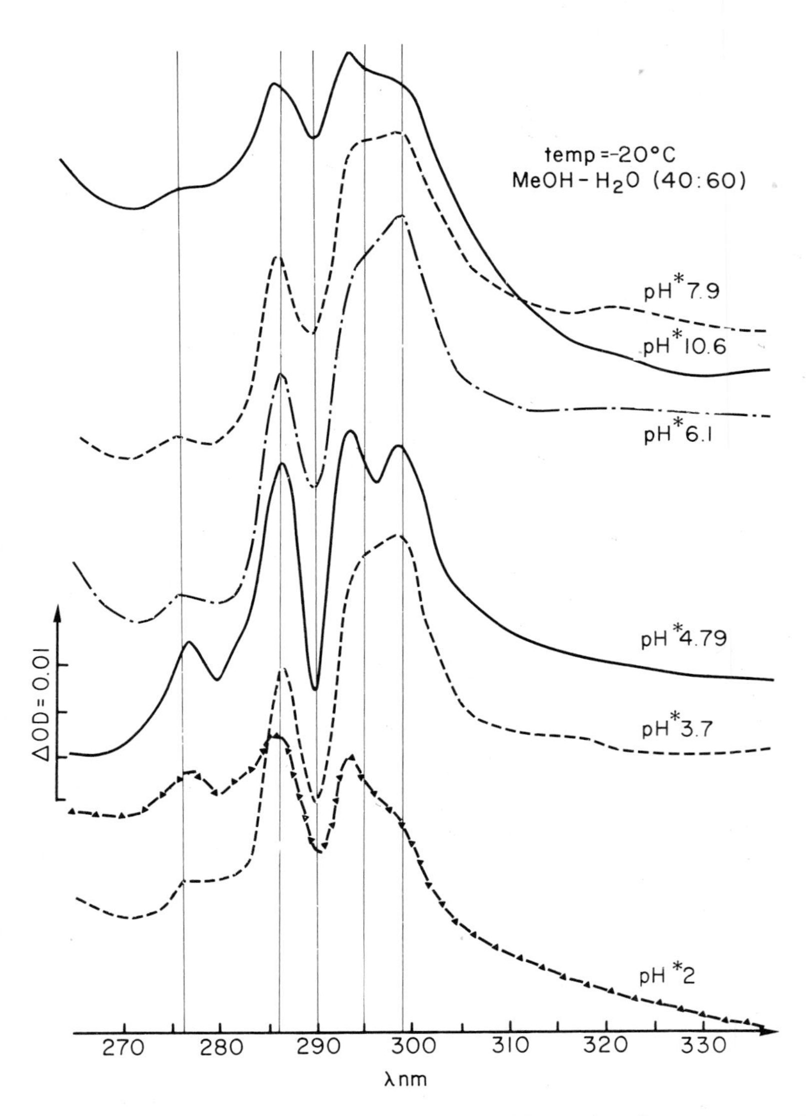

FIG. 35. Difference spectra of lysozyme + $(GL_cNA_c)_6$ against lysozyme at various *values* of pH* in 40% methanol at −20°C.

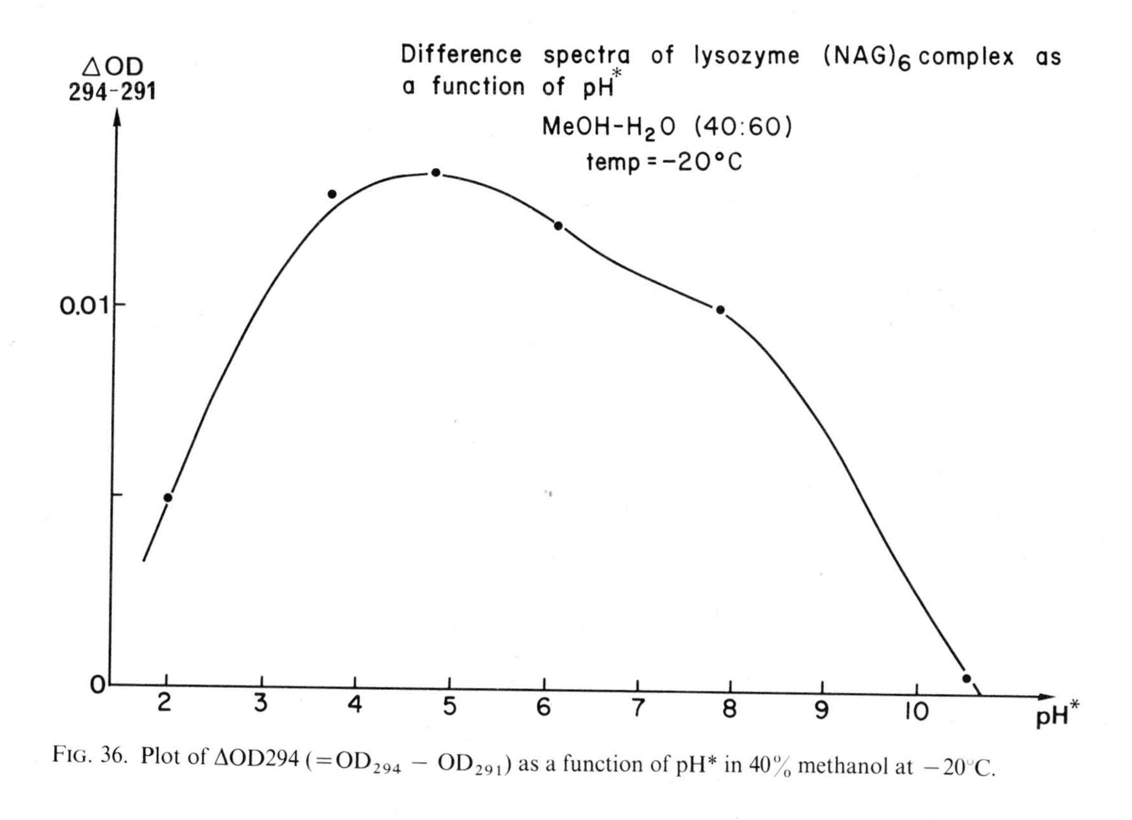

FIG. 36. Plot of ΔOD294 ($=OD_{294} - OD_{291}$) as a function of pH* in 40% methanol at −20°C.

2. Lowering the temperature decreases the reaction rates markedly, and the reaction is "quenched" at the level of the E—S complex between −20°C and −25°C.

3. Conditions of pH corresponding to the maximum of catalytic activity must be chosen to obtain an affinity maximum between E and S.

These studies suggest that the crystallographic determination of the structure of a productive enzyme-substrate complex is feasible for human lysozyme and oligosaccharide substrates. They also provide the information on pH, temperature and solvent effects on activity which are necessary to choose the best conditions for the crystal structure work. The system of choice for human lysozyme is mixed aqueous–organic solvents at −25°C, pH* 4.7. Data obtained by us in this and previous investigations on the dielectric constant, viscosity, and pH* behavior of mixed solvents enable these conditions to be achieved with precision, and should be valuable in studies of other proteins at subzero temperatures.

Membrane-bound multienzyme systems

Normal enzyme assay in membrane multienzyme systems is often difficult, due to the inherent shielding of the enzymes from their substrates, and also because one often has to use specific chain inhibitors as well as "exogenous" substrates different from the "endogenous" ones, in order to perform the assay of a single-step enzymic reaction.

On the other hand, the kinetic behavior of multienzyme systems is much more difficult to analyze quantitatively than the kinetics of single enzymes. Considerable progress could therefore be achieved under conditions of medium and temperature in which the kinetics can be more easily resolved.

The study of membrane-bound multienzyme systems in a fluid medium over a broad range of subzero temperatures allows one to influence selectively the activity of each enzyme and eventually to interrupt, in a temporal sense, sets of reactions at given levels, allowing the assay of single-step enzymic reactions without the addition of exogenous substrates and (or) specific chain inhibitors. However, such conditions can introduce other foreseeable difficulties which must be controlled or at least analyzed in order to interpret the results correctly. Keeping in mind the influence structural integrity may have on enzymic activity, and the possible influence of the solvent on this integrity, it is essential to check the eventual effect of the macroaddition of any organic "antifreeze" solvent upon such activity at normal temperatures and to select the solvent that is found to be least disruptive. The best solvents should be those which are known as protective additives towards organized systems in cryobiology, i.e. polyols (ethylene glycol, glycerol).

Working at subzero temperatures, one must also bear in mind that enzyme catalyzed reactions, to be effective, must occur in the correct temporal and spatial sequence and that a decrease in temperature might upset the delicate balance of the system, especially the sequence of events, and thus alter the proper functioning of the system. Such an imbalance can be anticipated since it is well known that the temperature coefficients and activation energies of the enzymes in the system are not all identical. Thus it is necessary to check for eventual perturbations of the normal reaction pathways as a result of these abnormal conditions. Up to now, a few quite promising investigations of membrane-bound multienzyme systems have been carried out in fluid mixed solvents at subzero temperatures. Let us examine preliminary observations obtained in several systems.

ENZYME ASSAY IN THE ENDOPLASMIC RETICULUM OF RAT LIVER

The drug-metabolizing multienzyme system of the endoplasmic reticulum of hepatocytes (rat liver microsomes) already intensively studied under normal conditions and which still raises a number of experimental problems has been investigated in this laboratory (77).

The system can be schematized as follows: the dashed-arrows indicate uncertainties in the sequential mechanism of the two electron process leading to the final drug metabolizing reaction by the terminal oxidase cytochrome P_{450}, where the sequence of substrate and oxygen interaction has yet to be definitely established (see Fig. 37). Enzyme assays can be performed in various ways using exogenous substrates (such as cytochrome c to test the activity of FP_1 and FP_2), specific chain inhibitors (such as PCMB to inhibit the FP_1 sequence, SKF to inhibit the binding site of cyt P_{450}, etc.).

The goal of the preliminary work has been to improve where possible our ability to study the single-step and sequential reactions and to find out new experimental conditions which will bring us nearer to understanding the complete functional mechanism of the system.

Solvent effect at normal temperatures

The rates of reduction of endogenous cytochrome P_{450} and exogenous cytochrome c by NADPH cytochrome c reductase are markedly decreased as the

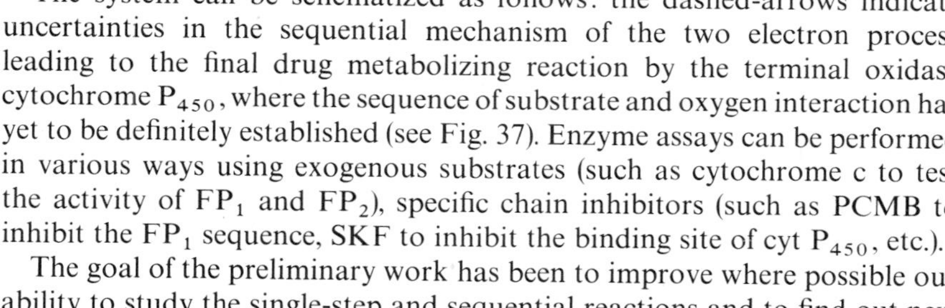

$$\mathrm{NADPH} \longrightarrow \mathrm{FP_1} \longrightarrow \text{“X”} \longrightarrow \mathrm{P_{450}} \left(\begin{array}{l} O_2 + SH \\ H_2O + SOH \end{array} \right.$$

$$\mathrm{NADH} \longrightarrow \mathrm{FP_2} \longrightarrow \text{cyt. } b_5$$

(dashed arrows: $FP_1 \dashrightarrow$ cyt. b_5; cyt. $b_5 \dashrightarrow P_{450}$)

FIG. 37. FP_1 represents the flavoprotein NADPH-cytochrome P_{450} reductase; FP_2 the flavoprotein NADH cytochrome b_5 reductase; cytochrome b_5; P_{450} is the cytochrome oxidase involved in the hydroxylation of substrates; X is an intermediate "carrier" of the electron transport chain, the existence of which is deduced from kinetic recordings.

concentration of added ethylene glycol to the aqueous microsomal suspensions at 27°C is increased.

A similar, though slightly smaller, decrease occurs when cytochrome c is reduced by the soluble NADPH cytochrome c reductase. In both cases, the solvent effect is totally reversible by dilution with water. This solvent effect was detected in many other experiments with soluble flavoproteins and hemoproteins performed in our laboratory and might be due to a reversible modification of the tertiary or quaternary structure of flavoproteins and hemoprotein that "cripple" enzyme activity. A similar partial "inhibition" was also reported in the mitochondrial respiratory chain in the presence of various organic solvents; this inhibition was explained in terms of the accessibility of membrane cytochromes to the solvent, which displaced water at the active center. The same conclusion was reached for the microsomal enzyme assembly, since the electron transfer from NADPH cytochrome c reductase to cytochrome P_{450} and to exogeneous cytochrome c is influenced to the same extent by the solvent.

However, the solvent effect on the reduction of cytochrome c is similar when NADPH cytochrome c reductase is either membrane bound or solubilized. It is more difficult to determine how the solvent might affect the enzyme activity. Moreover, the above results do not indicate that the solvent viscosity was one of the causes of the decrease in reaction rates.

Absorption spectra at subzero temperatures

At room temperature, addition of ethylene glycol to aqueous buffered suspensions of microsomes causes a slight decrease in the usual light scattering, indicating that there is no precipitation of microsomes. For instance, the ΔOD 450–700 nm of an oxidized suspension of 3 mg/ml of protein decreases by 0.05 in the presence of 50% ethylene glycol. In addition, cooling to temperatures below 0°C induces an expected sharpening and hyperchromicity of the absorption maxima, increasing the precision of recordings and kinetic measurements. Of course, exogenous compounds, such as cytochrome c, used to test the functioning of microsomal enzymes, similarly benefit under the same recording conditions.

Thus, the macroaddition of ethylene glycol as well as study of the suspensions at temperatures below 0°C can facilitate the recording of absorption spectra and kinetics. On another hand, these conditions alter the speed of the reactions and can greatly slow down some of them as reported below.

Enzyme assay at subzero temperature

The individual functioning of the two flavoproteins, their action upon cytochrome c and cytochrome b_5, and the reduction of cytochrome P_{450} between +20° and −45°C were recorded spectrophotometrically with the following results: the reduction rate of exogenous cytochrome c by both

NADPH and NADH reductases, already decreased in the normal range of temperatures by addition of ethylene glycol, is practically zero just above 0°C, whereas the reduction of cytochrome P_{450} is then unobservable (too slow) at about −5°C.

In the meantime, the reduction of cytochrome b_5 both by NADPH and NADH reductases is recordable down to −25°C, thus indicating that the interruption of the reduction of cytochrome P_{450} at about −5°C is due to the rate-limiting electron transfer from NADPH cytochrome c reductase to cytochrome P_{450}. The fact that NADPH and NADH reductases are still reactive towards cytochrome b_5 in a rather large span of temperatures below 0°C was used to record reduction kinetics. The reduction of cytochrome b_5 by NADPH cytochrome c reductase is a single first-order reaction, and the Arrhenius plot of the rate constant is a straight line between −7°C and −20°C.

Reduction of cytochrome b_5 by NADH cytochrome c reductase previously studied in normal conditions of media and temperature by stopped-flow experiments has been recorded in the above conditions and appears to be biphasic down to −21°C, where the Arrhenius plot of the higher rate constant shows a sharp discontinuity.

The biphasic reduction recorded down to −21°C may be due to the two different transfer processes already postulated from experiments in normal conditions (78), and the monophasic reduction occurring at −21°C and below may be due to a viscosity effect hindering one transfer process, as well as to a conformational change of the functional protein or (and) to a phase transition of the membrane similar to that reported by Steim and co-workers with certain membranes at temperatures below 0°C (79). On another hand, the slower phase of the reduction is surprisingly independent of temperature, and we cannot provide any satisfactory explanation.

P_{450} remained chemically reducible well below the temperature arresting its enzymic reduction. Moreover it was observed that the fixation of carbon monoxide upon ferrocytochrome was faster than the reduction of ferricytochrome, which remains rate limiting at each temperature (49). The fact that cytochrome P_{450} is chemically reducible down to −20°C permit us to predict useful experiments below 0°C which could provide new information about the intermediates of the reaction cycle of P_{450}. It can be seen that the normal behavior of several reactions is still exhibited under the present unusual conditions of medium and temperature, except the sudden kinetic change of the reduction of cytochrome b_5 by FP_2 at about −21°C.

Reversible suspensions of the hydroxylation, and storage conditions. Since it has been established that flavoproteins of the hydroxylating multienzyme do not react in fluid ethylene glycol–water mixtures at temperatures below

$-15°C$, these mixtures have been used to cool and store our microsomal preparations (freezing point $-25°C$) (80).

Under such conditions, we compared the oxygen consumption, the hydroxylating activity and the decay of individual concentrations of cytochromes b_5 and P_{450} from preparations stored in the above conditions at $-15°C$, or at $+4°C$, and from aqueous buffered preparations stored at $+4°C$. These analytical investigations were performed after warming to room temperature and aqueous dilution.

The oxygen uptake of microsomes in the presence of NADPH and the absence of substrate has been measured at 28°C. The consumption is constant for 7 days when microsomes are stored at $-15°C$, while it decreases greatly when microsomes are stored in buffer at 4°C.

The protein concentration of the stored microsomal preparations was about 25 mg/ml and 100 μl of each sample was used. The substrate used was codeine, incubated 10 min at $+28°C$ in Tris buffer (pH 7.5) and KCl (0.15 M). The addition of ethylene glycol did not influence the initial hydroxylating power at room or at subzero temperatures, but cooling efficiently protected the multienzyme system against aging, since most of the initial activity is recovered after several days storage at low temperature. Cytochrome b_5 from the electron transport chain of liver microsomes is completely recoverable and reactive after 72 h storage in the three different samples. However, the terminal oxidase enzyme (cytochrome P_{450}) seems to be very sensitive to storage conditions. About 92% of its initial concentration is retained in samples cooled in antifreeze after 72 h, but only 57–62% is retained in samples stored at $+4°C$. On the other hand, we noted that the activity of flavoproteins, and thus of the whole electron transport system, is stopped at about $-15°C$ in microsomes.

After 7 days' storage, the peroxide content of the samples increased threefold when the microsomes were stored in buffer at $+4°C$, and only 1.5 times when stored in EGOH-buffer at $-15°C$. Thus, cooling but not freezing results in a total suspension of the metabolism of liver microsomes and thus an efficient protection against aging processes. This is mostly due to the fact that the enzyme activity of flavoproteins is almost stopped at about $-15°C$ under our experimental conditions. Heating and dilution cause the recovery of initial enzymic and multienzymic activities. Thus one can say that the storage was nearly perfect, protecting the enzyme oxidase cytochrome P_{450} against its normal decay. The moderate concentration of ethylene glycol as macroadditive as well as the moderate temperature of storage eases both the operations of conservation and of recovery.

Since it is not cold *per se* but freezing and thawing which destroy living systems, and since cooling without freezing can reversibly interrupt enzyme reactions, some aqueous–organic mixtures already used as protective agents against freezing might be used as "antifreeze" to protect biological systems against aging at subzero temperatures.

STUDY OF ELECTRON TRANSFER AND OF OXYGEN INTERMEDIATES IN MITOCHONDRIA

The tolerance of mitochondria to mixed solvents and temperature variations has been explored by several authors and more recently in this laboratory (80) and it has been found that the presence of 30% ethylene glycol can produce a 20% loss of respiratory activity; however, it has been observed that cooling to −15°C protects the organelles against aging, and their respiration is almost constant after 7 days of storage. The diminished respiration and rate of electron transport observed in the mixture at room temperature are fully reversible by infinite dilution. These processes are practically reversibly inhibited at −15°C in the fluid mixed solvent used.

It has been found (81) that increased ethylene glycol and repeated freeze-thaw cycles damage respiratory control and energy coupling of mitochondria, but do not alter electron transfer kinetics through cytochrome c in response to oxygen pulses.

In such conditions, the low temperature procedure can be applied to the investigation of the cytochrome system of mitochondria; the trapping of reactive intermediates by rapid freezing has been carried out by Chance and Spencer (82), who demonstrated that the steady state electron transfer components of a respiring yeast system could be chilled rapidly enough to bring about the trapping of the oxido-reduction states of the respiratory carriers.

The feasibility of measuring fast electron transfer reactions in such frozen materials has been demonstrated in various experiments involving cytochromes (83,84,85). On the other hand, flash photolysis and recombination of the carbon monoxide compounds of hemoproteins in frozen states have been found to occur over a wide range of subzero temperatures; however, the cytochrome oxidase compound showed no measurable recombination below about 120°K.

The triple trapping procedure

Recently, Chance and colleagues (54) undertook the study of cytochrome oxidase–oxygen intermediates by a very elegant and efficient low temperature procedure comprising three steps, which can be summarized as follows:

1. Mitochondrial preparations are cooled in the dark at −15° to −30°C in presence of ethylene glycol as "antifreeze". In these conditions the velocity of dissociation of CO from the heme iron of the oxidase is markedly decreased so that oxygen does not combine with cytochrome oxidase–CO to a significant extent. The samples of mitochondria or soluble oxidase are

suspended in 15% ethylene glycol, 85% buffer (0.25 M mannitol, 0.05 M sucrose, Tris) at a protein concentration of 5 to 30 mg/ml, and supplemented with 5 mM succinate and 1.2 mM carbon monoxide. An alternative to this reaction mixture is an increased concentration of ethylene glycol (up to 45%) with a corresponding decrease in mannitol–sucrose–Tris buffer (pH 7.4), which may be replaced with phosphate buffer because of the lower temperature coefficient of the latter. A preferred sequence of additions is first, succinate, to activate respiration and render the system anaerobic; second, ethylene glycol; finally, carbon monoxide. This suspension may be stored at $-20°$ to $-40°C$ for some days. Other studies (81) indicate that respiratory control and coupled phosphorylation are retained by the mitochondria suspended in up to 30% ethylene glycol. Above this level, succinate oxidation is slowed and ascorbate plus tetramethyl *para*-phenylenediamine are more useful substrates. The sample is oxygenated and cooled to $-18°C$. Vigorous stirring for 5–10 s, up to a total of 20–40 stirs, gives oxygen concentrations in the range from 200 μM to 300 μM, sufficient for rapid activation of the system in the subsequent flash photolysis. The temperature of the suspension is monitored by a thermo-couple incorporated into the coiled stirring rod. Stirring should not start prior to thermal equilibration of the contents of the tube.

2. A rapid transfer at the end of the stirring period to $-80°C$ freezes the suspension and traps the reduced components of the respiratory chain, the CO-inhibited oxidase, and the dissolved oxygen in a stable mixture. Such mixtures appear to be unreactive for several hours. A number of samples may then be accumulated for flash photolysis at low temperature.

3. Following this step, the sample is placed at the temperature at which flash activation of the reaction is proposed. The third step of trapping then occurs when the reaction has proceeded to an appropriate extent. In this case, the reaction of interest may have a half-time of about 10 s and the chemical composition is readily trapped by rapid cooling. The sample is maintained at the temperature of liquid nitrogen in readiness for slow spectroscopic studies from 77° to 4°K.

Several alternative procedures are possible, the most usual one being that the temperature of spectrophotometric observation can be the same as the temperature at which oxygen is added so that the procedure can start immediately at $-40°C$, for example, if sufficient ethylene glycol is present, or at a higher temperature if less ethylene glycol is present. Under such conditions the kinetics are recorded in the liquid rather than the solid state. The third stage of trapping is still feasible but the rate of temperature drop is less because the total temperature drop is larger, and because of the heat of fusion of the sample.

Study of the kinetics of intermediates of cytochrome oxidase and oxygen

The trapping of intermediates in the cytochrome oxidase–oxygen reaction is made possible in mitochondria by the triple trapping procedure. In the second stage of the procedure the samples are placed in the spectrophotometer cell, in this case a Dewar flask to which the light guides are coupled for transmission spectroscopy. After temperature equilibration at $-40°C$ for 2 min, the measuring light is turned on and the spectrophotometer adjusted. The laser flash photolysis activates the reaction with oxygen and the oxidation of the cytochrome components occurs for cytochromes a, c and copper. These kinetics may be used as a signal for the third stage of trapping. Then, the third stage involves rapidly transferring the sample tubes from the Dewar at $-40°C$ to a liquid nitrogen bath or, preferably, to an isopentane–propane mixture chilled to the temperature of liquid nitrogen. The samples chilled to this temperature can be kept for several days in a stable condition pending study by slow physical methods.

In these conditions, the intermediates accumulate because the energy of activation of the terminal reactions of the respiratory chain increase with the sequence of steps in the chain. The initial reactions with oxygen are of lower energy of activation, and lowering the temperature makes them relatively faster than the oxidation of later components of the chain such as cytochrome c, thereby allowing intermediates to accumulate. Thus, a sequence of enzymic reactions is susceptible to analysis by the trapped state method only when the sequence of energies of activation is favorable.

The reaction kinetics are observed in the frozen state in small diameter circular tubes suitable for rapid trapping and we will merely summarize the results obtained by Chance and his associates (86,87,88). Three classes of functional intermediate compounds of cytochrome a_3 and oxygen have been identified:

Compounds of type A formed between $-125°$ and $-150°C$ do not involve electron transfer to oxygen and include oxy-compounds which are spectroscopically similar to the ferricyanide-pretreated oxidase $Cu^{2+}a_3^{2+}—O_2$, or the completely reduced oxidase $Cu^{1+}a_3^{2+}—O_2$, oxygen being readily dissociated.

Compounds of type B are formed between $-95°$ and $-60°C$, and involve oxidation of the heme and copper components of cytochrome a_3 to form peroxy-compounds $Cu^{2+}a_3^{3+}—O_2^{--}$ or $Cu_2^{2+}a_3^{2+}—O_2H_2$.

Compounds of type C are formed above $-100°C$ from mixed valency states of the oxidase obtained by ferricyanide pretreatment and might involve higher valency states of the heme component $Cu^{2+}a_3^{4+}—O_2^{--}$. These components act as electron acceptors for the respiratory chain and as functional intermediates in oxygen reduction, whereas compounds of type B above $-60°C$ serve as effective electron acceptors for cytochromes a, c and c_1.

All these compounds are functional in oxygen reduction and afford a basis for understanding energy conservation at the molecular level in oxygen metabolism by membrane-bound cytochrome oxidase.

Discussion

As stated by Chance (54), the actual range of the method for the study of macromolecules in biological processes is not clearly defined and may be either very limited or useful in the broadest sense of biological kinetics. Fortunately, terminal bimolecular reactions such as that of cytochrome oxidase with oxygen allow accumulation of the oxygen compound and can readily be made more rapid than the intercarrier reaction steps.

On the other hand the method presents some difficulties largely related to the versatility which has been built into the triple trapping approach and which needs a number of improvements including faster trapping in the last stage. This method appears to be superior to the Bray "spray quenching" (89) as well as to the droplet spray method used by Sangster (90), and might be usefully extended to a variety of other photo-activable reactions, as well as to terminal bimolecular reactions such as those involved in monooxygenase and dioxygenase systems.

STUDY OF THE PHOTOSYNTHETIC ELECTRON TRANSPORT CHAIN

The use of a liquid medium has obvious advantages in the study of chloroplast reactions since it avoids the changes associated with freezing and allows experiments involving the diffusion of small molecules to the chloroplasts. The possibility of studying chloroplast reactions in media with freezing points below that of water is thus potentially of great interest.

The advantages of a liquid medium cannot be obtained without the possible problems associated with the addition of high concentrations of organic solvents.

However, Inoué and Nishimura (91) showed that chloroplasts could be suspended in 50% (v/v) ethylene glycol with at least 50% retention of the ability to photoreduce the electron acceptor dichlorophenol indophenol (DCIP).

A number of reports of experiments to investigate the behavior of the chloroplast electron transport chain at subzero temperatures have appeared. Most of these have concerned the behavior of chloroplasts in frozen suspensions, or in glasses or extremely viscous solutions such as those obtained with glycerol at low temperatures. The only experiments in fluid media are those of Amesz *et al.* (92) who reported light-induced absorbence changes at $-40°C$ in chloroplasts suspended in 50% (v/v) ethylene glycol. The viscosity of solutions containing ethylene glycol is considerably less than those containing glycerol, and ethylene glycol was used in this laboratory by Cox, in an investigation of the photoreduction of artificial electron acceptors by chloroplasts suspended in aqueous organic mixtures at subzero temperatures. The goal was to investigate the possibilities of applying this technique to the study of photosynthetic electron transport, and to provide some information about the effect of solvents and temperature on chloroplast reactions under these conditions. Most of the previous experiments involving chloroplasts suspended in high concentrations of polyols seem to have been motivated by the desire to obtain an optically clear medium. These new experiments were rather concerned with exploiting to the full the advantages of a liquid medium in extending the range of possible types of experiment. More recently, Cox has studied the behavior of individual components of the photosystems in chloroplasts under these conditions (93).

Reduction of artificial electron acceptors by chloroplasts

The reduction of ferricyanide and 2-6-dichlorophenolindophenol (DCIP), of oxidized 2-3-5-6-tetramethyl-*p*-phenylenediamine (DAD), methyl purple, and the study of the effects of the inhibitors 3(3-4-dichlorophenyl)-1-1-dimethylurea (DCMU) and 2-5-dibromo-3-methyl-6-isopropyl-*p*-benzoquinone (DBMIB), have been carried out in ethylene glycol–water mixtures, glycerol–water and methanol–water mixtures in various volume ratios down to their freezing points and have shown the following major results: The chloroplast electron transport chain is relatively resistant to high concentrations of polyols. Electron flow can be observed at temperatures as low as $-40°C$ in 50% (v/v) ethylene glycol.

In any case, on the basis of the evidence obtained, it seems reasonable to suppose that electron transport from the donor side of Photosystem II to the acceptor side of Photosystem I can occur in chloroplasts suspended in 50% (v/v) ethylene glycol at temperatures at least as low as $-25°C$. Evidence for donation before Photosystem II was provided by inhibition of methyl purple reduction by DCMU and by the reduction of DAD; evidence for reduction after Photosystem I was provided by the biphasic kinetics of methyl purple reduction and the effect of DCMU on this, and also by the inhibition of methyl purple reduction by DBMIB.

It has been found that the presence of ethylene glycol causes an inhibition of the reduction of artificial electron acceptors by the chloroplast preparation. According to Cox (93), the inhibition caused by the solvent seems more likely to be due to a general alteration of the membrane structure than to inhibition at a specific site. The degree of inhibition is less than might be expected for a system involving a number of proteins reacting to the presence of the solvent in different ways. Possibly the small effect is due to many of the components being in a hydrophobic environment in the interior of the membrane and thus more protected from the medium than isolated enzymes. The "activation energies" for the various reactions obtained from the Arrhenius plots were found to be rather variable from sample to sample, and there is probably little significance in any case in the actual values in such a complex situation. Apparent "break points" in the Arrhenius plots similar to those observed here were reported by Shneyour *et al.* (94) for the reduction of various acceptors above 0°C, and interpreted as the result of phase changes in the membrane lipids.

It has been observed that the methyl purple reduction is biphasic, and that under suitable conditions, the slow phase but not the fast phase can be abolished in the presence of DCMU at lower temperatures (−20° to −30°C). This observation is a striking illustration of the potential value of the low temperature procedure, since a phenomenon which could previously only be observed by fast reaction techniques has now been brought into the time range where it can be studied with a commercial spectrophotometer.

The study of cytochrome f and P700 in chloroplasts

The above investigations open the way for the study of the behavior of individual chloroplast components using the low temperature procedure. Investigation of the properties of cytochrome f and P700 between 0° and −35°C has been carried out by Cox (95).

The relationship between these components is far from clear; in particular, the position of cytochrome f remains uncertain. The reduction of this component by Photosystem II via the plastoquinone pool, and its oxidation by Photosystem I, which is rapid, is well documented. There is a wealth of evidence that the oxidation of cytochrome f, but not its reduction, involves the participation of plastocyanin (96,97,98). However, kinetic results are difficult to reconcile with a simple series model: cytochrome f—plastocyanin —P700. Other results suggest that, in the dark following a sequence of flashes, the reduction of P700 occurs mostly by a pathway which does not involve cytochrome f, and recent results (99) suggest strongly that the component involved in transfer to P700 is plastocyanin.

Using the low temperature procedure, the reduction of P700 in the dark following a period of illumination, which would reduce the plastoquinone pool and oxidize components on the donor side of Photosystem I, has been investigated down to −35°C. Such an investigation provides a way of studying the flow of electrons through the rate-limiting step between Photosystems I and II.

Light-induced absorbence changes were measured in an Aminco–Chance spectrophotometer equipped for side illumination and fitted with an attachment enabling the temperature of the cuvette to be varied. Light from a 450 W Xenon arc lamp was passed through a 25 mm thickness of water and focussed on the cuvette. The optical path length of the cuvette was 10 mm and its width was 5 mm. Kinetic changes close to the response time of the recorder (0.3 s) were measured instead using a cathode ray oscilloscope.

The rate of the reaction is dependent on temperature and the kinetics can be easily measured at temperatures below −10°C. The reduction follows first-order kinetics. Three completely separate experiments with different preparations gave mean values for the activation energies of 80 ± 8 kJ mol^{-1} in the presence of NH_4Cl and 76 ± 7 kJ mol^{-1} in its absence. This value is similar to that obtained for the reduction of the electron acceptor, methyl purple, at low temperatures (below −10°C) (95). It may well represent a genuine activation energy for the rate-limiting step in the electron transport chain. Above −10°C, the reaction was too fast to measure accurately with the apparatus available.

At −20°C, the reduction is almost completely abolished by 80 μM 3-(3-4-dichlorophenyl)-1-1-dimethylurea. The cycle of oxidation and reduction could be repeated many times (at least 20) with the same kinetics. The half-time for reduction was independent of the length of the period of illumination even when this was not sufficient to oxidize P700 completely.

The effect of NH_4Cl suggests that the membranes still retain some of their impermeability towards protons under these conditions. The inhibition by dichlorophenyldimethylurea confirms that the reduction is dependent on Photosystem II and is not merely the result of an electron returning to P700 by a cyclic pathway around Photosystem I. Hence, it appears that electron transport from the donor side of Photosystem II to the acceptor side of Photosystem I can occur at −35°C.

The lower limit for the flow of electrons from the donor side of Photosystem II to the acceptor side of Photosystem I can now be reduced to −35°C. There is no evidence for any qualitative change in the behavior of the chloroplasts down to this temperature: at about −35°C, in contrast, there are suggestions of a change in the properties of the photosynthetic electron transport chain. Thus, experiments performed at temperatures lower than this may be more difficult to relate to the usual situation.

The Arrhenius plot for the reduction of cytochrome f in the presence of NH_4Cl in 50% (v/v) ethylene glycol medium between −5° and −30°C gives an activation energy of 72 ± 10 kJ mol^{-1} in three separate experiments with different preparations of spinach or lettuce chloroplasts. This is similar to the value obtained for the reduction of P700 in the same medium. A comparison of the plots, however, suggests that the half-time for the reduction of cytochrome f exceeds that for P700 and this is supported by the values of the time for half-reduction of these components.

From room temperature observations, Haehnel's conclusion (99) was that cytochrome f is not a component of the electron transport pathway between plastoquinone and P700. The results obtained with cytochrome f confirm the experimental observations of Haehnel without the use of a repetitive flash technique. His conclusion that cytochrome f does not lie on the direct pathway of electron transport between plastoquinone and P700 depends on the assumption that the spectrum of the initial portion of the first-order reduction is solely that of cytochrome f. Otherwise, a small degree of sigmoidicity could be masked by other changes. The present experimental accuracy was not sufficient to preclude a series model: cytochrome f—plastocyanin—P700 with a low equilibrium constant between cytochrome f and P700.

However more data will be needed before it is possible to decide between the large number of models, but the experiments summarized above provide an example of the flexibility which the use of cooled fluid media brings to investigations of the photosynthetic systems at subzero temperatures. A number of observations demonstrate that the major part of the electron transport chain can function in chloroplasts suspended in mixed media, and suggest that the membrane is still able to function specifically and efficiently.

References

1 S. Freed. *Science* **150**, 576 (1965).
2 B. Chance. *Arch. Biochem. Biophys.* **41**, 404 (1952).
3 P. Douzou, R. Sireix and F. Travers. *Proc. Nat. Acad. Sci. U.S.A.* **66**, 787 (1970).
4 F. Leterrier and P. Douzou. *Biochim. Biophys. Acta* **220**, 338 (1970).
5 S. Nakamura, K. Yokota and I. Yamazaki. *Nature* **222**, 794 (1969).
6 H. Degn. *Biochim. Biophys. Acta* **180**, 271 (1969).
7 H. Degn and D. Meyer. *Biochim. Biophys. Acta* **180**, 291 (1969).
8 P. Douzou. *Biochimie* **53**, 17 (1971).
9 P. Douzou. *Biochimie* **53**, 307 (1971).
10 J. W. Hastings and Q. H. Gibson. *J. Biol. Chem.* **238**, 2537 (1963).
11 Q. H. Gibson, J. W. Hastings, G. Weber, W. Duane and J. Massa. In *Flavins and Flavoproteins*, E. C. Slater, Ed., p. 341. Elsevier Publishing Company, Amsterdam (1966).
12 J. W. Hastings, Q. H. Gibson and C. Greenwood. *Proc. Nat. Acad. Sci. U.S.A.* **52**, 1529 (1964).
13 A. Eberhard and J. W. Hastings. *Biochem. Biophys. Res. Commun.* **47**, 348 (1972).
14 F. Lavelle, J. P. Henry and A. M. Michelson, *C. R. Acad. Sci. Paris* **270**, 2126 (1970).
15 J. W. Hastings, K. Weber, J. Friedland, A. Eberhard, G. W. Mitchell and A. Gunsalus. *Biochemistry* **8**, 4681 (1969).
16 L. G. Whitby. *J. Biochem.* **54**, 437 (1953).
17 G. W. Mitchell and J. W. Hastings. *J. Biol. Chem.* **244**, 2572 (1969).

18 E. A. Meighen, M. Ziegler-Nicoli and J. W. Hastings. *Biochemistry* **10**, 4062 (1971).
19 J. Lee. *Biochemistry* **11**, 3350 (1972).
20 E. A. Meighen and J. W. Hastings. *J. Biol. Chem.* **246**, 7666 (1971).
21 T. Watanabe and T. Nakamura. *J. Biochem.* **72**, 647 (1972).
22 F. McCapra and D. W. Hysert. *Biochem. Biophys. Res. Commun.* **52**, 298 (1973).
23 A. Gunsalus Miguel, E. A. Meighen, M. Ziegler-Nicoli, K. H. Nealson and J. W. Hastings. *J. Biol. Chem.* **247**, 398 (1972).
24 G. W. Mitchell and J. W. Hastings. *Anal. Biochem.* **39**, 243 (1971).
25 J. W. Hastings and G. Weber. *J. Opt. Soc. Am.* **53**, 1410 (1963).
26 G. Hui Bon Hoa and P. Douzou. *J. Biol. Chem.* **248**, 4649 (1973).
27 M. Nicoli. *Active center Studies on Bacterial Luciferase*, Ph.D. Thesis. Department of Biological Chemistry, Harvard University (1972).
28 C. Balny, G. Hui Bon Hoa and F. Travers. *J. Chim. Phys.* **68**, 366 (1972).
29 V. Massey, F. Müller, R. Feldberg, M. Schuman, P. A. Sullivan, L. G. Howell, S. G. Mayhew, R. G. Mathews and G. P. Foust. *J. Biol. Chem.* **244**, 3999 (1969).
30 W. H. Walder, P. Hemmerich and V. Massey. *Helv. Chim. Acta* **50**, 2269 (1967).
31 C. R. Jefcoate, S. Ghisla and P. Hemmerich. *J. Chem. Soc.* **C**, 1689 (1971).
32 P. Hemmerich, S. Ghisela, U. Hartman and F. Müller. In *Flavins and Flavoproteins, Third International Symposium*, H. Kamin, Ed., p. 83. University Park Press, Baltimore, (1971).
33 T. Spector and V. Massey. *J. Biol. Chem.* **247**, 5632 (1972).
34 O. Shimomura, F. H. Johnson and Y. Kohama. *Proc. Nat. Acad. Sci. U.S.A.* **69**, 2086 (1972).
35 G. A. Michaliszyn, D. K. Dunn, E. A. Meighen. *Fed. Proc.* **32**, 627abs (1973).
36 M. J. Cormier and J. Totter. *Biochim. Biophys. Acta* **25**, 229 (1957).
37 J. W. Hastings, C. Balny, C. Le Peuch and P. Douzou. *Proc. Nat. Acad. Sci. U.S.A.* **70**, 3468 (1973).
38 L. G. Augenstein. *Progr. Biophys. Chem.* **13**, 1 (1963).
39 J. W. Hastings, Q. H. Gibson and C. Greenwood. *Photochem. Photobiol.* **4**, 1227 (1965).
40 P. Rogers and W. D. McElroy. *Arch. Biochem. Biophys.* **75**, 87 (1958).
41 J. W. Hastings, Q. H. Gibson, J. Friedland and J. Spudich. In *Bioluminescence in Progress*, F. H. Johnson and Y. Haneda, Eds., pp. 151–186. Princeton University Press, Princeton, New Jersey (1966).
42 E. H. White and R. B. Brundrett. In *Chemiluminescence and Bioluminescence*, M. J. Cormier, D. M. Hercules and J. Lee, Eds., pp. 231–244. Plenum Publishing Co., New York (1973).
43 G. W. Mitchell and J. W. Hastings. *Biochemistry* **9**, 2699 (1970).
44 T. W. Cline and J. W. Hastings. *Biochemistry* **11**, 3359 (1972).
45 M. Z. Nicoli, E. A. Meighen and J. W. Hastings. *J. Biol. Chem.* **249**, 2385 (1974).
46 E. A. Meighen and R. E. MacKenzie. *Biochemistry* **12**, 1482 (1973).
47 J. W. Hastings and C. Balny. *J. Biol. Chem.* **250**, 7288 (1975).
48 C. Balny and J. W. Hastings. *Biochemistry* **14**, 4719 (1975).
49 P. Debey, G. Hui Bon Hoa and P. Douzou. *FEBS Letters* **32**, 227 (1973).
50 P. Debey, C. Balny and P. Douzou. *FEBS Letters* **35**, 86 (1973).
51 A. L. Fink. *Arch. Biochem. Biophys.* **155**, 473 (1973).
52 A. L. Fink. *Biochemistry* **12**, 1736 (1973).
53 E. Kraiscovitz and P. Douzou. *Biochimie* **55**, 1007 (1973).
54 B. Chance, N. Graham and V. Legallais. *Anal. Biochem.* **67**, 552 (1975).
55 B. W. Griffin and J. A. Peterson. *Biochemistry* **11**, 4740 (1972).

56 R. Lange, G. Hui Bon Hoa, P. Debey and I. C. Gunsalus. (In press).
57 B. W. Low, C. C. H. Chen, J. E. Bergher, L. Singman and I. F. Pletcher. *Proc. Nat. Acad. Sci. U.S.A.* **56**, 1746 (1966).
58 D. J. Haas. *Acta Cryst.* **B24**, 604 (1968).
59 D. J. Haas and M. G. Roosmann. *Acta Cryst.* **B. 26**, 998 (1970).
60 F. G. Parak, R. L. Mössbauer and W. Hoppe. *Ber. Bunsenges. Phys. Chem.* **74**, 1207 (1970).
61 V. F. Tomanek, F. Paraf, R. L. Mössbauer, R. I. Formanek, P. Schwager and W. Hoppe. *Acta Cryst.* **29**, 263 (1973).
62 G. Petsko. *J. Mol. Biol.* **96**, 381 (1975).
63 D. C. Phillips. *J. Sci. Instrum.* **41**, 123 (1964).
64 D. W. Banner. Ph.D. Thesis, University of Oxford, England (1972).
65 P. R. Evans. Ph.D. Thesis, University of Oxford, England (1973).
66 G. Petsko and A. Tsernoglou. Manuscript in preparation.
67 D. G. Marsh and G. Petsko. *J. Applied Cryst.* **6**, 76 (1973).
68 D. C. Phillips. *Scient. Amer.* **11**, 119 (1966).
69 P. E. Koshland and K. E. Neet. *Ann. Rev. Biochem.* **37**, 359 (1963).
70 J. A. Rupley and V. Gates. *Proc. Nat. Acad. Sci. U.S.A.* **57**, 496 (1967).
71 N. Sharon. *Proc. Roy. Soc. London. B.* **167**, 402 (1967).
72 D. C. Phillips. *Proc. Nat. Acad. Sci. U.S.A.* **57**, 484 (1967).
73 S. M. Banyard. Ph.D. Thesis, University of Oxford, England (1973).
74 K. Hagashi, K. Imoto and M. Funatsu. *J. Biochem.* **54**, 381 (1963).
75 S. K. Banerjee and J. A. Rupley. *J. Biol. Chem.* **248**, 2117 (1973).
76 C. C. Blake, L. N. Johnson, G. A. Mair, A. C. North, D. C. Phillips and V. R. Sarma. *Proc. Roy. Soc. London. B.* **167**, 378 (1967).
77 P. Debey, C. Balny and P. Douzou. *Proc. Nat. Acad. Sci. U.S.A.* **70**, 2633 (1973).
78 P. Strittmatter. In *Flavins and Flavoproteins*, E. C. Slater, Ed., p. 325. Elsevier Publishing Company, Amsterdam (1966).
79 J. M. Steim, M. E. Tourtelotte, J. C. Reinert, R. N. McElhanev and R. L. Rader. *Proc. Nat. Acad. Sci. U.S.A.* **63**, 104 (1969).
80 C. Balny, P. Debey and P. Douzou. *Cryobiology* **9**, 465 (1972).
81 M. Chance, A. Salcedo and B. Chance. *Analyt. Biochem.* (in press).
82 B. Chance and E. L. Spencer. *Farad. Soc. Disc.* **27**, 200 (1959).
83 B. Chance and M. Nishimura. *Proc. Natl. Acad. Sci. U.S.A.* **46**, 19 (1960).
84 T. Kihara and B. Chance. *Biochim. Biophys. Acta* **189**, 116 (1969).
85 B. Chance and W. D. Bonner Jr. In *Photosynthesis Mechanisms in Green Plants*, 66. N.R.C., NAS Washington DC. (1963).
86 B. Chance, C. Saronio and J. S. Leigh. *Proc. Nat. Acad. Sci. U.S.A.* **72**, 1635 (1975).
87 B. Chance, C. Saronio and J. S. Leigh. *Fed. Proc.* **33**, 1289 (1974).
88 B. Chance, C. Saronio and J. S. Leigh. *Fed. Proc.* **34**, 515 (1975).
89 R. C. Bray. In *Rapid Mixing and Sampling Techniques in Biochemistry*, 159, B. Chance, R. Eisenhardt, Q. Gibson and K. K. Lonberg-Holm, Eds. Academic Press. New York (1964).
90 M. Sangster. In *Rapid Mixing and Sampling Techniques in Biochemistry*, B. Chance, R. Eisenhardt, Q. Gibson and K. K. Lonberg-Holm, Eds., p. 193. Academic Press, New York (1964).
91 H. Inoué and M. Nishimura. *Plant Cell Physiol.* **12**, 137 (1971).
92 J. Amesz, M. P. J. Pulles and B. R. Velthuys. *Biochim. Biophys. Acta* **325**, 472 (1973).
93 R. Cox. *Biochim. Biophys. Acta* **387**, 588 (1975).

94 A. Shneyour, J. K. Raison and R. M. Smillie. *Biochim. Biophys. Acta* **292**, 152 (1973).
95 R. Cox. *Eur. J. Biochim.* **55**, 625 (1975).
96 G. Hindt. *Biochim. Biophys. Acta* **153**, 225 (1966).
97 M. Plesnicar and D. S. Bendall. *Eur. J. Biochim.* **34**, 483 (1973).
98 S. Izawa, S. Kraeyenkof, E. K. Ruuge and D. Devault. *Biochim. Biophys. Acta* **314**, 328 (1973).
99 W. Haehnel. In *Proceedings of the Third Internat. Congress on Photosynthesis*, M. Avron, Ed., p. 537. Elsevier Publishing Company, Amsterdam (1975).

6

General Conclusions

There are several conditions essential for the safe and successful investigation of enzyme systems in mixed fluid media at subzero temperatures which must and can be fulfilled. Solubility must be preserved and denaturation avoided by finding suitable values for dielectric constant, ionic strength and proton activity in selected mixtures prepared by a special simple and efficient sampling procedure; higher enzyme concentrations are needed to maintain sufficient enzyme activity at low temperatures; and, finally, a drastic reduction of reaction rates can be obtained but must leave the mechanism essentially unaltered.

Once prepared at subzero temperatures, the samples of a large number of enzyme systems can be studied at leisure; cooling $\rightleftharpoons$ heating cycles often allow a temporal resolution of reactions, step by step, while direct spectroscopic investigation gives details of reaction mechanisms from the detection and analysis of stabilized enzyme-substrate intermediates.

In these conditions, the low temperature combined with fast kinetic techniques open the way for a kinetic study of pre-stationary states and the elementary stages of enzyme reactions, bringing greater precision and uncovering new details of their mechanisms. Some reaction steps involving very short-lived species can be analyzed by the freeze-trapping procedure described in chapter 5 which represents one of several predictable improvements of the cooling procedure

Protein crystallography in mixed mother liquors at subzero temperatures represents a very promising field for the application of the technique: X-ray crystallography presents several spectacular improvements and opens the way to the three-dimensional structure studies of stabilized "productive" enzyme-substrate compounds.

Attractive though these results may be, it must be stressed that knowledge of enzyme-substrate intermediates is just part of the understanding of reaction mechanisms which can be obtained by the recording of kinetic and thermodynamic parameters, corresponding to the energy barriers and transition states comprising any given reaction. Since, as we have shown repeatedly, cosolvents used as "antifreeze" induce reversible changes in enzyme specific activity, sometimes casting doubt on the significance of data, it is essential to make substantial progress in the study of such artificial processes and even, if possible, to search for new media.

Since the bulk of this book was completed, we have followed new experimental trends in cryoenzymology, first by investigations in pure supercooled water, and secondly by "shielding" enzymes with polyelectrolytes solubilized in mixed solvents. We will only summarize these procedures, which are to be published soon: supercooled aqueous solutions of enzymes can be obtained by suitable suspensions of water in oil already described in the literature (1) and allow enzyme-catalyzed reactions to be carried out down to $-40°C$. On the other hand, it is possible to protect enzymes from cosolvents used as antifreeze, by causing interactions with soluble polyelectrolytes of opposing charges, and then by exploiting the selective solvation by water molecules resulting from the strong electrostatic potential developed by any polyelectrolyte, thus inducing an electrostatic "sorting out" of solvent molecules.

These new procedures are, of course, not free from problems but are indicative of the kind of progress which can be achieved with low temperature procedures using cooled fluid media, whose sole aim is to slow down enzyme reactions, but which can also introduce unwanted side-effects.

Cryoenzymology is just part of cryobiochemistry, which includes the major problem of protein fractionation. Low temperature column chromatography used to isolate an enzyme-substrate complex and, more recently, to purify a membrane-bound enzyme normally thermodynamically unstable in the test tube at room temperature, is quite representative, in the author's opinion, of the kind of progress which can be achieved with other current techniques of biochemical analysis if adapted to low temperatures. Recently, other adsorption chromatographic techniques, besides electrophoresis and isoelectric focusing, have been similarly adapted to resolve protein fractionation problems. In chapters 2 and 3, when dealing with the various cosolvent and temperature effects on enzyme-specific activity, we repeatedly mentioned some obvious effects on protein structure and bonding as well as their possible incidence in protein association and dissociation. These observations are essential to further developments in protein fractionation now under close investigation in this laboratory.

Protein fractionation at subzero temperatures involves changes in solvent composition, controlled variation in pH* values and the modification of the separation patterns based on a selected type of interaction for separating proteins. Some of these changes will lead to new, favorable or unwanted separation patterns, which, when suitably analyzed, could lead to a better understanding of the types of forces involved in protein associations. Thus use of mixed solvents and subzero temperatures should create a new dimension in protein chemistry.

GENERAL CONCLUSIONS

In recent years, cosolvent studies on a number of enzyme systems have been initiated by the observation that polyols present in storage buffer preparations increased the overall activity, could synergize the action of inducers and antagonize the effect of repressors of enzyme reactions (2). It is obvious that knowledge and control of physico-chemical parameters would prevent misleading observations.

Enzyme systems involving effectors which appear to be necessary to enable reactions to go to completion could be studied in this way and help determine the kinds of rearrangements that proteins undergo under their influence, and then gain information about the macromolecular basis of catalysis.

There is increasing evidence that dissimilar cosolvents at low concentration have similar stimulating effects on DNA and RNA synthesis *in vitro*. Patterns of such effects are identical to that reported in the case of bacterial luminescence: the overall synthetic reaction is increasingly stimulated up to a "critical" molar concentration of cosolvent and then moves back down and becomes inhibited as the concentration is increased. Further investigations have shown that cosolvents can stimulate, or act as substitutes for some protein factors involved in transcription (3) and it has been assumed that both the cosolvent and the factors act by producing local perturbations in the structure of the DNA template.

Cosolvents have been used to study the association equilibrium of vacant ribosomes (4) as well as perturbing agents in a number of the initiation and elongation reactions in the synthesis of polypeptides *in vitro* (5). Cosolvents such as methanol (10–20% v/v) have been used as "substitutes" for the aminoacyl tRNA of ribosomal particles during the catalytic action of some protein elongation factors. These different effects are somewhat difficult to understand and analyze in terms of parameters which could give a satisfactory explanation of both the cosolvent action and the normal reaction mechanisms.

Lack of the kind of data we have gathered concerning changes in the physico-chemical properties of solvents and solutes upon addition of increasing amounts of cosolvent, as well as a lack of any precise kinetic investigation (often impossible to carry out in such complex processes, where any recordable parameter will be the sum of various unequal or opposing contributions from several constituent reactions) can be very misleading and possibly without any precise significance, and further investigations need to be resumed on a new physico-chemical basis.

Thus the use of cosolvents and temperature variations as tools for investigating the effects of cofactors on enzyme systems, which can undergo conformation changes, in terms of their magnitude and effects on mechanisms is a promising field of study. Investigation through kinetic analysis, spectrophotometric and potentiometric titrations, recording of thermodynamic data and their interrelationships could provide a wealth of information about a number of biochemical mechanisms related to the macromolecular basis of enzyme activity.

Cryobiochemistry is quite different from cryobiology even though the composition of some of the media used are derived from the latter. Their goals are in fact opposite since cryobiology is faced with the necessity of keeping cells dormant (but alive) whereas cryobiochemistry has to preserve the reactivity of biological compounds. Nevertheless, knowledge and control of the variation of a number of physico-chemical properties, when the composition and temperature of media are changed, might lead to new attempts to protect soluble enzyme systems, membrane-bound multienzyme systems and organelles currently stored in so-called cryoprotective liquors. Such might be the immediate contribution of cryobiochemistry to cryobiology.

The purpose of this book has been to sum up a team's work over several years and to give an introduction to low temperature biochemistry. In spite of the quantity of data and results obtained, we are aware that such a task is only a very humble beginning but, we hope, sufficiently convincing and stimulating to prompt a number of colleagues to use and improve this new technique in their respective fields of interest.

References

1 D. H. Rasmussen, M. N. Macarslay and A. T. Mackenzie. *Cryobiology* **12**, 328 (1975).
2 J. W. Hastings, Q. H. Gibson and C. Greenwood. *Proc. of the Nat. Acad. Sci. U.S.A.* **52**, 1529 (1964).
3 E. Brody and J. Leautey. *Eur. J. Biochem.* **36**, 347 (1973).
4 A. Spirin. *FEBS Letters* **14**, 414 (1971).
5 A. Crepin and F. Gros. *Proc. of the Nat. Sci. U.S.A.* **72**, 333 (1975).

Index